# ACOUSTICS AND NOISE CONTROL HANDBOOK FOR ARCHITECTS AND BUILDERS

by
Leland K. Irvine and Roy L. Richards

**KRIEGER PUBLISHING COMPANY**

MALABAR, FLORIDA
1998

Original Edition 1998

Printed and Published by
**KRIEGER PUBLISHING COMPANY**
**KRIEGER DRIVE**
**MALABAR, FLORIDA 32950**



**Library of Congress Cataloging-In-Publication Data**

Irvine, Leland K., 1909–
Acoustics and noise control handbook for architects and builders / by Leland K. Irvine and Roy L. Richards. — Original ed.
p. cm.
Includes bibliographical references and index.
ISBN 0-89464-922-1 (hardcover : alk. paper)
1. Architectural acoustics. 2. Noise control. I. Richards, Roy L., 1928– . II. Title.
NA2800.I78 1998
690'.2—dc21 96-47818
CIP

10 9 8 7 6 5 4 3 2

# Contents

# List of Illustrations

***Chapter 3 (Sound Transmission Loss is abbreviated TL)***

# List of Tables

# Acknowledgments

The authors are indebted to a number of people who provided valuable assistance during the preparation of this handbook. Martha Mitchell was very helpful with typing, collecting acoustical data from manufacturers, performing calculations and serving as "eyes" for Lee Irvine when his vision began to fail. Janine Boyer-Richards helped keep the numerous files in good order and directed production of the many graphics. Additional graphics production work was performed by Byron Canfield. Herb Chaudiere and Jim Fullmer were major contributors to Chapter 6 on Electroacoustic Systems.

We wish to acknowledge the following manufacturers (most of whom are listed in the handbook's Appendix, Section A.12), that provided acoustical data, figures and tables:

- BBN Laboratories, Inc. for permission to use absorption tables for plywood panels developed by this organization.
- Industrial Acoustics Company, Inc. for permission to use figures from their Noise Control Reference Handbook.
- The estate of Mike Rettinger for permission to use the table "Sones as a function of Sound-Pressure Level," from his book *Acoustics, Room Design & Noise Control* (1968).
- State of California—Department of Health Services from their *STC/IIC Ratings for Wall/Ceiling Assemblies*.
- Synergetic Audio Concepts/Consulting-Seminars for material from their "Tech Topic" Vol. 12, Number 4.
- United McGill Corporation for information on the fundamentals on HVAC duct system acoustics design from their Acoustical Engineering Reports (AERs).
- United States Gypsum Company for providing sound transmission loss data.
- The Wenger Corporation for drawings of an acoustical shell (Figure 5-4).
- Panelfold Company for data on sound reduction of operable partitions (Figure 3-22).
- RPG Diffusor Systems, Inc., for the drawing of patterns of reflected sound (Figure 2-24).

In particular, we are grateful to the American Society of Heating, Refrigerating and Air-Conditioning Engineers (ASHRAE) for their permission to reprint versions of figures and tables from the 1991 *Applications* and 1993 *Fundamentals Handbooks*.

This handbook was sponsored by the University of Utah's Graduate School of Architecture. Dean William Miller and his staff gave early encouragement, and later, Professor Robert Young provided an excellent review. Finally, the authors are deeply indebted to Krieger Publishing Company for its patience as the authors failed to meet one deadline after another.

# About the Authors

The authors of this Handbook are Leland K. Irvine and Roy L. Richards. The sponsor is the Graduate School of Architecture at the University of Utah, Salt Lake City, Utah.

Mr. Irvine is a registered Professional Engineer in Acoustical and Electrical Engineering as well as an Adjunct Professor of Architecture at the University of Utah. He was the principal of the firm Acoustical Engineers, Inc. from 1947 to 1992.

Mr. Irvine graduated from the University of Utah in 1932 with a Bachelor of Science in Electrical Engineering. He attended summer sessions on acoustics at the Massachusetts Institute of Technology and the Illinois Institute of Technology. His first acoustical consulting project was in 1933, involving the acoustics and noise control for a broadcast booth for the Church of Jesus Christ of Latter Day Saints (Mormon) Tabernacle on Temple Square in Salt Lake City. Since forming his consulting firm, Mr. Irvine has been involved in over 2700 projects throughout the United States. Most of his architectural projects involve broadcast and school facilities, as well as considerable work in industrial and community noise control.

In his teaching position at the University of Utah Graduate School of Architecture, Mr. Irvine presented lectures on architectural acoustics and noise control for 40 years. Much of the material in this handbook is taken from the many references and notes developed for his lectures on acoustics. Mr. Irvine has presented numerous seminars to schools, industrial organizations and allied professional societies.

Organizations Mr. Irvine has been a member of include the Acoustical Society of America, the Institute of Noise Control Engineers, the National Council of Consulting Engineers, the American Society of Heating, Refrigeration and Air Conditioning Engineers, and the American Society for Testing and Materials.

Mr. Richards' career in Acoustics began in 1947 at Riverbank Acoustical Laboratories located in Geneva, Illinois. Riverbank is a world-renowned laboratory for testing the acoustical properties of architectural building materials. He was responsible for conducting tests of sound absorption, sound transmission loss and acoustic impedance, as well as light reflectance and flame spread ratings. While at Riverbank he participated in research projects of ballistic noise studies, propagation from high altitude sound sources and a number of architectural noise studies. He was instrumental in developing a new sound transmission loss measuring facility at the laboratory.

During his employment at Riverbank, Mr. Richards majored in Physics at the Illinois Institute of Technology, completing his studies in 1958.

In 1960 Mr. Richards joined Robin M. "Buzz" Towne in Seattle, Washington, to begin the first acoustical consulting firm in the Pacific Northwest. The firm currently operates under the name of Towne, Richards & Chaudiere, Inc., where Mr. Richards serves as vice president and senior consultant. The firm provides services in the fields of architectural acoustics and environmental noise. Mr. Richards has provided acoustical design recommendations for hundreds of building projects of all types, specializing in educational facilities and performing arts centers. He has directed environmental noise studies involving airports, freeways and industrial noise sources.

Mr. Richards has written numerous articles published in technical journals, has lectured at several universities and appeared as an expert witness on many occasions. He is a member of the Acoustical Society of America, the Institute of Noise Control Engineers and the National Council of Acoustical Consultants.

# Introduction

The acoustical qualities of the environments we work, live and play in are important to our productivity, health and well being. Buildings that are designed with care to the acoustic environment can make our lives more comfortable and productive. Apartments can become private, offices quieter and musical performances enhanced. This handbook will give architects or builders the ability to make their buildings function much better.

*Acoustics and Noise Control Handbook for Architects and Builders* is not a complete textbook on acoustics, but more a "what to do" and "what not to do" in the design profession. Most architects will never have the opportunity to design a major concert hall or opera house, where acoustical concerns dictate the highest level of acoustical analysis. For this reason, very little is devoted to projects of this type.

The emphasis in the handbook is on the categories of projects with which most architects and builders are involved: Educational and medical facilities, office buildings, multifamily residences, multipurpose auditoriums, churches and similar projects. The first four chapters cover acoustical fundamentals, acoustical properties of building materials, and the behavior of sound in enclosed spaces. This prepares the reader for Chapter 5, Designing Acoustical Spaces, which is a guide for good acoustical design in a broad cross section of architectural spaces. The final four chapters provide supporting material on noise and vibration control in building mechanical systems, electroacoustics covering sound reinforcement and enhancement systems and open plan masking systems, an account of current issues concerning community noise, and methods of industrial noise control.

The purpose of this handbook is to provide a guide for lecture courses for architectural (and engineering) students at either the undergraduate or graduate level. Practicing architects will also find it a handy reference as well as a reminder to seek professional help when appropriate. The handbook will be useful for mechanical, heating and ventilating engineers and their contractors. Much of the material will aid building contractors and owners who have no background in acoustics or building noise control. The chapter on sound reinforcement will be helpful to sound system designers, specification writers and contractors. Finally, acoustical consultants will find a wealth of realistic acoustical performance data on building materials, conveniently organized by type of material or construction. Most such data are presented in a graphical format, thus enabling the reader to quickly grasp the essential differences in performance, a task made much more difficult with only numerical tables. All dimensions are given in both U.S. and metric units for the convenience of the reader.

The Appendix contains reference and conversion tables, calculation procedures, a discussion of acoustical measurements, numerous lists including agencies issuing acoustical standards and manufacturers of acoustical products.

The contents of this handbook are a compilation of materials gathered over a combined total of eighty years of consulting, teaching and delivering professional seminars by the authors. More information on their backgrounds can be found in the section About the Authors.

The acoustical performance data presented in this handbook have been obtained from many sources, such as laboratory and field measurements performed by product manufacturers, and by acoustical consultants, including the authors. Because the performance of most materials depends on the environment in which it was both tested and ultimately installed, the authors cannot guarantee that the indicated performance will be achieved in actual practice. The handbook stresses, however, methods by which the highest level of performance may be obtained. In critical situations, it is always advisable to employ the services of a qualified acoustical consultant.

# Chapter 1

# Sound Fundamentals

Sound in air is produced by the vibration of a surface, such as a loudspeaker diaphragm, moving the air molecules back and forth. This causes an alternating condensing and rarefaction of the air molecules, creating pressures slightly above and below atmospheric pressure. These pressure fluctuations create a wave which travels through the air, similar to waves moving across the surface of water, illustrated in Figure 1-1. The basic characteristics for describing these waves are *frequency*, *amplitude*, and *temporal*, the latter describing how the first two may vary with time. These characteristics are explained in the following sections.

## 1.1 Frequency— Wave Length

The number of pressure peaks that pass a given point in one second is termed *frequency*, and is stated in *Hertz*, symbolized Hz, formerly cycles per second. Sounds may consist of a single frequency, or pure tone, such as produced by a tuning fork; a combination of several frequencies such as from the instruments of an orchestra all playing together; or many frequencies occurring at random intervals such as from a mountain stream or the ocean surf. Most sounds are a complex combination of frequencies, but each source has a unique pattern, permitting the listener to distinguish between the myriad sound sources of our world.

The range of audible frequencies is from about 20 Hz to 20,000 Hz. The upper limit decreases as people become older. A piano keyboard is a good frequency scale, with the low frequencies on the left and the high frequencies on the right. Middle C is at a frequency of 256 Hz. See Figure 1–2.

A graphic representation of a single frequency produces the familiar "sine wave," as shown in Figure 1–3. The distance between two adjacent peaks is the *wavelength* of the sound wave, symbolized by the Greek letter lambda, symbolized λ. Sound travels through the air at a speed of about 1,130 ft/sec (344 m/sec) at room temperature and normal atmospheric pressure. The wavelength of a given frequency is determined by dividing the distance traveled by sound in one second by the frequency in Hz. For example, the wavelength at a frequency of 256 Hz is given by

$$L = \frac{1130 \text{ ft/s}}{256 \text{ 1/s}} = 4.41 \text{ ft}\left(\frac{344 \text{ m/s}}{256 \text{ 1/s}} = 1.34 \text{ m}\right)$$

Listed in Table 1-1 are the wavelengths for the octave band center frequencies commonly used in architectural acoustics and noise control (see Section 1.1.1).

Wavelengths at the extremes of the audible frequency range are 56.5 ft (17.2 m) at 20 Hz, and about 2/3 of an inch (17 mm) at 20,000 Hz. Wavelengths in water are more than 4 times what they are in air, because the speed of sound in water is about 4887 ft/sec (1490 m/sec). Wavelength is an important consideration in the reflection and diffraction of sound (see Section 1.4.2).

Architectural acoustics and noise control are concerned with frequency and wavelength with regard to sound absorption, transmission, reflection, diffraction, reverberation and loudness. Noise and vibration control in structures are directly related to the frequencies involved. The quality of sound amplification systems is also a function of frequency response.

### 1.1.1 Frequency Bands

Frequency is important in architectural acoustics because the acoustical characteristics of building materials change with frequency (refer to Chapters 2 and 3). Also, human ears are more sensitive to certain frequencies, as explained in Section 1.5.1.

Architectural acoustics usually deals with frequencies from about 50 to 8000 Hz, though frequencies below 50 Hz must occasionally be considered. It is customary to divide this frequency range into octave bands, similar to the octaves on a piano, as shown in Figure 1-2. These bands are referred to by their center frequencies of 63, 125, 250, 500, 1000, 2000, 4000 and 8000 Hz. Notice the frequency doubles for each octave band. Recall that middle C on the piano is 256 Hz. C above middle C is exactly one octave higher, 512 Hz, and C below middle C is 128 Hz, and so forth.

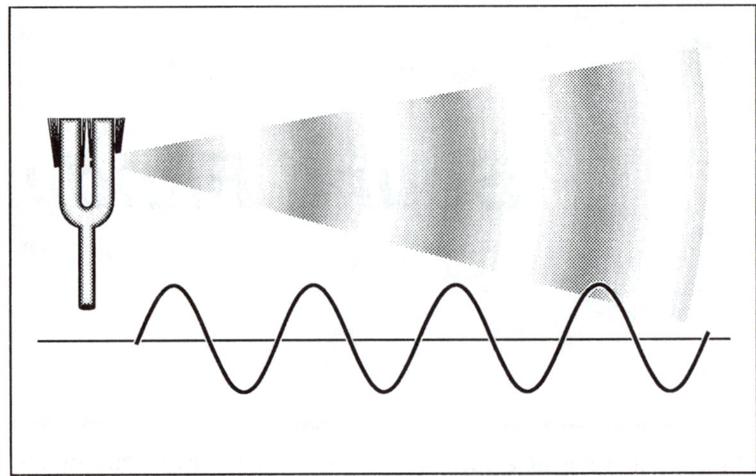

**Figure 1-1.** Sound is a sequence of pressure waves generated by a vibrating object.

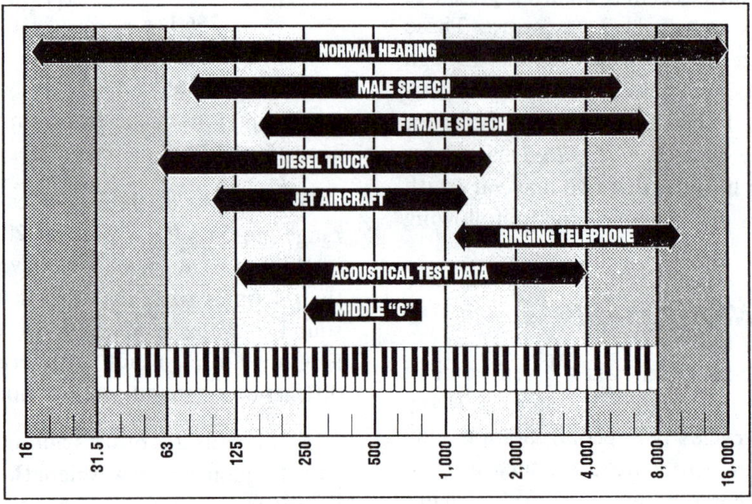


**Figure 1-2.** The frequency range of human hearing includes up to 10 octaves, and can be represented by a piano keyboard.

Sometimes greater frequency resolution is required for certain problems, where the sound may be divided into one-third octave or even narrower bands. Center frequencies for the one-third octave series are listed in Appendix A.2. A musical instrument playing a single note consists of a pure tone or "fundamental frequency" plus harmonics of this frequency that are 2, 3, 4, etc. times the fundamental.

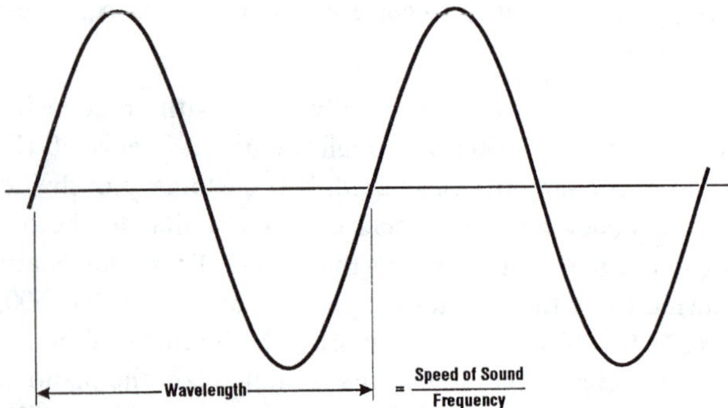


**Figure 1-3.** The distance sound travels for one complete cycle is the wavelength.

**TABLE 1-1.** Octave Band Center Frequencies and Wavelengths

| Frequency, Hz | Wavelength, ft | (m) |
|---|---|---|
| 31.5 | 35.9 | (10.9) |
| 63 | 17.9 | (5.46) |
| 125 | 9.04 | (2.76) |
| 250 | 4.52 | (1.38) |
| 500 | 2.26 | (0.69) |
| 1000 | 1.13 | (0.34) |
| 2000 | 0.57 | (0.17) |
| 4000 | 0.28 | (0.09) |
| 8000 | 0.14 | (0.04) |

These frequencies (and their relative amplitudes—see Section 1.2) determine the quality of the sound, and are different for a violin, for example, compared to those for a horn. These unique frequency and amplitude patterns distinguishes the sounds from the different musical instruments.

### 1.1.2 Noise

In acoustics, the term *noise* has two meanings. In the area of sound perception, noise is considered to be unwanted sound. This aspect of noise is discussed in Section 1.8.

The other definition of noise relates to the physical characteristics of broad-band (including many frequencies) sounds with specified frequency and amplitude distributions. Within this definition, there are two commonly referred to "colors" of noise:

1. "White" noise, which has equal energy per cycle. Because the number of cycles increases in each successive unit bandwidth (such as an octave), the amplitude of a white noise signal increases with frequency.
2. "Pink" noise, which has equal energy per unit bandwidth. The sound level remains constant with increasing frequency.

There are other types of noise with special characteristics for applications such as sound masking and acoustical measurements. See Chapter 5, Section 5.3.2, Chapter 6, Section 6.3, and Appendix A.10.

## 1.2 Pressure—Intensity—Power

A second important characteristic of sound is *amplitude*, which relates to the magnitude of the pressure of the sound wave. Two sound waves with the same frequency but with different amplitudes are shown in Figure 1-4. This pressure can be measured, but the measurement does not give any information on the direction the sound is traveling. Another measure of amplitude is sound intensity, a measure of the energy per unit area contained in a sound wave, but traveling in a specified direction. A third

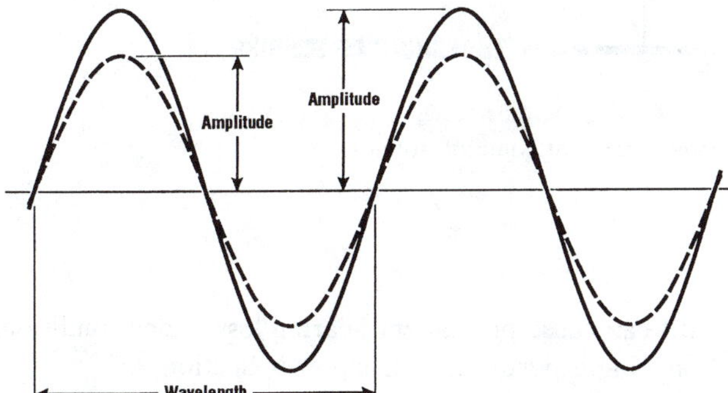


**Figure 1-4.** Two sounds may have equal wavelengths (frequencies) but different amplitudes.

measure of amplitude is sound power, the rate at which a source produces acoustic energy. All three of these characteristics relate to the physical measures of sound amplitude.

### 1.2.1 Decibels—Levels

The range of acoustic energy involved in human hearing is extremely large—more than 1 to 10,000,000,000,000, from the quietest sound we can detect to a deafening roar. This extreme range of numbers becomes unmanageable in working everyday acoustical problems.

By taking the *logarithm* to the base 10 (log) of this energy range, the numbers become more manageable. For example, the log of 1 is 0, while the log of 10,000,000,000,000 is 13, resulting in a range of only 13 units.

However, acousticians went one step further. The human ear is able to distinguish about 100 discrete steps across this entire range of sound amplitudes. If the logarithms are multiplied by 10, the range now extends from 0 to 130, and each step corresponds approximately to the smallest change in sound amplitude that the human ear can detect. This procedure forms the basis of the *decibel* scale. The symbol commonly used for decibel is dB.

Zero decibels, approximating the threshold of hearing, corresponds to a sound pressure of 20 micropascals (or $2 \times 10^{-5}$ Newtons/sq m). Micropascals is symbolized μPa. Earlier publications used a pressure reference value of $2 \times 10^{-4}$ dynes/sq cm, which is the same as the reference pressure 20 μPa. This exceedingly small pressure is referred to as the reference pressure. To illustrate just how sensitive the human ear is, the sound pressure represented by 0 dB corresponds to a weight of only 0.6 oz (17 grams), spread uniformly over a typical city block that is 300 ft (91.4 m) on a side! Ten times the logarithm of the ratio of a specific sound pressure to this reference pressure, quantity squared, results in the sound pressure level, Lp. Notice that the sound pressure level is proportional to the square of the pressure, and that the level expression is dimensionless since it is based on a ratio. *It is standard practice to add the word* level *when the units are in decibels.* Figure 1-5 lists both the sound pressure and sound pressure level for a variety of common sound sources.

The sound pressure level is calculated by the expression:

$$\begin{aligned} Lp &= 10 \log (p / 20\ \mu\text{Pa})^2 \text{ db} \\ &= 20 \log (p / 20\ \mu\text{Pa}) \text{ db} \end{aligned}$$

For example, assume the sound pressure $p$ of a truck driving by is 200,000 μPa. Then

$$\begin{aligned} LP &= 20 \log (200{,}000 / 20) \\ &= 20 \log (10{,}000) \\ &= 80 \text{ dB} \end{aligned}$$

Sound pressure levels below 20 dB would be considered

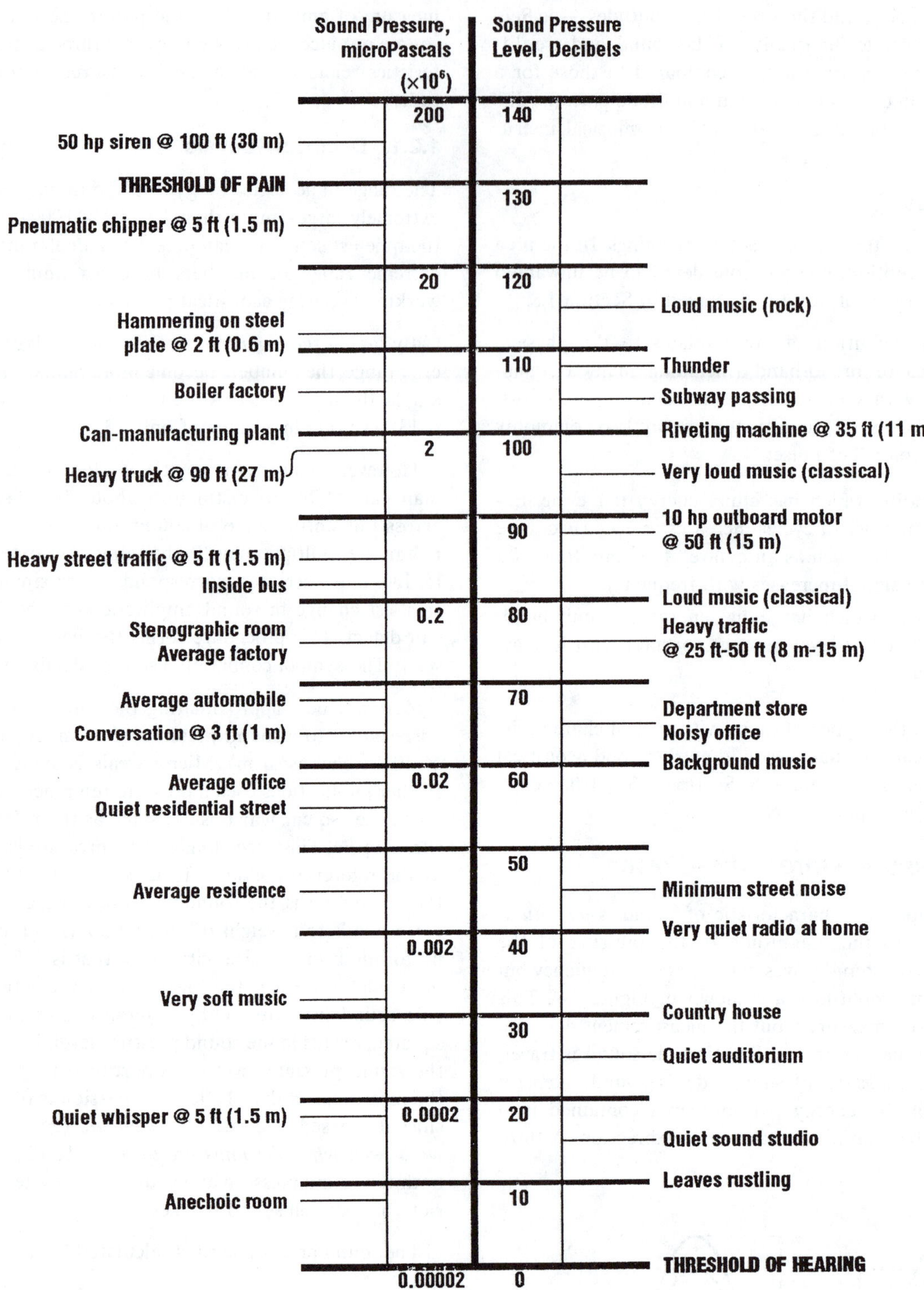


**Figure 1-5.** Sound pressure and sound pressure level of common sounds.

extremely quiet, while levels above 100 dB as extremely loud. Most people living in urban areas spend their lives experiencing sound pressure levels in the range of about 30 to 90 dB, though some industrial noise and rock music can produce levels well above 100 dB. Levels above about 85 dB can cause permanent hearing loss under conditions of ongoing exposure (see Chapter 9, Section 9.1).

A change in sound level of one decibel is the smallest change that the human ear can detect, and even then only under ideal listening conditions. A 3 dB change is readily detectable, and a 10 dB change will be judged as twice (or one-half) as loud. See Section 1.5.1.

Decibels can also be used to express sound intensity levels and sound power levels. Over the years several different reference quantities have been used for these calculations, so it is mandatory to state the *reference value* for intensity and power decibel levels on charts and tables.

The intensity of sound at the threshold of hearing is $10^{-16}$ watt/sq cm, a value used as a reference intensity for computing sound intensity levels (Li), determined in a manner similar to the calculation of sound pressure levels (Lp) as explained earlier. For the calculation of sound power levels (Lw), the reference acoustic power used is $10^{-12}$ watt, though $10^{-13}$ watt was used in earlier publications. As stated earlier, it is important to state the reference quantity for these decibel level calculations. Table 1-2 lists the sound power and sound power level for a number of representative sources.

It is interesting to note that when a sound power source of $10^{-12}$ watt is located at the center of a sphere whose area is one sq m, the resulting intensity on the surface of the sphere is $10^{-16}$ watt/sq cm; both are reference quantities as explained earlier.

For example, if the sound power of a source is one watt, the sound power level is obtained by

$$\text{Lw} = 10 \log\ (1/10^{-12})\ \text{dB Re: } 10^{-12}\ \text{watt}$$
$$= 120\ \text{dB Re: } 10^{-12}\ \text{watt}$$

**TABLE 1-2.** Typical Sound Power and Sound Power Level*

| | *Approximate Power Output* | |
|---|---|---|
| Source | Watts, w | Decibels re: $10^{-12}$w |
| Saturn Rocket | $10^8$ | 200 |
| Turbojet engine with afterburner. | $10^5$ | 170 |
| Jet aircraft at takeoff, four engines. | $10^4$ | 160 |
| Turboprop at takeoff | 1000 | 150 |
| Propeller aircraft at takeoff, four engines. | 100 | 140 |
| Large pipe organ | 10 | 130 |
| Small aircraft engine | 1 | 120 |
| Blaring radio | 0.1 | 110 |
| Automobile at highway speed | 0.01 | 100 |
| Voice, shouting | 0.001 | 90 |
| Garbage disposal unit | $10^{-4}$ | 80 |
| Voice, conversation level | $10^{-5}$ | 70 |
| Electronic equipment ventilation fan | $10^{-6}$ | 60 |
| Office air diffuser | $10^{-7}$ | 50 |
| Small electric clock | $10^{-8}$ | 40 |
| Voice, soft whisper | $10^{-9}$ | 30 |
| Rustling leaves | $10^{-10}$ | 20 |
| Human breath | $10^{-11}$ | 10 |
| Threshold of hearing | $10^{-12}$ | 0 |

*Reprinted by permission of the American Society of Heating, Refrigerating and Air-Conditioning Engineers, Atlanta, Georgia, from the 1993 *ASHRAE Handbook—Fundamentals*

The sound intensity on the surface of the sphere is $10^{-4}$ watt/sq cm, therefore the sound intensity level is obtained by:

$$\text{Li} = 10 \log\ (10^{-4}/10^{-16})\ \text{dB Re: } 10^{-16}\ \text{watt/sq cm}$$
$$= 120\ \text{dB Re: } 10^{-16}\ \text{watt/sq cm}$$

### 1.2.2 Adding Decibels

Since any level expressed in decibels is based on the logarithm of a ratio, it is not possible to add levels arithmetically. If two trucks each produce a sound pressure level of 80 dB, their combined level is 83 dB—definitely *not* 160 dB!

Decibels are added by converting the level back to the original *energy* ratio. (Note that these ratios must *always* be proportional to the energy content of the quantity being expressed.) This is accomplished by taking the antilog of each level, summing the antilogs, and reconverting to a level, as follows:

$$\text{Lc} = 10 \log\ (10^{L1/10} + 10^{L2/10} + \ldots 10^{Ln/10})\ \text{dB}$$

where Lc = the combined level, dB

Ln = individual source level, dB

For example, the resultant level by adding two sounds of 40 dB each is determined as follows:

$$\text{Lc} = 10 \log\ (10^{40/10} + 10^{40/10})\ \text{dB}$$
$$= 10 \log\ (10{,}000 + 10{,}000)\ \text{dB}$$
$$= 10 \log\ (20{,}000)\ \text{dB}$$
$$= 43\ \text{dB}$$

The curve in Figure 1-6 may be used for combining two levels. If there are more than two levels to be added, a level for the first two has to be calculated, and that result then added to the next level, and so forth.

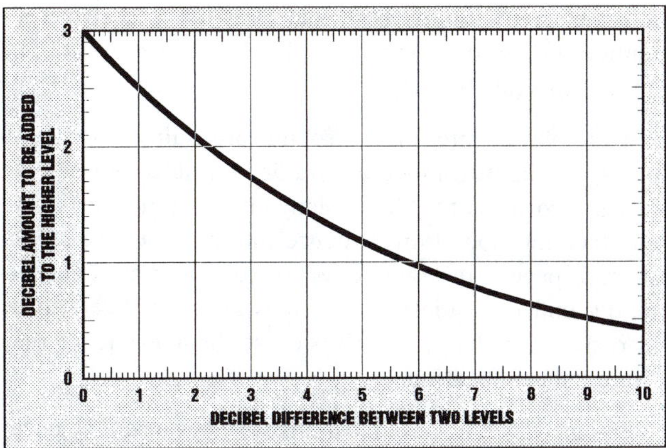


**Figure 1-6.** Chart for adding decibel levels.

For example, if the levels 40, 40 and 43 dB are to be added, 40 + 40 = 43, and 43 + 43 = 46 dB. That is, increasing the acoustic energy by two times equals a 3 dB increase, and by four times, a 6 dB increase.

Decibel subtraction is accomplished in the same manner, except the antilogs are subtracted instead of added. One application of the subtraction of decibels involves measuring a noise source in the presence of background noise. The level of the noise source is obtained by subtracting the background noise from the total noise, as follows:

$$Ls = 10\log(10^{Lc/10} - 10^{Lb/10})\ \text{dB}$$

where Ls = source level, dB

Lc = combined level, dB

Lb = background level, dB

For example, the background noise level in a shop is 40 dB and the total level with the compressor operating is 45 dB. The level of the compressor alone is determined by

$$\begin{aligned} Lc &= 10\log(10^{45/10} - 10^{40/10})\ \text{dB} \\ &= 10\log(31{,}623 - 10{,}000)\ \text{dB} \\ &= 10\log(21{,}623)\ \text{dB} \\ &= 43.3\ \text{dB} \end{aligned}$$

## 1.3 Temporal

How a sound varies with time can be important in how it will affect the listener. The time-varying—or temporal—characteristics of sound can be classified into the general categories of steady, intermittent and impulsive. There are special sound descriptors that are used to help define these characteristics.

### 1.3.1 Steady—Intermittent—Impulsive

The level of a steady sound changes very little with time. Refer to Figure 1-7. Examples would be the noise from a fan or blower, or a waterfall. The noise produced by heavy traffic on a freeway can be fairly steady, particularly at distances of several hundred feet. People are usually able to tolerate steady sounds as they are less likely to attract attention, unless the sound is so loud as to cause interference with some activity.

Examples of intermittent sounds are trains, aircraft flyovers, and light to moderate traffic with lulls between vehicles or groups of vehicles. Many machines, industrial and construction operations produce intermittent sound. Music and speech are examples of sounds which are intermittently loud and quiet. Such sounds are more likely to be intrusive than steady sounds because the interference they cause keeps repeating.

Impulsive sounds are most intrusive because they occur without warning and may cause a startle reaction. Examples are gunshots, fire crackers and automotive backfires. Some industrial and construction operations produce impulsive sounds, such as a punch press, hammering and pile driving. Some impulsive noises occur so rapidly that they may be treated as a combination of intermittent and steady, such as a jack hammer, riveting and chipping.

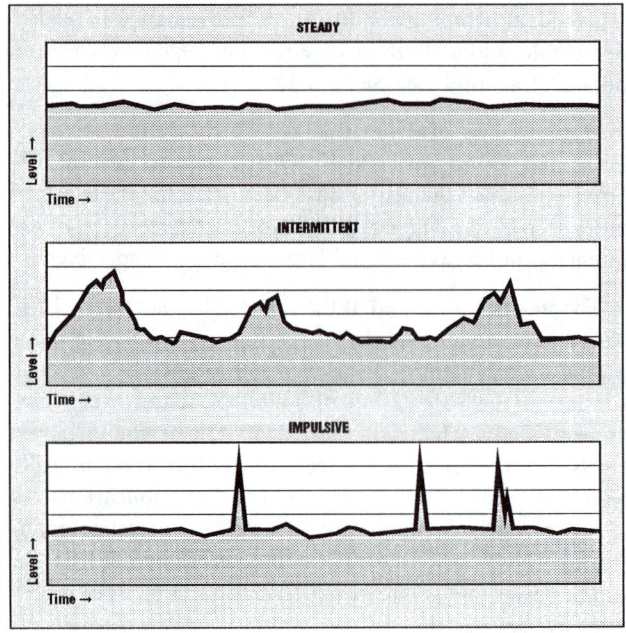


**Figure 1-7.** Graphical representations for steady, intermittent and impulsive sounds.

### 1.3.2 $L_{eq}$—Percentiles—$L_{max}$—$L_{min}$

The temporal characteristics of a sound may be defined by selecting an appropriate sound measure or *descriptor*. The most commonly used descriptors currently in use are defined below:

$L_{eq}$. Equivalent Energy Level. This is the sound level which, if held constant for a specified time period, yields the same total energy as contained in the actual sound. The Leq is probably the most useful single number for describing the level of sound, and is used extensively for sound specification purposes. It is affected most by the highest sound levels occurring during the time period, since they contain the most energy, hence it is an appropriate descriptor for rating noise impacts.

$L_{max}$. This is the maximum level that occurs during a specified time period. It is usually of very short duration, possibly caused by an impulsive sound. It is particularly useful for evaluating sleep interference.

$L_{min}$. This is the minimum level that occurs during a specified time period. It is useful in identifying the background sound level in the absence of intermittent higher sound levels.

$L_x$. This descriptor, or percentile, specifies that sound level which is exceeded "X" percent of the time for a specified time period. For example, the $L_{50}$ is that sound level exceeded 50% of the time, the $L_{10}$ 10% of the time, and so forth. The $L_1$ or $L_{0.1}$ is usually close to the $L_{max}$, and the $L_{99}$ close to the $L_{min}$. The $L_{eq}$ is usually between the $L_{50}$ and the $L_{10}$. For a steady sound, the $L_1$ and $L_{99}$ will be close together. The further apart they are, the more the sound fluctuates.

## 1.4 Sound Propagation

It is necessary to understand how sound travels from source to receiver in order to predict what the sound characteristics will be at the receiver. Also, most of the opportunities for controlling sound occur along the transmission path, such as absorption, reflection, diffusion, etc. The ability to impose such controls requires an understanding of how sound travels through the air, and how it behaves when it encounters typical architectural elements.

### 1.4.1 Distance—Inverse Square Law

Sound radiating from a point source will spread as spherical waves. Since the area of the sphere increases as the *square* of the distance from the source, the intensity (watts/sq cm) decreases as the square of the distance. That is, each time the radius of a sphere is doubled, the surface area increases by 4 times, so the intensity decreases to 1/4th, corresponding to a reduction of 6 dB (10 log 1/4). See Figure 1-8. This characteristic follows the well known inverse square law, applicable to all types of wave radiation from a point source.

If the dimensions of the sound source are large compared to the propagation path, the decrease in sound level with distance will not follow the inverse square law. For example, a heavily traveled freeway acts as a line source, and the sound expands in a cylindrical manner. The surface area of a cylinder doubles with twice the radius (not the square of the radius), and the reduction in sound level is 3 dB per double distance. This phenomenon is illustrated in Figure 8-3 of Chapter 8. However, as the distance to the freeway becomes greater in length compared to the segment of freeway exposed, the decrease in sound level approaches the point source rule.

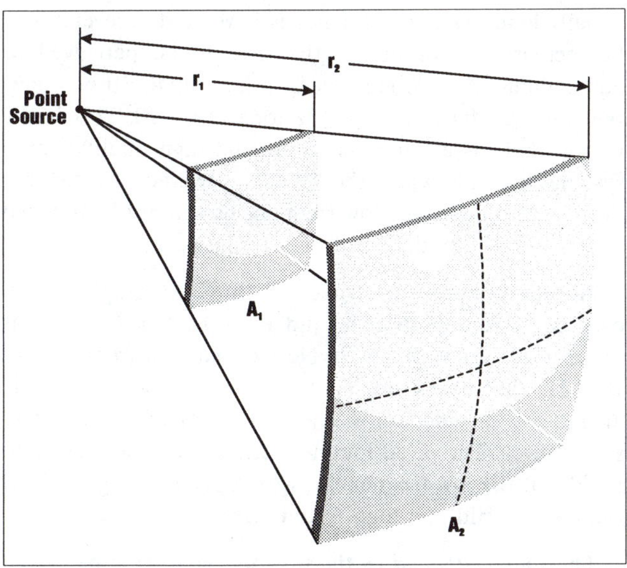


**Figure 1-8.** Sound from a point source expands spherically. When the distance is doubled, the area increases by 4 times, resulting in a 6 dB drop in sound pressure level (10 log 1/4).

Other examples of sound sources that may be large compared to the propagation path are machinery enclosures, open doors to noisy shops, and large arrays of loudspeakers such as may be used in a sports arena. The propagation characteristics of such sound sources must be separately evaluated in order to accurately predict the sound level at some distance from the source.

### 1.4.2 Reflection—Diffraction—Refraction

When sound is reflected from a smooth, plane surface, the wave continues to move in a direction following the rule that the angle of reflection equals the angle of incidence, as shown in Figure 1-9. In tracing the direction of propagation following reflection, it is convenient to construct a ray which traces the direction the sound is moving. An image source is located behind the reflecting surface on a line passing through the true source, perpendicular to the reflecting surface, with both sources equidistant from the surface. This method of tracking the direction of sound reflections is called *geometrical* acoustics.

If the reflecting surface is convex, the sound will be dispersed more rapidly than if the surface is plane. Convex reflective surfaces are often used in auditoriums to improve sound distribution. A concave surface tends to concentrate or focus the sound, resulting in poor sound distribution. Concave surfaces can cause serious problems in architectural acoustics, and generally should be avoided, or

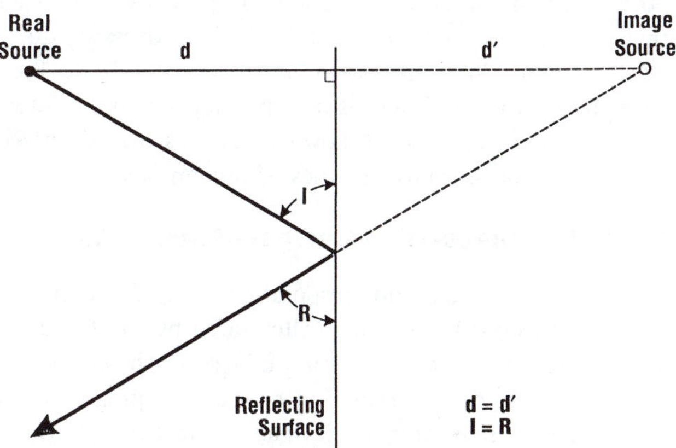


**Figure 1-9.** The angle of incidence (I) equals the angle of reflection (R).

used with great caution. The subject of reflections in rooms is treated in greater detail in Chapters 4 and 5.

Geometrical acoustics fails to adequately define a sound reflection if the reflecting surface is about the same size or smaller compared to the wavelength of the sound. That is because the sound "sees" the edge of a reflecting surface, and some of the sound arriving at or near the edge of the surface spills into the shadow zone behind the surface. This phenomenon is called *diffraction*, and is typical of all forms of wave motion. If the reflecting surface is very small compared to the wavelength of the sound, the sound continues on virtually undisturbed. This phenomenon is displayed in Figure 1-10. Tracing of rays from reflecting surfaces, therefore, is frequency-dependent, and a reflection pattern that is accurate at high frequencies may be unreliable at low frequencies.

In much the same manner as described above, sound diffraction also takes place at the boundary of two surfaces with different absorption characteristics. More sound tends to strike the surface having greater absorption, a condition affecting sound absorption measurements and techniques for installing acoustical materials (see Chapter 2, Section 2.4, and Chapter 5, Section 5.2.4).

Sound *refraction* occurs when a wave encounters air of a different temperature (density). Because the speed of sound is faster in warm air compared to cold, a wave front tends to bend in a direction away from the warmer air. This phenomenon is particularly important in sound propagation out-of-doors. A diagram of sound refraction is contained in Chapter 8, Figure 8-4.

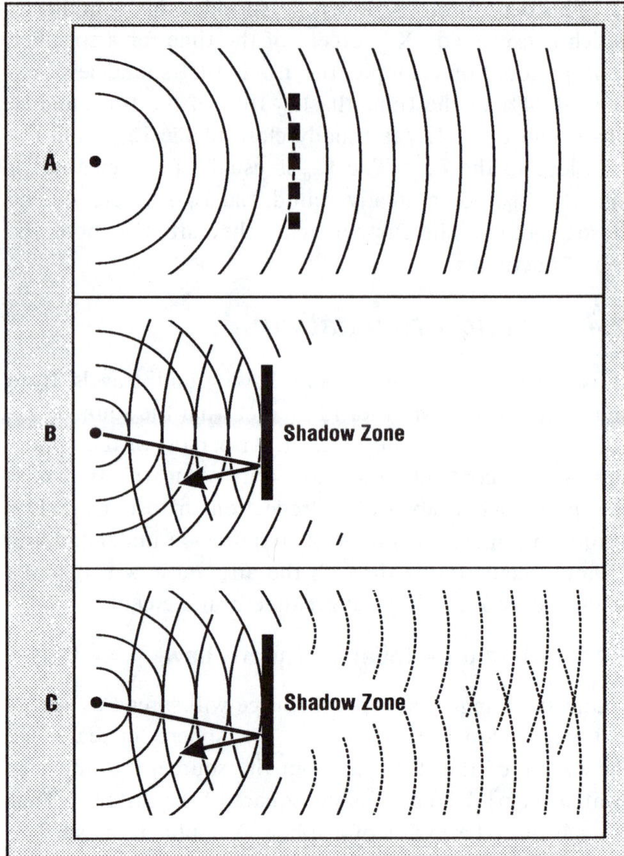


**Figure 1-10.** A: Sound diffracts around objects which are small compared to the wavelength. B: Sound reflects from objects which are large compared to the wavelength. C: Because of diffraction at the edges of the object, some sound enters the shadow zone.

## 1.5 Hearing

Eventually most problems in architectural and environmental acoustics must consider the phenomenon of human hearing. When sound is received by a person there is a transformation from the physical aspects of sound to the psychological: How sound is perceived by humans. It is important to understand how sound is perceived in order that design criteria may be established—addressing questions such as: Will the sound be clear; will it be loud enough—or too loud; will it be annoying, or will it interfere with sleep? In the following sections are discussed some of the basic characteristics of human hearing.

### 1.5.1 Loudness—F-M Curves—Sones—Phons

Human hearing does not respond linearly to either frequency or sound level. This is illustrated by the Fletcher-Munson (F-M) curves of equal loudness, as shown in Figure 1-11. These curves, representing loudness judgments of pure tones (single frequencies) as a function of sound level, illustrating a basic characteristic of human hearing, that the ear is most sensitive to frequencies around 4000 Hz. Each solid line in the graph represents tones that are equally loud. As the frequency is lowered, the level has to be increased in order for the tone to be perceived as equally loud. A 1000 Hz tone at a level of 40 dB will have the same loudness as a 50 Hz tone at 65 dB. This is the reason there is a loudness control on high quality music systems, so that when the level is lowered (by turning down the volume), the low frequencies will not be reduced as much as the middle and high frequencies.

Another interesting aspect of the F-M graph is that each curve represents tones judged to be twice as loud as those in the curve below it. Notice that at a frequency of 1000 Hz the curves are separated by exactly 10 dB, while at 100 Hz the separation is less, particularly at lower sound levels. Perceived loudness, therefore, increases more rapidly at the low frequencies with a given change in level than at the mid and high frequencies.

The *sone* is defined as the unit for measuring perceived loudness. The tones represented by each curve in the F-M graph have the same sone value. By definition, a 1000 Hz tone at a level of 40 dB has a value of one sone. The curve

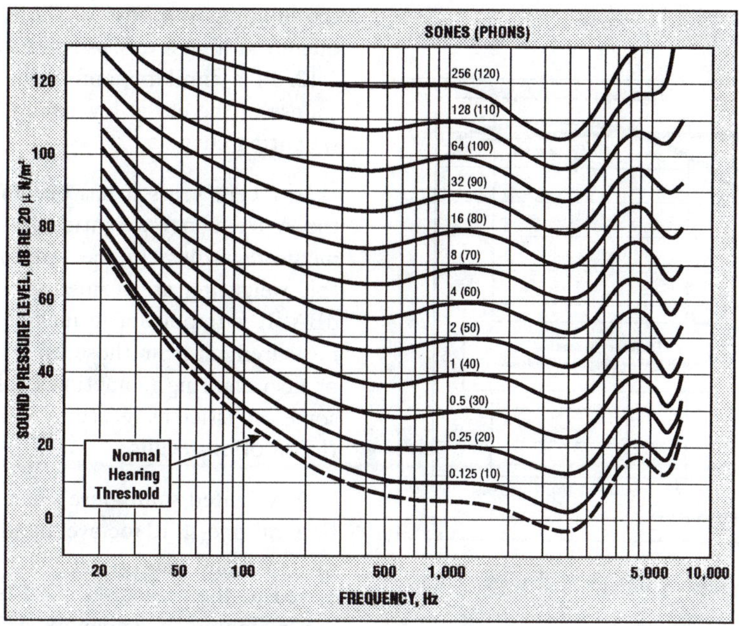


**Figure 1-11.** Loudness and loudness level contours for pure tones. Reprinted by permission of the American Society of Heating, Refrigerating and Air-Conditioning Engineers, Atlanta, Georgia, from the 1993 *ASHRAE Handbook—Fundamentals.*

above it, which is twice as loud, has a value of two sones, and the curve above that four sones, and so forth. Each time the sone value doubles, the loudness doubles.

The F-M equal loudness contours shown in Figure 1-11 are for single frequencies. Equal loudness contours have also been developed for octave bands of random noise. The loudness function is somewhat different for bands of noise, but still exhibits the same basic characteristic of the ear's reduced sensitivity to low frequency sound.

The method for calculating the total loudness of a complex sound is explained in Appendix A.8, including a sample calculation.

The loudness reduction achieved by correcting a noise problem is determined by comparing the number of sones before and after. Since the sone scale is linear, the percentage difference in loudness can be calculated. If the initial noise has a value of 2 sones and is reduced to 1 sone, it is a 50% reduction in loudness. Another example calculation can be found in Chapter 9, at the end of Section 9.3.2. Differences in loudness cannot be as easily determined from decibel levels as they are on a logarithmic scale and cannot be divided.

It is sometimes more convenient to convert the loudness value in sones to a loudness level. The unit for designating loudness level is the *phon*. The conversion from sones to phons is explained in Appendix A.8. In Figure 1-11, each contour is given a sone and phon value. Note that a 10 dB change in phon level is a doubling or halving of perceived loudness.

### 1.5.2 Directionality

How people determine the direction from which a sound arrives has important applications in architectural acoustics and the design of sound reinforcement systems. Obviously it is desirable to have the sound produced by a speaker or musical instrument be perceived as coming from the source. In concert halls, however, sound arriving from the sides of the hall also provides an important attribute for musical enhancement. See Reflectors in Chapter 5, Section 5.1.3.

Exactly how human hearing determines directionality is a complex psychoacoustic phenomenon, but it is related to the arrival time of the sound, the phase relationship, and comparative loudness at each ear. (Two sounds are said to be in phase if they are identical and arrive at a specified point or points with exactly the same time alignment.) Sound coming from straight ahead arrives at both ears at exactly the same time, and is exactly in phase. The sound is thereby perceived as coming from straight ahead. Sound arriving from the side is not the same at each ear, permitting a judgment as to where the source is located.

The situation becomes even more complex if the listener is inside a room, where the first arriving sound, or direct sound, may be from straight ahead, but then followed by reflected sound from the side walls or ceiling overhead. Even if the reflected sound (or sound from a loudspeaker) from another direction is slightly higher in level and the delay is not excessive, the listener will locate the direction of the source correctly on the basis of the first arriving direct sound. (See Chapter 4, Section 4.1.) This characteris-

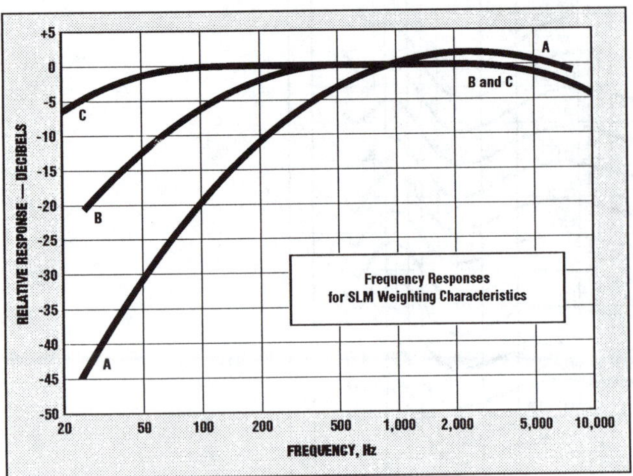


**Figure 1-12.** A, B, and C frequency weighting characteristics.

tic of human hearing also has important applications in the design of electroacoustic systems.

## 1.6 dBA -B -C

As demonstrated by the Fletcher-Munson equal loudness curves (Figure 1-11), the ear does not respond to the low frequencies the same as at higher frequencies. If several sounds have different frequency characteristics or have different levels, a sound level meter with equal response at all frequencies will not give a true comparison of the sounds as the ear perceives them. For this reason, frequency-weighting networks were developed so that the meter readings would approximate the F-M loudness curves.

Three frequency-weighting networks were originally developed, A, B and C, for measuring low, moderate and high level sounds, respectively. The three response curves are shown in Figure 1-12. Referring to Figure 1-11, notice the loudness response of the ear is nearly "flat" with frequency at high levels, as is the C-weighting curve in Figure 1-12. In a similar manner, the poorer low frequency response of the ear at low sound levels is reflected in the A-weighting curve.

In recent years it has become common practice to use the A frequency-weighting for most noise level measurements regardless of the level. Noise measurements using this weighting are termed A-weighted levels, symbolized dBA. It is important to include the A to differentiate the measurement from those using other frequency weightings or no weighting (sometimes called linear). Unfortunately some acousticians refuse to make this distinction and confusion often results.

A-weighted sound levels can be calculated from full octave or one-third octave band measurements. Each frequency band has an associated weighting factor to be added to the measured band level. The weighting factors are presented in Appendix A.2. After the weighting factors have been applied, the adjusted levels are added to determine the A-weighted level. An example calculation is shown in Table 1-3 (see Section 1.2.2 for the procedure for addition of decibels).

Notice the adjusted low frequency levels contribute very little to the A-weighted level. A-weighted levels are frequently used to specify maximum permissible noise levels for both background noise in rooms caused by mechanical systems, and for environmental noise. A-weighted room noise criteria are presented in Appendix A.4, and environmental noise standards are discussed in Chapter 8.

## 1.7 Noise Criteria (NC) and Room Criteria (RC) Curves

The prevalent method for defining the maximum permissible background noise generated by mechanical systems in buildings has been developed by the American Society for Heating, Refrigerating, and Air-Conditioning Engineers (ASHRAE). The method consists of comparing the

**TABLE 1.3.** Determining the A-Weighted Level from Octave Band Data

| Octave Band Center Frequency, Hz | Unweighted sound level, dB | A-Weighting, dB | Adjusted Level, dB |
|---|---|---|---|
| 31.5 | 78 | −39 | 39 |
| 63 | 76 | −26 | 50 |
| 125 | 72 | −16 | 56 |
| 250 | 82 | −9 | 73 |
| 500 | 81 | −3 | 78 |
| 1000 | 80 | 0 | 80 |
| 2000 | 80 | +1 | 81 |
| 4000 | 73 | +1 | 74 |
| 8000 | 65 | −1 | 64 |
| SUM | (unweighted): 88 dB | | (A-weighted): 85 dBA |

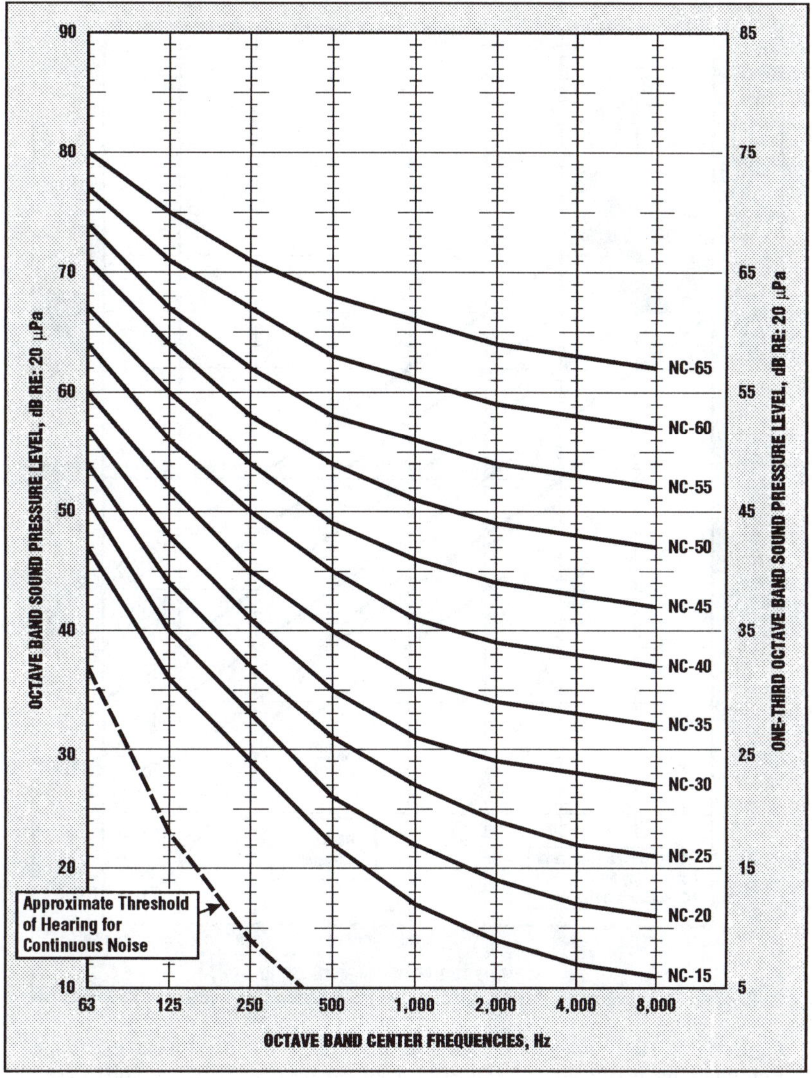


**Figure 1-13.** NC curves for specifying design level in terms of maximum permissible sound pressure level for each frequency band. Reprinted by permission of the American Society of Heating, Refrigerating and Air-Conditioning Engineers, Atlanta, Georgia, from the 1993 *ASHRAE Handbook—Fundamentals.*

octave band sound pressure levels of a noise with a family of curves known as NC (Noise Criteria) or RC (Room Criteria). These sets of curves are shown in Figures 1-13 and 1-14, respectively. Notice the shape of the curves compensate for the loudness response of the ear, permitting higher levels at the low frequencies.

RC curves are slightly more restrictive than the NC curves at the lowest and highest frequencies, and are preferred by many mechanical system designers because the slope of the contours presumably matches more closely a "well-balanced, bland-sounding" spectrum. Another set of curves designated NCB were recently issued in ANSI (American National Standards Institute) S12.2—1995, "Criteria for Evaluating Room Noise", and have similar characteristics to the RC and NC curves. This ANSI standard is an excellent reference for methods of evaluating room noise.

Noise levels are classified by the NC or RC curves by plotting the octave band levels (or one-third octave band levels if single frequency components are present in the spectrum) on a graph similar to Figures 1-13 and 1-14. The highest level on the curves determines the rating of that noise. For example, the rating of the dashed line shown in Figure 1-14 is RC-32, determined by the sound pressure level in the 250 Hz octave band intersecting the RC-32 contour (interpolated between the RC-30 and 35

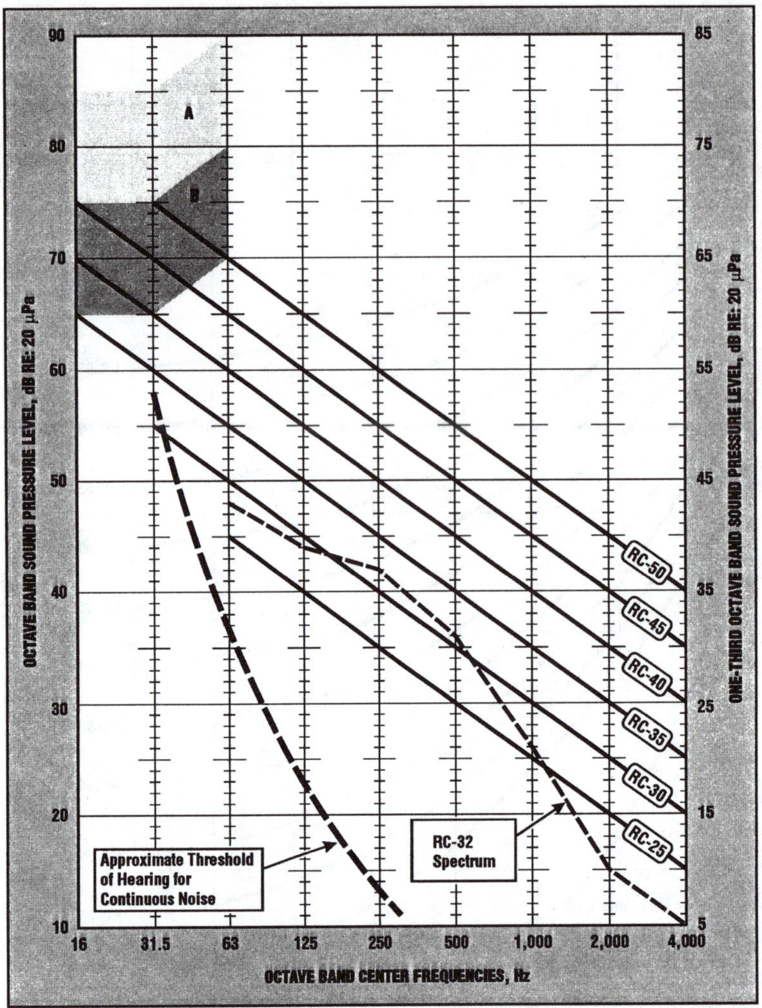


**Figure 1-14.** RC curves for design level in terms of balanced spectrum shape. Reprinted by permission of the American Society of Heating, Refrigerating and Air-Conditioning Engineers, Atlanta, Georgia, from the 1991 *ASHRAE Handbook—Applications.*

contours). A satisfactory rating for a classroom is in the range RC-30 to 35. Recommended NC and RC levels for different use categories are listed in Appendix A.4.

While the NC and RC methods were developed primarily for rating noise from mechanical systems (see Chapter 7), they are frequently used to classify other noise sources as well.

## 1.8 Effects of Noise

In Section 1.1.2 two definitions of noise were given, followed by a description of the physical characteristics of two types of noise. The second meaning of the term defines noise as unwanted sound because of the adverse impacts that noise can have on human health and activities.

The more important and best defined effects of noise on people relate to interference with speech, sleep, study or other activities, and, at higher levels, permanent loss of hearing. A consequence of these interfering effects is annoyance. Other psychological and physiological effects are often attributed to noise, but the causal relationship is less clear. The aforementioned effects are so pervasive, however, that further discussion is warranted. The attributes of speech and sleep interference, and annoyance, are presented in the following sections of this chapter, and throughout the handbook. Damage to hearing is presented in Chapter 9.

There are some applications where the presence of noise is beneficial. Low level background noise is frequently used in open plan offices for masking intruding conversations and business machine noise. See Section 5.3.2 in Chapter 5. Background noise also is useful in masking confidential conversations transmitted from one room to another. Typical applications would be counseling

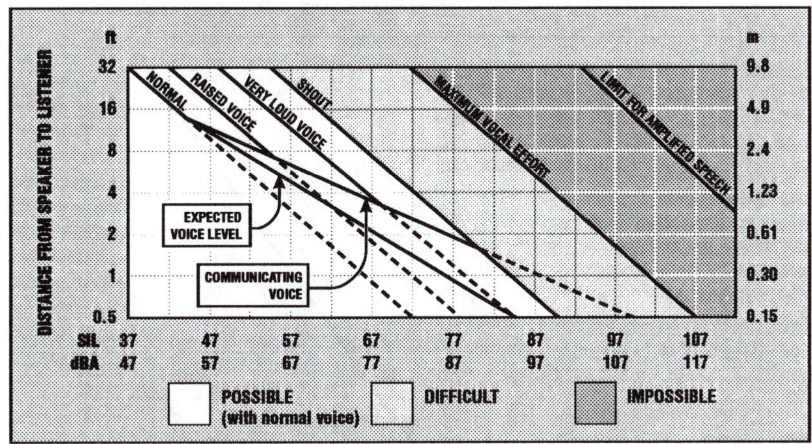


**Figure 1-15.** Rating chart for determining speech communication capability from speech interference levels (SIL) and A-weighted levels (dBA).

or medical exam rooms adjacent to reception or waiting areas.

### 1.8.1 Speech Interference

One of the most important negative attributes of noise is interference with speech communication. Because noise can be so devastating of the ability to understand speech, several rating methods have been developed specifically for this purpose. In architectural acoustics these descriptors enable one to determine the degree of acoustical privacy at work stations, in classrooms or any other location where speech communication is important.

Speech intelligibility is conveyed mainly by the mid and high frequencies, the consonant sounds that are formed in the mouth by shaping the tongue and lips. The lower frequency vowel sounds are formed in the throat and chest cavity, and while not important for understanding speech, do make the speech more pleasant to listen to, and help to identify the speaker. It is probably more than coincidental that speech intelligibility is a function of the frequencies where the ear is most sensitive.

### 1.8.2 Speech Interference Level—Articulation Index—RASTI

Because speech intelligibility is conveyed primarily by the mid and high frequencies, the frequency components of a noise which interferes with, or *masks* the speech, are in the same frequency range. An appropriate noise descriptor for rating speech interference, therefore, will emphasize the mid and high frequencies.

Understanding speech is essentially a signal-to-noise problem. The level of speech is determined by the voice level (whisper, normal, raised or shouting) and the distance separating speaker and listener. Intelligibility is a function of comparing the speech level at the listener to the background noise, or masking level.

The Speech Interference Level (SIL) is a well established measure for determining the speech interfering effects of masking noise. SIL is defined as the arithmetic average of the sound levels in the three octave bands with center frequencies of 500, 1000 and 2000 Hz (occasionally the 4000 Hz octave band is also included). Figure 1-15 shows the relationship between the SIL and the ability to understand speech at various voice levels and speaker to listener distances. If a space is very large, or so noisy that a raised or shouting voice level is required and the background noise cannot be reduced, it may be necessary to utilize a sound reinforcement system.

Another more complicated method of determining the intelligibility of speech in the presence of noise is the Articulation Index (AI). This method compares the level of speech with the masking noise level in one-third octave frequency bands from 200 to 5000 Hz, each band being assigned a weighting to give it equal importance to speech intelligibility. This weighting is applied to the *difference* in each band between the speech and masking levels, and the AI is the sum of these weighted level differences. The AI may have values from zero to 1.0. A value of zero indicates that speech is totally unintelligible, and a value of 1.0 indicates that virtually all words spoken are understood, even if the words are read from lists and are unrelated. An AI of 0.3 is required for about 80% of sentences unfamiliar to a listener to be understood. AI values below 0.3 indicate an unsatisfactory condition, while higher values are usually no cause for concern. More information on AI can be obtained from ASTM (American Society for Testing and Materials) Designation: E 1110, "Standard Classification for Determination of Articulation Class."

The understanding of speech within rooms is affected by reverberation. Long reverberation times mask speech syllables, causing interference in a manner similar to the effect of high background noise levels. In halls to be used

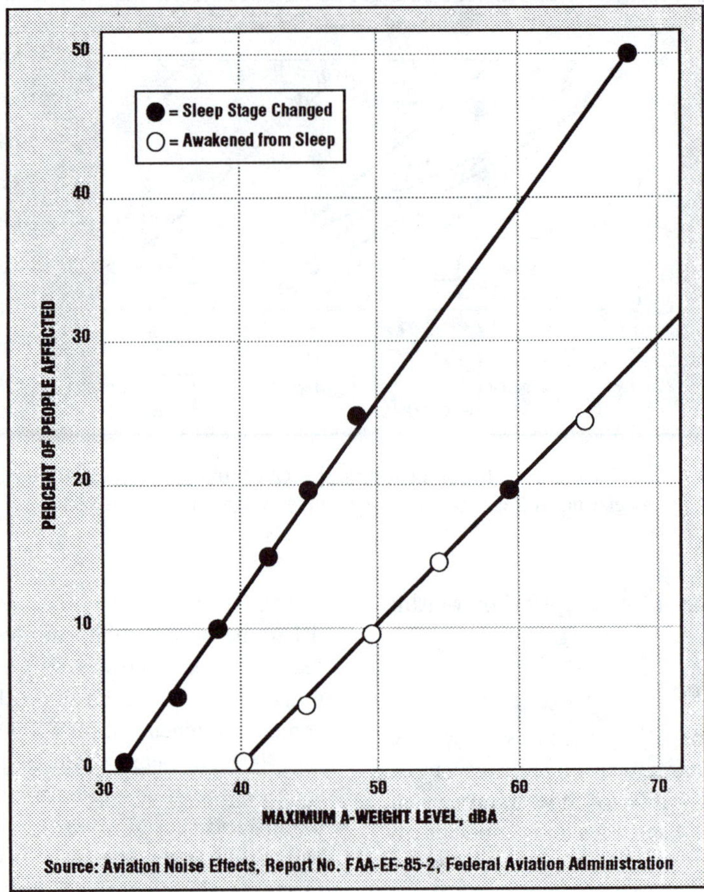


**Figure 1-16.** Relationship between noise level and sleep interference.

for both speech and music, where some reverberation is desirable to enhance the quality of the music, speech intelligibility can be improved by providing strong early reflections, passively by reflectors, or actively by electronic sound reinforcement. These subjects are thoroughly discussed in Chapters 5 and 6.

A more recently developed method for rating speech intelligibility is called RASTI, or Rapid Speech Transmission Index. This method involves determination in several frequency bands of a modulation transfer function between two locations in an enclosed space. The transfer function is dependent on the acoustical characteristics of the space, such as reverberation, the pattern of reflections and background noise.

Probably the simplest procedure for evaluating the speech masking effects of noise, yet still quite reliable for many types of noise, is the A-weighted sound level (see Section 1.6). This involves a single calculation or measurement with no need for a frequency analysis. Figure 1-15 gives A-weighted levels as well as SIL values for evaluating speech intelligibility at various voice levels and speaker to listener distances. This is a very useful figure for identifying a potential voice communication problem.

### 1.8.3 Sleep Interference

The relationship between noise and sleep interference is complicated by the fact that there are various stages of sleep (ranging from light to deep), and the level of noise which causes awakening is different for each stage. There is also a process of "accommodation," wherein some people are able to adjust to a noisy environment with continuing exposure. Other people, however, tend to become more sensitive with continuing exposure.

Since awakening caused by noise is usually the result of a single noisy event, most studies of sleep interference utilize a single event noise descriptor such as the maximum level or the single event level (see Chapter 8, Section 8.4.2). Figure 1-16 shows the relationship between the maximum A-weighted noise level indoors of an event and a change in the stage of sleep, and awakened from sleep.

### 1.8.4 Annoyance

Because noise can interfere with human activities, a probable reaction to such interference is annoyance. Unlike the effects of noise discussed in the preceding sections, an individual annoyance reaction is not something which can be reliably predicted from a physical measurement of the

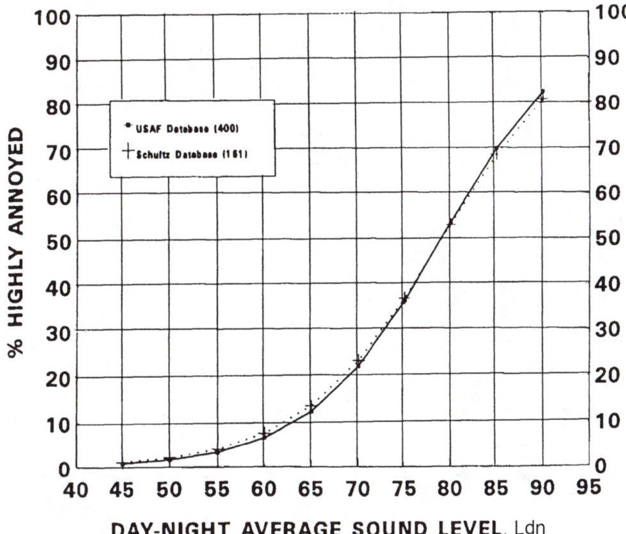


This curve was derived from numerous community annoyance social surveys assessing the impacts of general transportation noise on people.

**Figure 1-17.** Percent of residential community highly annoyed by noise exposure measured in Ldn.

offending noise. Each of us places different values on the same sound. The noise from a departing jet may be very annoying to a mother trying to get her baby to sleep, but reassuring to a pilot leaving for work. The opera may be music to the ears of some, but an intolerable cacophony to others.

While recognizing the inability to accurately predict *individual* annoyance reactions to noise, there is a clear trend of increased *community* annoyance to increasing noise exposure. Figure 1-17 shows the percent of highly annoyed people as a function of the day-night average sound level (Ldn). (See Chapter 8, Section 8.2.1). Furthermore, it is probably safe to assume that a community with a significant number of highly annoyed people is not a desirable or healthy condition, and the possibility of other adverse health effects, even if not well correlated, cannot be dismissed.

In an attempt to more accurately predict the annoying aspects of a noise, a method was developed which closely follows the method for determining loudness and loudness level, except the subjective attribute was defined as "noisiness" instead of loudness. The unit of noisiness is called the "noy," equivalent to the sone for loudness, and the unit for expressing the total noisiness is called perceived noise level (PNL), equivalent to the loudness level. PNL is expressed in perceived noise level decibels (PNdB), equivalent to phons for loudness level. At the present time, the use of PNL is mostly limited to use by the FAA for jet aircraft noise certification purposes (see Section 8.4 in Chapter 8).

### 1.8.5 Sound Quality

A recent development in the perception of different sounds is that of Sound Quality (SQ). The sounds produced by many consumer products are often a significant factor in the consumer's perception of the quality or effectiveness of the product. While this Section of the handbook appears under the NOISE heading, SQ covers desirable sounds as well as undesirable sounds, the latter previously defined as noise.

An example of SQ is the sound of a closing car door. A solid thump is perceived as strength and durability, while a tinny sound is considered cheap and poorly constructed. Door chimes can be both effective and pleasant in announcing the arrival of a guest, but alternatively can be unduly startling and obnoxious if harsh or too loud. A vacuum cleaner that is too quiet may be perceived as lacking suction and cleaning power.

Research on SQ is currently focusing on such issues as setting goals and criteria for product sound, achieving the right sound by design, manufacturing quality assurance for sound, and the economics of good sound.

## *1.9 Acoustical Measurements*

Whenever possible it is useful to make acoustical measurements in existing buildings or communities with problems, to better define the nature of the problems and to assist in designing solutions. Such measurements could include sound isolation between spaces, reverberation time, background noise, mechanical system noise and vibration, and performance of sound systems. Also, measurements are often made of environmental noise sources, such as adjacent to freeways and around airports, to develop base line conditions for environmental impact statements, or to assist in the design of exterior construction or barriers for buildings located in noisy areas. Such measurements and other types are discussed in Appendix A.10.

# Chapter 2

# Acoustical Materials—Sound Absorption

All building materials are, in a sense, acoustical materials, in that they all have acoustical properties. Common building materials such as wood paneling, gypsum board, carpet and glass all absorb, reflect or transmit sound in different ways. Even the presence of furniture and people must be considered to accurately predict sound behavior in rooms.

In the customary sense of the term, however, acoustical materials are those materials used specifically for sound absorption. The various types of acoustical materials and other building materials are described in following sections of this Chapter, including their sound absorption (and transmission) characteristics, and methods of use.

## 2.1 Absorption—Sabins—NRC

Sound absorption is defined as all the sound that strikes a surface that is not otherwise reflected back toward the source. This includes the sound actually absorbed within the material as well as that transmitted *through* it. An open window is an excellent sound absorber because no sound is reflected back. However, the open window is also a very poor sound barrier; it readily transmits sound from one space to another. The separate functions of *absorbers* and *barriers* in architectural acoustics are often confused, resulting in frequent misapplication of acoustical materials. Chapter 3, Building Noise Control—Sound Transmission, includes a comprehensive discussion of the sound insulating properties of building materials.

The sound absorption of a material is expressed as a coefficient of absorption, ranging between zero (a perfect reflector) and 1.0 (a perfect absorber). It is symbolized by the Greek letter alpha ($\alpha$). The absorption of most materials varies with frequency. Coefficients are measured in laboratories at standard frequencies of 125, 250, 500, 1000, 2000 and 4000 Hertz. The test method is specified in ASTM Designation: C 423, "Sound Absorption and Sound Absorption Coefficients by the Reverberation Room Method."

The Noise Reduction Coefficient (NRC) is frequently used by architects for specification purposes. The NRC is the arithmetic average of the coefficients at the 4 middle test frequencies, reported to the nearest 0.05. For example, assume the coefficients for a ceiling lay-in panel have the following values:

| FREQUENCY, Hz | 125 | 250 | 500 | 1000 | 2000 | 4000 |
|---|---|---|---|---|---|---|
| COEFFICIENT, $\alpha$ | 0.22 | **0.45** | **0.67** | **0.74** | **0.82** | 0.69 |

The NRC is calculated by

$$\text{NRC} = \frac{(0.45 + 0.67 + 0.74 + 0.82)}{4}$$

$$= \frac{2.68}{4}$$

$$= 0.67, \text{ or } 0.65 \text{ reported to the nearest } 0.05.$$

As this handbook is in preparation, a revised draft of this test method is under consideration. The draft provides for the NRC being replaced by the Average Absorption Coefficient (AAC), and is defined as follows: "A single number rating, the average, rounded off to the nearest multiple of 0.01, of the sound absorption coefficients of a material, measured according to this Test Method and rounded off to the nearest 0.01, for the twelve one-third octave bands from 200 through 2500 Hz, inclusive, with the provision that the exact midpoints be reported as the next higher multiple of 0.01. For example, 0.625 would be reported as 0.63."

(Authors' note: Since most acoustical laboratories are now measuring sound absorption coefficients in one-third octave bands, it is deemed appropriate to incorporate the added data into the single number rating. Because the recommendation is in draft form, however, this handbook will present sound absorption coefficients at octave band frequencies only, and will report the single number rating as the NRC.)

The total absorption of a surface may be expressed in units of "perfect" absorption (having a coefficient of 1.0) called a *sabin*. One U.S. sabin is the equivalent of 1.0 sq ft with a coefficient of 1.0; the equivalent metric sabin would be 1.0 sq m. The total absorption in sabins for a surface is the product of the coefficient times the area. For example, if the area of the ceiling for the lay-in panel coefficients given above is 194 sq ft (18.0 sq m), the total absorption at a frequency of 500 Hz. is 194 × 0.67 = 130 sabins (18 × 0.67 = 12 metric sabins). The total absorption in a room at a given frequency may be obtained by adding the sabins of all absorbing elements in the room. This procedure is necessary to calculate reverberation times (see Chapter 4, Section 4.6.3).

The NRC may properly be used as a specification for general noise control applications, such as in corridors, shops or other noisy or noncritical locations. However, the NRC alone is inadequate where a space is to be used for critical speech communication, music rehearsal or performance. The design of such spaces requires reverberation time analyses at all six standard frequencies, and acoustical material specifications should define an acceptable range of sound absorption coefficients for each frequency. Where there is a noise problem, noise should be analyzed in octave bands and acoustical treatment prescribed to match the noise content in the different frequency bands.

It is not unusual to see sound absorption data with coefficients greater than 1.0. This seeming paradox (How can a material absorb more sound than strikes it?) is caused by sound diffraction that takes place at the boundaries of the finite-sized test sample (see Chapter 1, Section 1.4.2). The effect can be quite substantial with highly absorbent materials, and permits more efficient use of acoustical materials through ingenious methods of installation. For example, an acoustical material can be placed in patches which then takes advantage of the higher absorption values. Otherwise, areas solidly covered with acoustical materials will have coefficients less than 1.0 because the "edge effect" will not be present.

The ASTM test method referenced above requires that the acoustical material be tested in a reverberation chamber. Within the chamber the sound undergoes many reflections, creating what is called a diffuse sound field. The sound absorption coefficients measured under these conditions are "random incidence," meaning the sound strikes the surface of the material from all directions. Test samples are typically 48 to 72 sq ft (4.5 to 6.7 sq m), hence the edge effect is present and can result in measured coefficients greater than 1.0.

Another test method, using an acoustic tube, measures sound absorption at normal (90°) incidence, as specified in ASTM Designation: C 384, "Impedance and Absorption of Acoustical Materials by the Impedance Tube Method." The acoustic tube method is used mostly for product development, though normal incidence coefficients may occasionally be encountered in product literature. Random incidence coefficients are usually slightly higher.

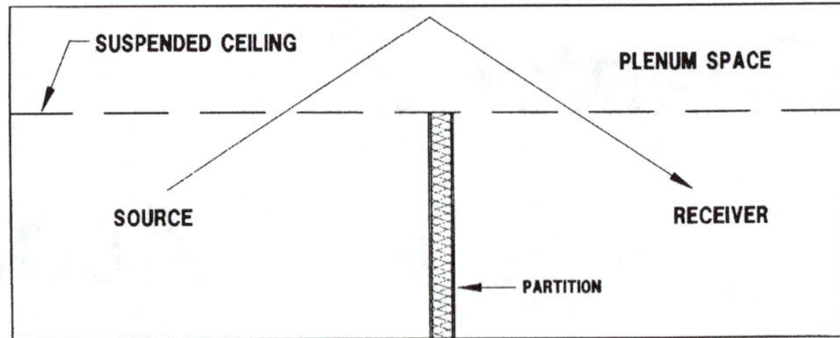


The suspended ceiling may be continuous over the partition, or the partition may protrude several inches into the plenum space.

**Figure 2-1.** Ceiling sound transmision path between rooms.

In some applications, it is useful to know sound absorption at specific angles of incidence. An example would be where sound undergoes a single reflection between a defined source and receiver locations, as between two adjacent work stations in an open plan office. For this application there are several more ASTM test methods under the classification of "interzone attenuation," which are discussed in Chapter 5, Section 5.3.2.

## 2.2 Suspended Ceiling Sound Attenuation between Rooms

Suspended acoustical ceilings are often rated for airborne sound attenuation between rooms. The airborne sound attenuation is a measure of the sound reduction between two adjacent rooms with identical suspended ceilings. The partition between the rooms either extends to the plane of the ceiling or protrudes several inches through it, but not all the way to the deck above. The two rooms, therefore, share a common space above the ceiling, and the sound path makes two passes through the suspended ceiling, as illustrated in Figure 2-1.

The test method for measuring the sound attenuation between rooms is specified in ASTM Designation: E 1414, "Airborne Sound Attenuation Between Rooms Sharing a Common Ceiling Plenum," which closely follows a test method developed many years ago by the Acoustical Materials Association, designated as AMA-I-II. The current ASTM test method requires that the sound attenuation (or *ceiling* attenuation as specified in the standard) be measured in contiguous one-third octave bands with center frequencies from 125 to 4000 Hz. Representative ceiling attenuation data for five types of suspended acoustical ceilings are shown in Figures 2-2 and 2-3.

The single number rating for specifying ceiling attenuation data is called the Ceiling Attenuation Class (CAC), and is determined in accordance with ASTM Designation: E 413, "Rating Sound Insulation." The test method and single number rating are analogous to the more familiar airborne sound transmission loss test method and Sound Transmission Class (STC) rating for walls and partitions (see Chapter 3, Section 3.2.1). Indeed, CAC was formerly

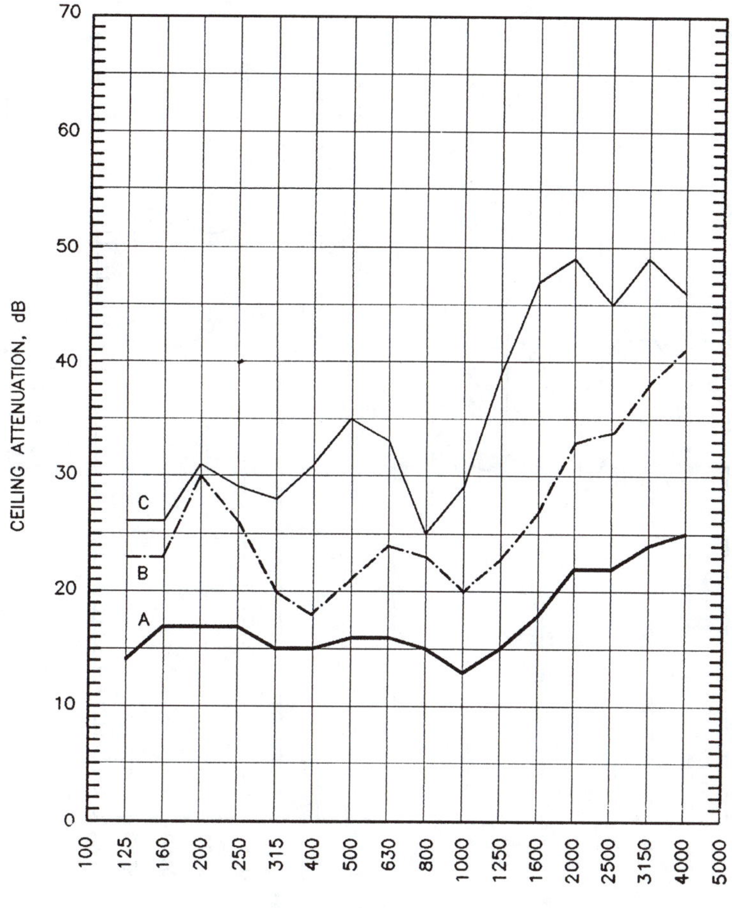


1/3 OCTAVE BAND CENTER FREQUENCY, Hz

CEILING ATTENUATION, dB

| CURVE | 100 | 125 | 160 | 200 | 250 | 315 | 400 | 500 | 630 | 800 | 1000 | 1250 | 1600 | 2000 | 2500 | 3150 | 4000 | 5000 | CAC RATING |
|---|---|---|---|---|---|---|---|---|---|---|---|---|---|---|---|---|---|---|---|
| A | -- | 14 | 17 | 17 | 17 | 15 | 15 | 16 | 16 | 15 | 13 | 15 | 18 | 22 | 22 | 24 | 25 | -- | 18 |
| B | -- | 23 | 23 | 30 | 26 | 20 | 18 | 21 | 24 | 23 | 20 | 23 | 27 | 33 | 34 | 38 | 41 | -- | 24 |
| C | -- | 26 | 26 | 31 | 29 | 28 | 31 | 35 | 33 | 25 | 29 | 39 | 47 | 49 | 45 | 49 | 46 | -- | 31 |

———————— A) 1″ (25 mm) Nubby, nonfoil-backed, CAC 18
—·—·—·—·· B) 1″ (25 mm) Nubby, foil-backed, CAC 24
———————— C) 1 1/2″ (38 mm) Nubby, foil-backed, CAC 31

Both panel thickness and foil-backing contribute to higher ceiling attenuation values.

**Figure 2-2.** Ceiling attenuation, glass fiber lay-in ceiling panels, mounting E-400, ceiling continuous at top of partition.

expressed as an STC value, and will so appear in older texts and literature from manufacturers.

In some applications it is desirable to know the sound attenuation provided by a *single* pass of sound through a suspended ceiling. An example would be where noisy mechanical equipment, such as fan-powered air mixing boxes, are located above a suspended ceiling in a noise-sensitive area. Single pass noise reduction data on several types of suspended ceilings are presented in Chapter 7, Section 7.3.3, Table 7-2.

## 2.3 Types of Absorbers

There are four general types of acoustical absorbers, each different in the way that it absorbs sound, and each behaving differently as a function of frequency. The four types are as follows:

1. **Porous absorbers**, such as acoustical tile, carpets, blankets, etc., which absorb acoustical energy by resisting the movement of air molecules and converting the energy to heat as the acoustic wave penetrates the porous

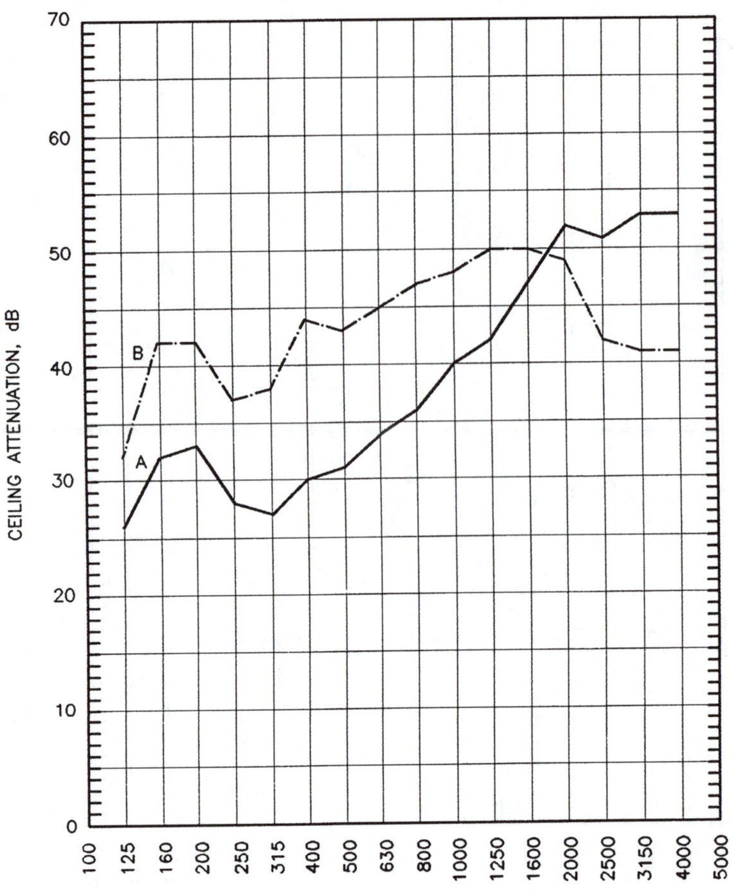


CEILING ATTENUATION, dB

| CURVE | 100 | 125 | 160 | 200 | 250 | 315 | 400 | 500 | 630 | 800 | 1000 | 1250 | 1600 | 2000 | 2500 | 3150 | 4000 | 5000 | CAC RATING |
|---|---|---|---|---|---|---|---|---|---|---|---|---|---|---|---|---|---|---|---|
| A | -- | 26 | 32 | 33 | 28 | 27 | 30 | 31 | 34 | 36 | 40 | 42 | 47 | 52 | 51 | 53 | 53 | -- | 37 |
| B | -- | 32 | 42 | 42 | 37 | 38 | 44 | 43 | 45 | 47 | 48 | 50 | 50 | 49 | 42 | 41 | 41 | -- | 45 |

———————— A) 5/8″ (16 mm) mineral fiber panel, CAC 37
—·—·—·—·· B) 1/2″ (13 mm) gypsum board, CAC 45

The decline in ceiling attenuation values for the gypsum board at the high frequencies is probably due to sound leakage between the panel and the T-grid.

**Figure 2-3.** Ceiling attenuation for lay-in suspended ceilings, mounting E-400, ceiling continuous at top of partition.

structure of the material. The absorption is affected by the porosity of the material, its thickness and how it is mounted. Many glass fiber insulation boards and blankets have excellent broad-band (all frequency) sound absorption characteristics.

2. **Panel absorbers**, such as thin plywood, hardboard or plastic with an airspace behind it. The acoustic wave pushes the panels backward and forward, and this flexural motion converts sound energy into heat. The absorption of panels occurs mostly at the low frequencies.
3. **Resonant absorbers**, which consist of a small opening into a larger hollow cavity. The resistance to air moving back and forth in the opening converts sound energy into heat. This type of absorber is commonly called a Helmholtz Resonator, and typically has maximum absorption at a single resonant frequency. Any material with perforations or slits leading into a larger space fall into this category.
4. **Active sound cancellation**, which technically, is not true sound absorption, but a method of canceling the sound

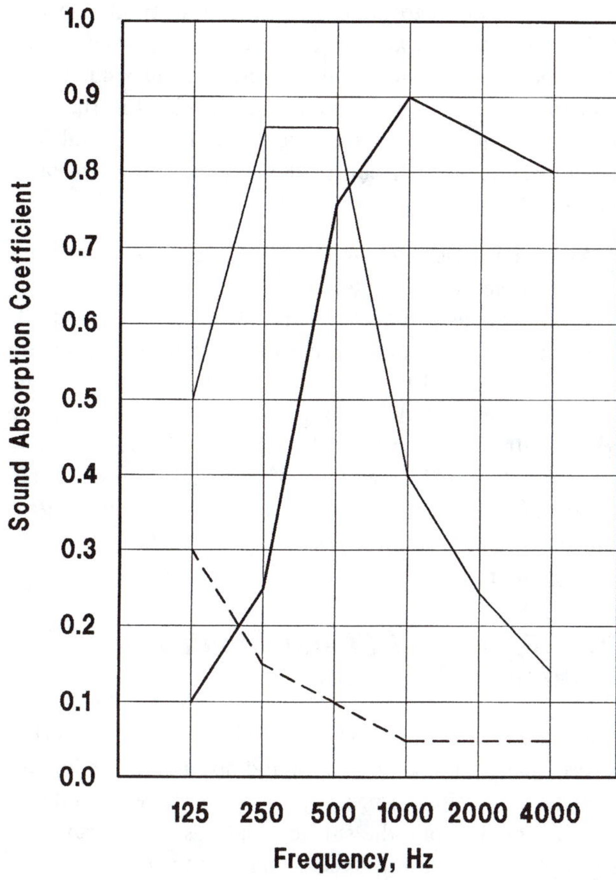


Porous—fissured mineral fiber tile, Type B MTG NRC 0.70

Panel—1/4″ (6.4 mm) plywood, airspace behind NRC 10

Resonant—pegboard over 2″ (51 mm) glass fiber blanket NRC 0.60

Each basic type has very different absorption characteristics.

**Figure 2-4.** Typical sound absorption characteristics of basic types of acoustical materials.

by a process called "destructive interference." One application of this method is in air ducts with a sampling microphone followed by a loudspeaker which is fed by the sampled noise signal but 180° out of phase, which cancels the sound in the duct. See Chapter 7, Section 7.4.4.

The absorption characteristics as a function of frequency for absorber types 1-3 are shown in Figure 2-4. All building materials have some acoustical absorption, but the mechanism of absorption is always one of the four types described above.

### 2.3.1 Unit Absorbers

Not all sound absorption takes place at the surfaces of a room. Objects within the room can also have significant absorption, such as furniture and people. There are several products available, sometimes called "space absorbers," that are intended to be suspended within a room to provide additional absorption (see Section 2.5.3).

Because it is sometimes difficult to determine the area of a unit absorber (such as a sofa), the sound absorption is usually given as sabins per unit (in either U.S. or metric units) rather than as coefficients. Unit absorbers are often very efficient absorbers, because the sound strikes them from behind as well as from the front.

Another advantage of unit absorbers is that they can be added to a room and not risk moisture condensation problems. This problem can occur when an acoustical material is added to the inside surface of an exterior wall. The material acts as a heat insulator and may lower the temperature of the wall surface to a temperature below the dew point, and moisture will condense. The problem cannot be solved with a vapor barrier over the acoustical material, as this will cause a loss of absorption, particularly at the higher frequencies.

*Caution*: In a composite construction, the vapor barrier must *always* be on the warm side of the dew point. If more material is added to the warm side of the construction, the dew point may shift to the warm side of the barrier, and condensation can occur at the barrier. The solution is to add more insulation to the outside of the construction, so that the temperature at the barrier is once again above the dew point. Alternatively, install the acoustical material in panels which are spaced out from the wall surface (using vertical furring for walls), thus providing air movement behind the material and raising the wall or ceiling temperature.

Manufacturers (see Appendix A.12): 1, 7, 10, 14, 19, 23, 25, 33, 40, 42, 47, 49, 58, 65, 68 & 74.

## 2.4 Mountings—ASTM (American Society for Testing and Materials)

How acoustical materials are installed can have an effect on their sound absorption at different frequencies. Typically, increasing the airspace behind a material will improve the absorption at the lower frequencies. The improvement in low frequency absorption, compared to a material with no back airspace, is shown in Figure 2-5. Placing a fibrous blanket in the air space in back of the material will further increase the absorption. For panel-type absorbers, it is essential to provide an airspace in back of the panel, with or without glass fiber insulation.

The more common mounting systems are shown in Figure 2-6, and are specified in ASTM Designation: E 795, "Mounting Test Specimens During Sound Absorption Tests". The nomenclature was changed sometime back, so

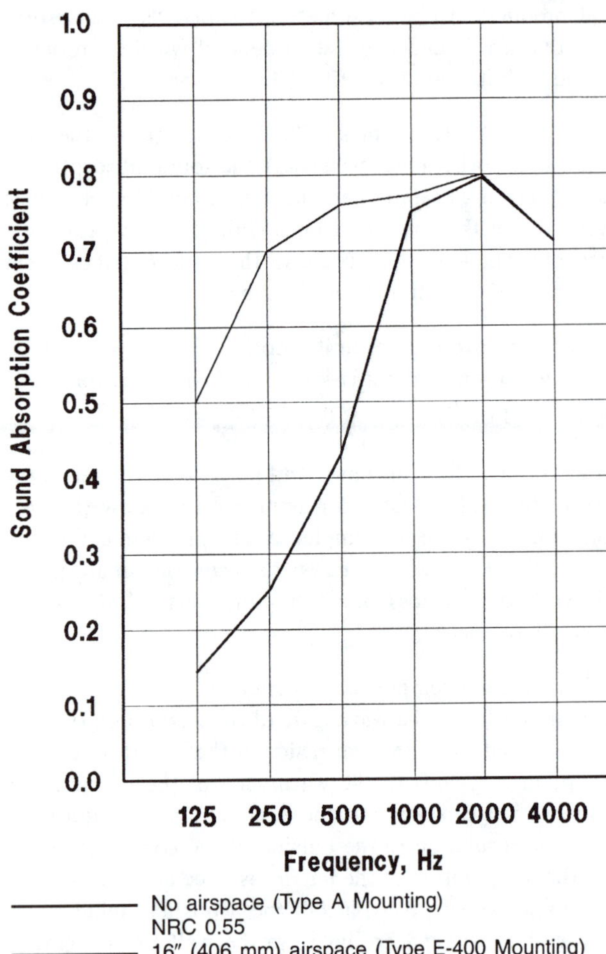


**Figure 2-5.** Effect of back air space on sound absorption.

both old and new designations are shown. Current specifications should use only the newer designations.

By far the most common method of installing acoustical ceilings is with a suspended grid system, which may be exposed, recessed or concealed. Many lighting fixtures and air diffusers are manufactured to fit into these grid dimensions. Easy accessibility to the space above is accommodated by this type of ceiling, particularly for lay-in panels in an exposed T-grid. Typical grid dimensions are 2 by 2 ft (0.61 m by 0.61 m) or 2 by 4 ft (0.61 by 1.22 m). Grids with dimensions greater than 4 ft. (1.22 m) are available, but must be used with some caution because of possible sagging of the material installed in the grid.

Suspended acoustical ceilings may be selected not only for their sound absorption and ceiling attenuation, but also for their fire rating, flame spread rating and light reflectance (see Sections 2.6 and 2.7). There are many ASTM standards relating to nonacoustical characteristics of material suspension systems, as listed in Appendix A.9. Additionally, structural requirements for suspended ceilings, including seismic restraints, are specified in the Uniform Building Code (UBC) Standard: 47:18, "Metal Suspension Systems for Acoustical Tile and for Lay-in Panel Ceilings."

As explained earlier in Section 2.1, the sound absorption of an acoustical material is enhanced when it is surrounded by a hard, reflective surface because of sound diffraction at the boundary. The total absorption of a square patch of acoustical tile, 20 ft (6.1 m) on each side, can be increased substantially by separating it into four square patches, 10 ft (3.05 m) on each side, with 1 ft (0.30 m) or so spacing between. This method of installing acoustical materials has the added advantage of increasing sound diffusion, a useful attribute for many spaces such as music rehearsal rooms.

## 2.5 *Types and Characteristics of Acoustical Materials*

There are many different types of acoustical materials whose characteristics such as sound absorption, ceiling attenuation and fire ratings, vary with the type of material they are made from, the surface coatings and texture, the density, porosity and thickness, and how they are installed. Also, similar products will vary between manufacturers. New, specialized acoustical materials are constantly being introduced to the market, often times by new companies as well. These companies should be able to provide the same basic acoustical performance data presented in this handbook. Products change; basic acoustical principles affecting sound absorption do not.

Manufacturers of acoustical materials publish test data on these characteristics of their products. The sound absorption values (either coefficients or sabins per unit—be sure to distinguish between U.S. and metric sabins) are listed at the 6 standard test frequencies, along with the NRC, as defined in Section 2.1. Ceiling attenuation data plus the single-number CAC are presented in Section 2.2. Fire ratings include both endurance and flame spread, as discussed in Section 2.6.

Listed in the following sections are the main categories of acoustical and building materials with acoustical performance characteristics, a discussion of other nonacoustical characteristics and typical applications. For most categories, sound absorption values are shown in graphs, with representative coefficients at the 6 standard test frequencies. The intent is to show the general absorption characteristics, which may be used for most applications. Specific product data are not always given because of frequent design or manufacturing changes. For current per-

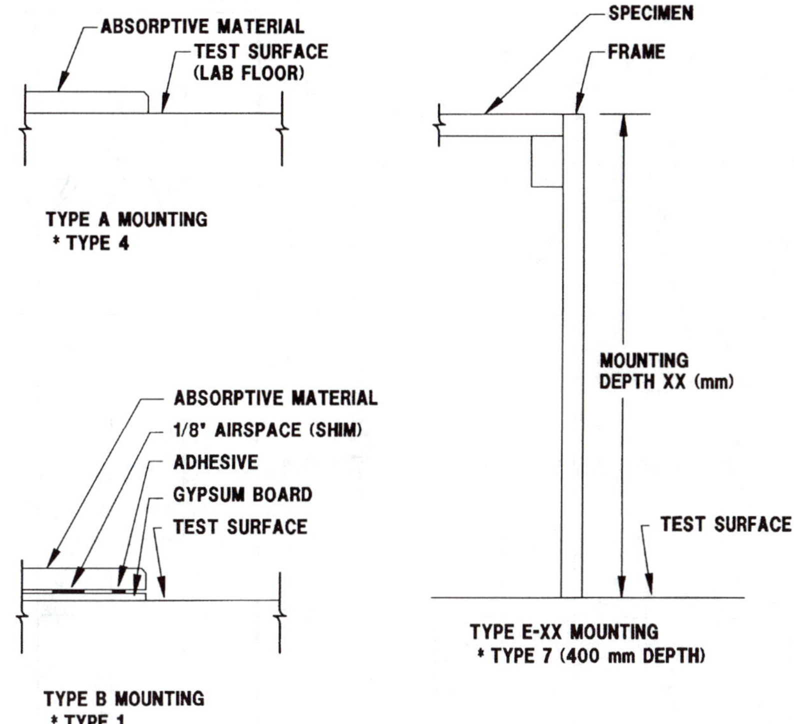


**Figure 2-6.** ASTM designated test mountings (*designates old mounting type).

formance data, consult the manufacturer's catalogs. Each material category lists a number or series of numbers identifying manufacturers of that type of material. The manufacturers and associated numbers are shown in Appendix A.12.

*Caution*: Painting acoustical ceiling materials must be done properly to avoid diminishing the sound absorbing properties. Painting should be avoided if possible, but when absolutely necessary, the following method should be followed.

If paint is applied improperly to porous acoustical materials, the surface will be sealed and much of the sound absorption will be lost. A light coating of a nonbridging, flat latex paint can be used. The paint should be spray-applied, never with a brush or roller. There is some question as to whether air-type or airless spray application should be used. One argument is that the air pressure will keep the paint particles from penetrating into the porous structure of the material. The applicator must carefully observe which method causes the least sealing of the surface. With spray-on materials (see Figure 2-12), it is preferred to respray with the same material rather than paint.

Painting or other surface treatment can be used to seal the surface of highly porous concrete masonry units to improve sound isolation. Figure 3-12 in Chapter 3 shows the reduction in transmitted sound achieved by applying 2 coats of paint to a porous block wall.

### 2.5.1 Ceiling Systems

*Mineral Fiber Lay-in Panels (Figure 2-7).* Probably the most common type of suspended acoustical ceiling, and lower in cost than most other types. Panels are available in many surface patterns, including fissured, grooved, finely perforated and many more. There are patterns that match up so the panel edges are not discernible. Some tiles are manufactured with very small pin-type perforations which may not be consistent, causing significant variations in sound absorption. It is suggested that these tiles be avoided where a definite amount of absorption is required at different frequencies.

Manufacturers: 6, 7, 10, 11, 13, 25, 33, 40, 47, 52, 65, 74 & 79.

*Glass Fiber Lay-in Panels (Figure 2-8).* Glass fiber panels are more expensive than mineral fiber, but the sound absorption is significantly higher. It is the preferred ceiling for open plan offices (see Chapter 5, Section 5.3.2), or where a highly absorptive ceiling is required.

A disadvantage of glass fiber panels is that they have poor ceiling attenuation performance. To improve this characteristic, panels may be covered on the back with a heavy foil which retards the transmission of sound through it (see Figure 2-2). Be aware, however, that such a backing may reduce the sound absorption. Where a high degree of speech privacy is required between offices with a common plenum space above the ceiling, even the backed panels

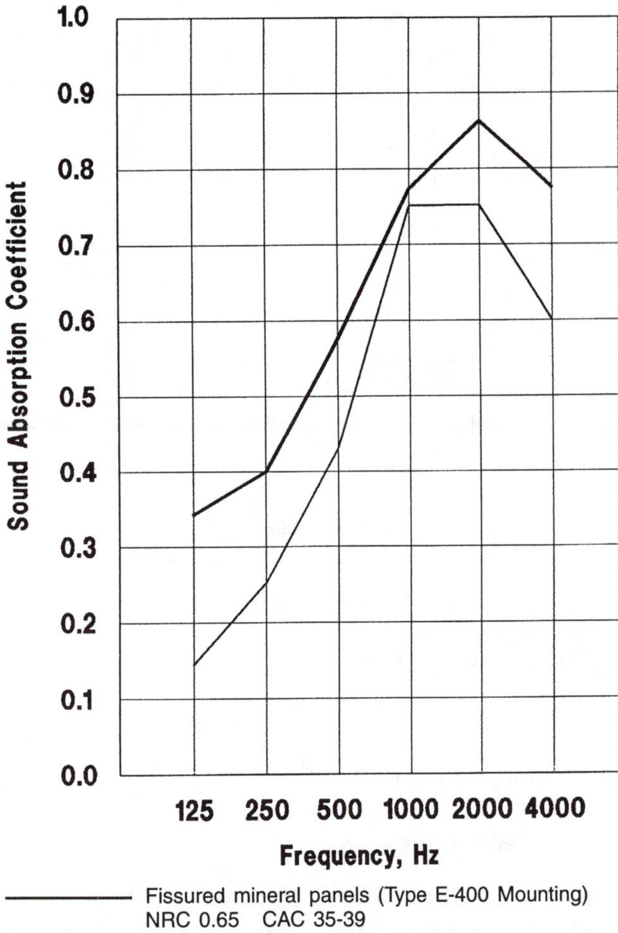


Fissured mineral panels (Type E-400 Mounting) NRC 0.65 CAC 35-39

Fire-rated mineral panels (Type E-400 Mounting) NRC 0.55 CAC 35-39

This type of material is probably the most common of suspended acoustical ceilings.

**Figure 2-7.** Mineral fiber lay-in panels.

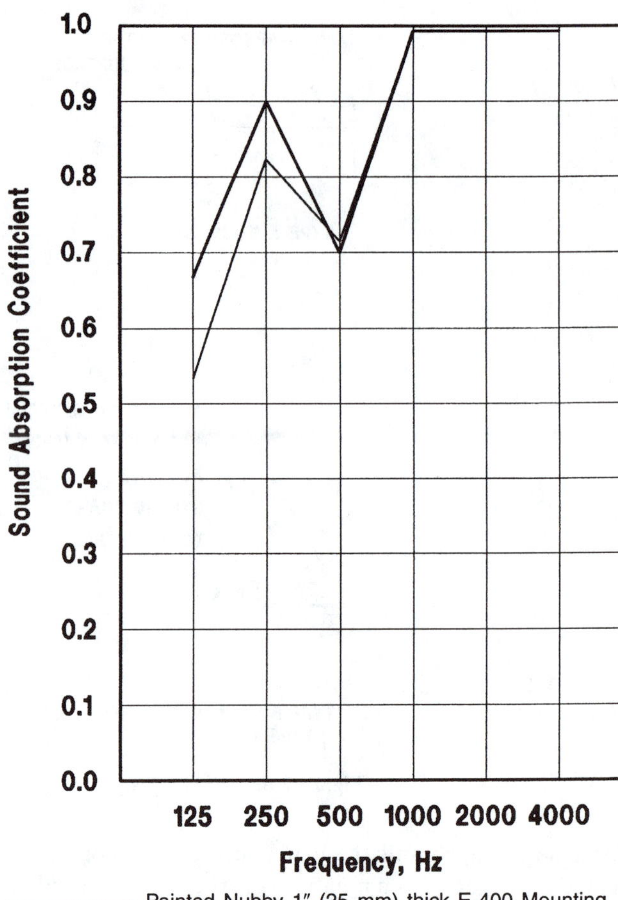


Painted Nubby 1″ (25 mm) thick E-400 Mounting NRC 0.95 CAC 15-19

Same foil-backed NRC 0.95 CAC 20-24

The foil-backed material loses a small amount of absorption at the low frequencies. Notice the characteristic dip for this material at 500 Hz, a function of the 400 mm back air space.

**Figure 2-8.** Glass fiber lay-in panels.

will not provide the required sound isolation, and dividing partitions may have to be full height.

Because of its relatively light weight, hold-down clips may be required for this type of ceiling if the space above the ceiling is used as a return air plenum. The air pressure differential may cause the panels to "rattle" in the grid system.

Manufacturers: 6, 7, 10, 13, 15, 23, 25, 33, 40, 47, 49, 52, 60, 65, 66, 68, 70, 73, 74, 79 & 83.

*Glue-on Acoustical Tile (Figure 2-9).* This type of ceiling is not nearly as popular as it once was, but is useful for application to existing surfaces, or where there is insufficient height for a suspended ceiling. Most glue-on tiles are composed of mineral fiber, are one foot square, and often are installed with splines to maintain a flat surface. A variety of surface patterns are available, similar to those for mineral fiber panels.

Manufacturers: 5, 6, 7, 10, 11, 13, 25, 49, 52, 65, 79 & 83.

*Ceramic-type Tile (Figure 2-10).* These tiles are composed of a ceramic-type material which is incombustible and very resistant to moisture and chemical fumes. They are also very durable. Applications may include ceilings for laboratories, operating rooms and swimming pools. Ceramic tiles are available both as panels for suspended ceilings and glue-on tiles.

Manufacturers: 5, 6, 13 & 79.

*Shredded Wood Fiber Panels (Figure 2-11).* This material consists of finely shredded wood fibers held together with a binder. It is a high density, tough, very durable ma-

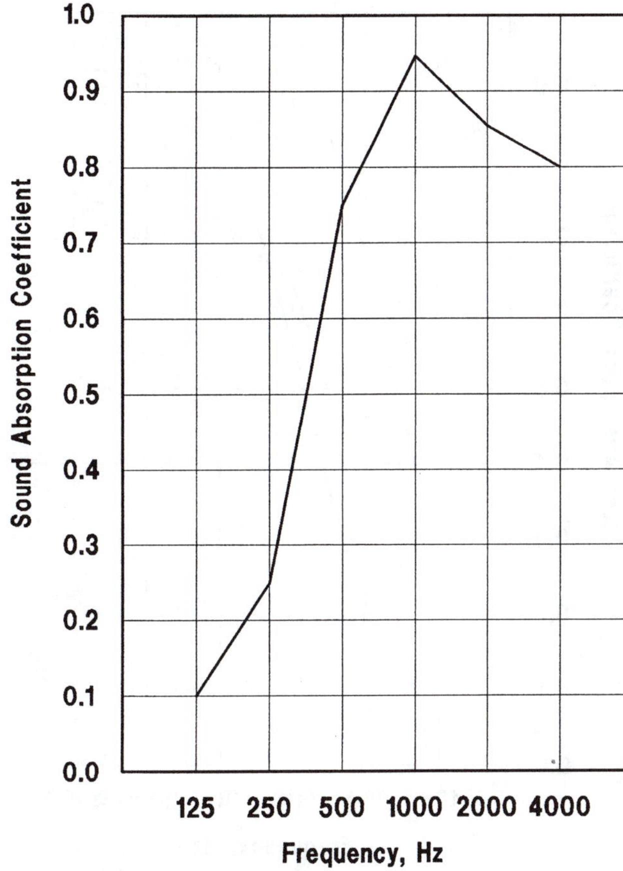


—— Fissured mineral fiber ascoustical tile (Type B Mounting)
NRC 0.70

The low frequency absorption is limited by lack of airspace behind the tile.

**Figure 2-9.** Glue-on acoustical tile.

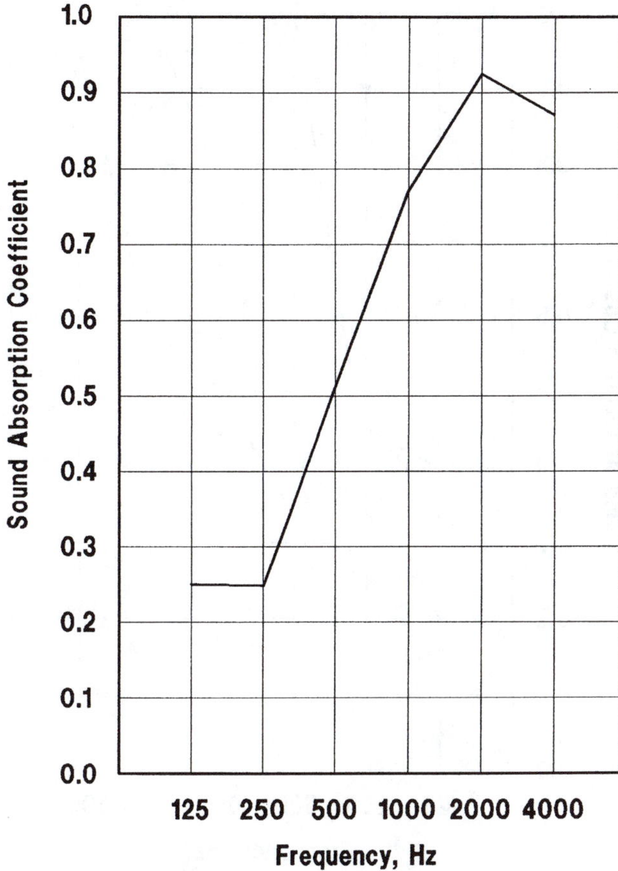


———— Ceramic panels, (Type E-400 Mounting)
NRC 0.60 CAC 40-44

The material is rated incombustible.

**Figure 2-10.** Ceramic-type acoustical tile.

terial that is suitable for areas subject to impacts and abuse. However, it is not suitable for out-of-doors applications where it would be subjected to weather, particularly moisture.

One of the few products of this type currently manufactured is marketed under the trade name of TECTUM™. It is available in panels of 1, 2 and 3 inches thick (25, 51 and 76 mm). The thicker panels may be used as a form board in roof systems, incorporating both acoustical treatment and roof deck in one material. Special versions are marketed primarily for wall application, one with a glass fiber board backing which increases sound absorption, and another with a carpet surface to enhance the appearance.

Manufacturers: 40, 75 & 79.

*Spray-on Material (Figure 2-12).* This category of acoustical material is relatively inexpensive and can be applied to existing surfaces or irregular surfaces which are curved or stepped. The surface has a textured appearance, which can be fairly rough or smooth, depending on the method of application. Thicknesses typically vary from about 3/4″ to 2″ (19 mm to 51 mm), the thicker treatments usually requiring more than one application.

Several companies market spray-on materials, ranging from low density cellulose fibers to higher density materials with gypsum or cement bases. The latter category can be very hard and durable, and suitable for out-of-doors applications. Another type of spray-on material, acoustical plaster, has had considerable use in the past but lost favor because of the inconsistency of sound absorption. There was also the problem of the applicator not putting on the specified thickness. This predicament will always exist and applications require close supervision.

*Caution*: There are a number of spray-on products that have the appearance and are marketed as an acoustical material, but consist of a thin, textured coating. Because

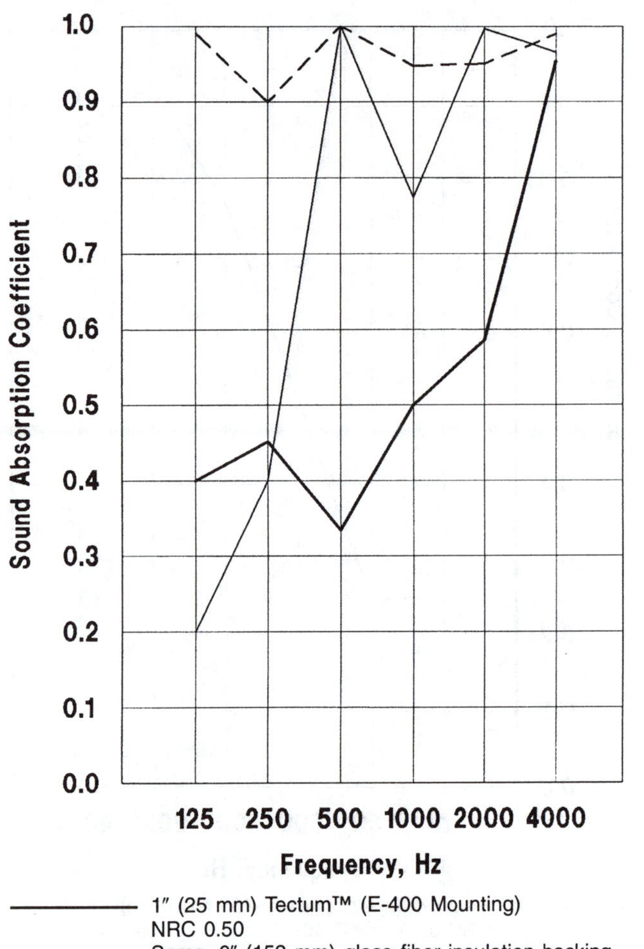


——— 1″ (25 mm) Tectum™ (E-400 Mounting) NRC 0.50

– – – – Same, 6″ (152 mm) glass fiber insulation backing (E-400 Mounting) NRC 0.95

——— 3″ (76 mm) Tectum™ roof deck (Mounting A) NRC 0.80

The absorption is greatly increased by backing the suspended panel (E-400 Mounting) with glass fiber insulation.

**Figure 2-11.** Shredded wood fiber panels.

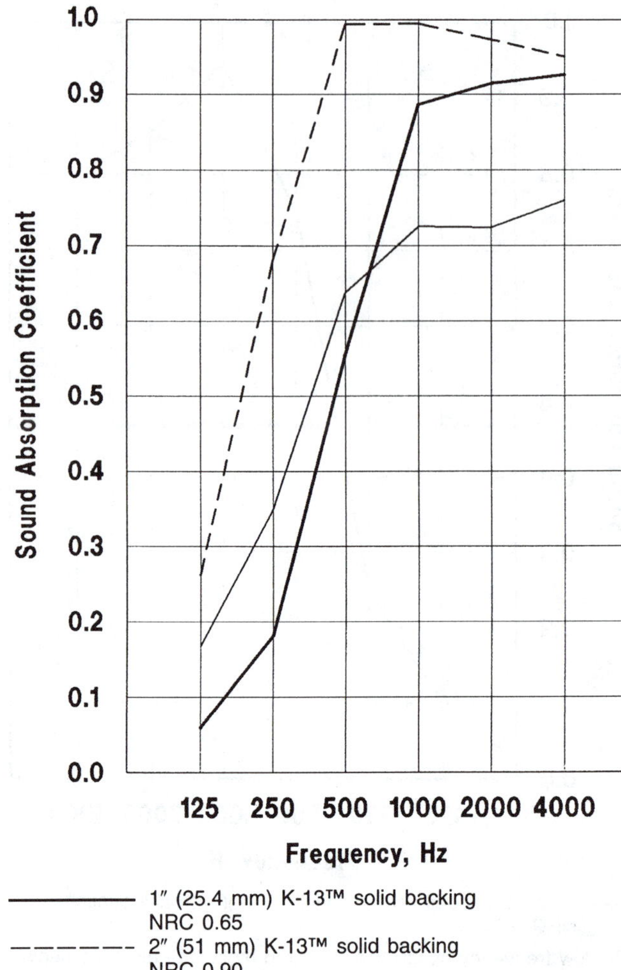


——— 1″ (25.4 mm) K-13™ solid backing NRC 0.65

– – – – 2″ (51 mm) K-13™ solid backing NRC 0.90

——— 1″ Pyrox™ solid backing NRC 0.60

Because the material is usually applied to a solid backing, the low frequency absorption is minimal.

**Figure 2-12.** Spray-on material.

these materials are so thin and often not porous, they do not add absorption to the surface on which they are applied.

Manufacturers: 26, 34, 35, 59, 68, 76 & 79.

*Perforated Metal or Hardboard (Figure 2-13).* This category of material consists of a perforated covering usually backed with glass fiber blanket or insulation board. It is very useful in applications, for both ceilings and walls, where a durable and cleanable surface is required. The sound absorption as a function of frequency can be controlled by the pattern of perforations, and by the thickness of insulation. A finely perforated material, with 20 to 30% open area and two or more inches of insulation behind will have excellent absorption across the frequency range. A surface with fewer perforations, such as common "peg board" with holes 1″ (25 mm) on centers each way will be tuned to have maximum absorption at 500 Hz., but much less absorption at 4000 Hz. This characteristic may be useful to balance the absorption in a room that would otherwise have too much high frequency absorption. (An excellent reference on this subject is "Acoustical Uses for Perforated Metal: Principles and Applications," available from the Industrial Perforators Association, 710 North Plankinton Avenue, Milwaukee, WI 53203. A summary report is also available.)

Manufacturers: 1, 4, 5, 6, 7, 8, 13, 16, 19, 22, 32, 33, 40, 42, 66, 71, 79 & 83.

*Linear Metal Panels (Figure 2-14).* The metal panels are typically 3″ to 4″ (76 mm to 102 mm) wide with about 1/2″ (13 mm) space between, and run the full length of the room in which they are installed. A 1″ or 2″ (25 mm or 51 mm) thick absorbent blanket is laid over the top. If the

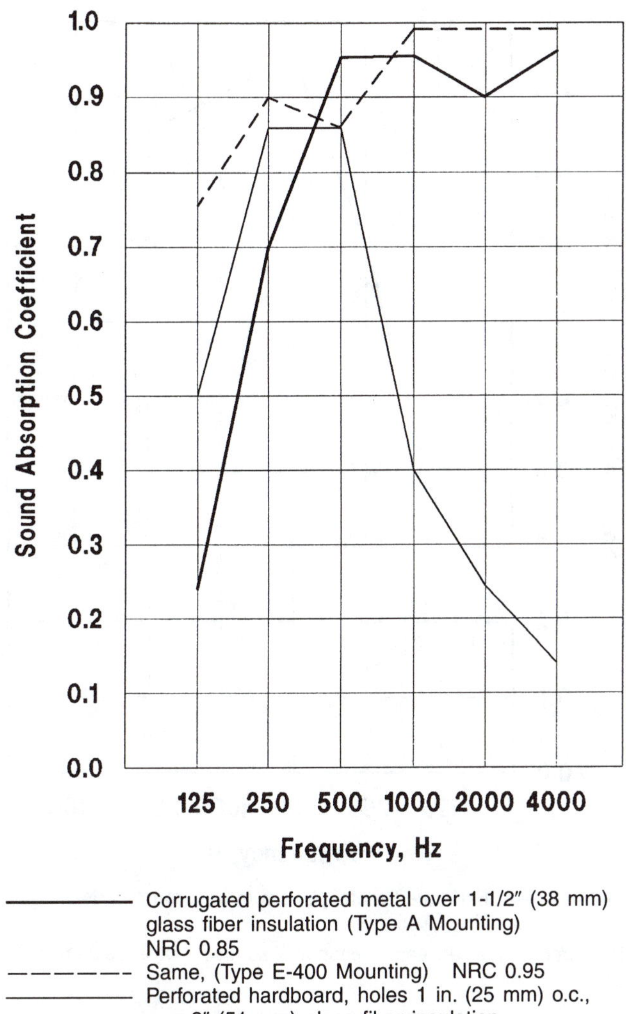


Corrugated perforated metal over 1-1/2″ (38 mm) glass fiber insulation (Type A Mounting) NRC 0.85

Same, (Type E-400 Mounting) NRC 0.95

Perforated hardboard, holes 1 in. (25 mm) o.c., over 2″ (51 mm) glass fiber insulation (Type A Mounting) NRC 0.60

The pegboard covering, with holes 1 in (25 mm) o.c., restricts high frequency absorption.

**Figure 2-13.** Perforated metal or hardboard.

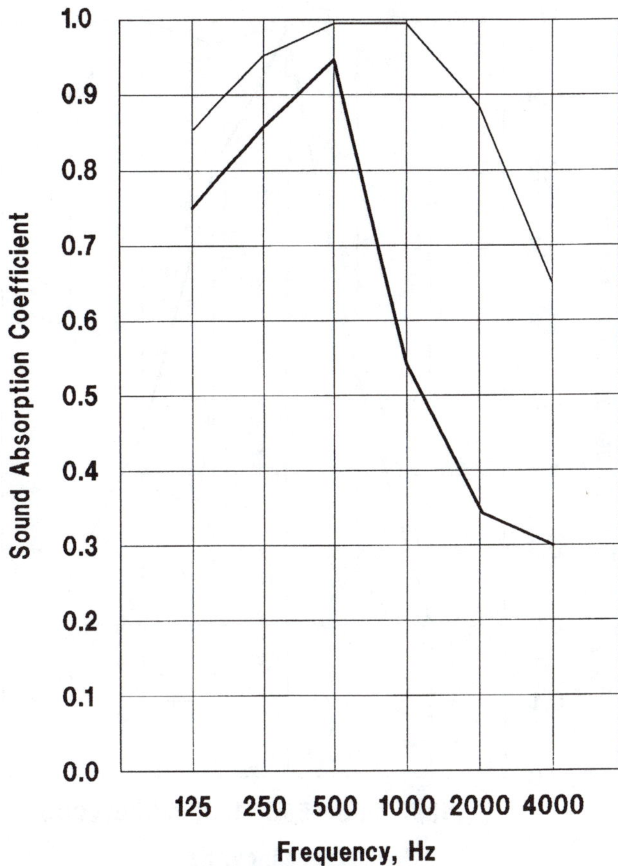


Paraline™ panels, not perforated 2″ (51 mm) glass fiber insultion (bagged) Type E-400 Mounting NRC 0.65

Same, panels perforated NRC 0.95

The 3-1/4 in (83 mm) wide nonperforated panels restrict high frequency absorption. Unbagged insulation would increase high frequency absorption but only for the perforated panels.

**Figure 2-14.** Linear metal panels.

panels are not perforated, the absorption falls off at the higher frequencies. By perforating the panels, this ceiling is an excellent broad-band absorber.

Manufacturers: 5, 8, 12, 13, 19, 21, 22, 32, 33, 40 & 79.

*Perforated Metal Roof Decks (Figure 2-15).* The utility of this material is that it combines acoustical treatment with part of a roof structural system. Several configurations of deck are available, differing in thickness and perforation pattern. Decks with spaced flutes that are perforated on the sides only and filled with glass fiber have moderate absorption at the low and mid frequencies and poor absorption at the high frequencies. Other patterns have flat bottoms which are almost completely perforated and have much higher absorption at all frequencies.

Manufacturers: 8, 22, 33 & 74.

*Perforated Metal Pans with Absorbent Pads (Figure 2-16).* Typical pans are 24″ × 48″ (610 mm × 1.22 m) that snap into an inverted metal "T" grid system with sound absorbent pads, usually wrapped in paper, placed in the pans with a wire spacer. The pan surface is considered cleanable and used in places where sanitation is a concern. There are other special types of these units to be used in "clean rooms" where the ceiling also acts as an air filter. This type of ceiling was once very popular but is seldom specified today.

Manufacturers: 1, 4, 5, 6, 7, 12, 13, 19, 21, 22, 32, 33, 37, 71 & 79.

*Fibrous Blankets and Boards (Figure 2-17).* Acoustical blankets may consist of glass fiber or mineral wool, and are available with insulation exposed or wrapped in a vinyl covering, in thicknesses ranging from 1″ to 6″ (25 mm

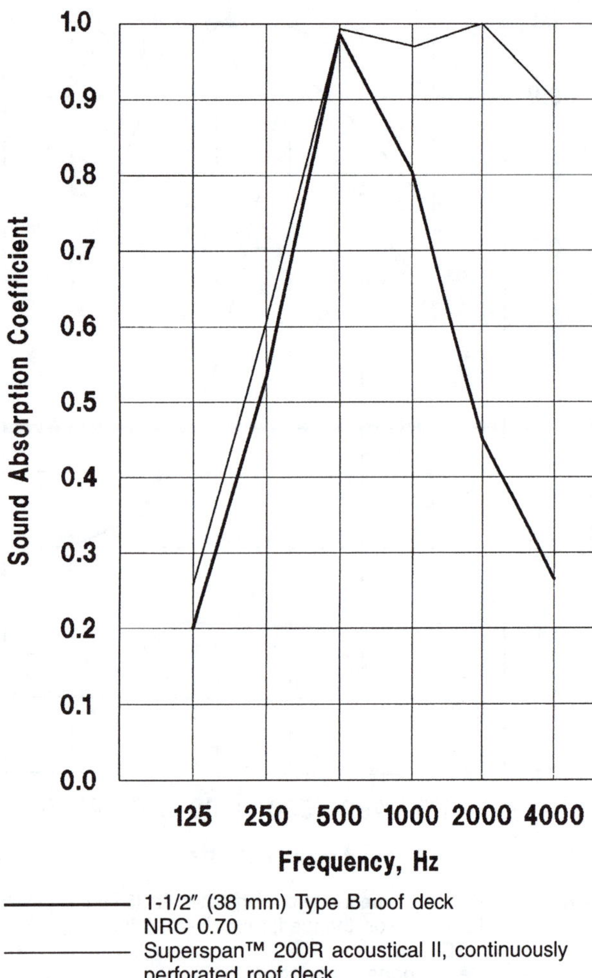


1-1/2″ (38 mm) Type B roof deck
NRC 0.70

Superspan™ 200R acoustical II, continuously perforated roof deck
NRC 0.90

The sound absorption and the higher frequencies depends critically on the extent of perforated areas. The Type B roof deck is perforated only on the sides of the flutes.

**Figure 2-15.** Perforated metal roof decks.

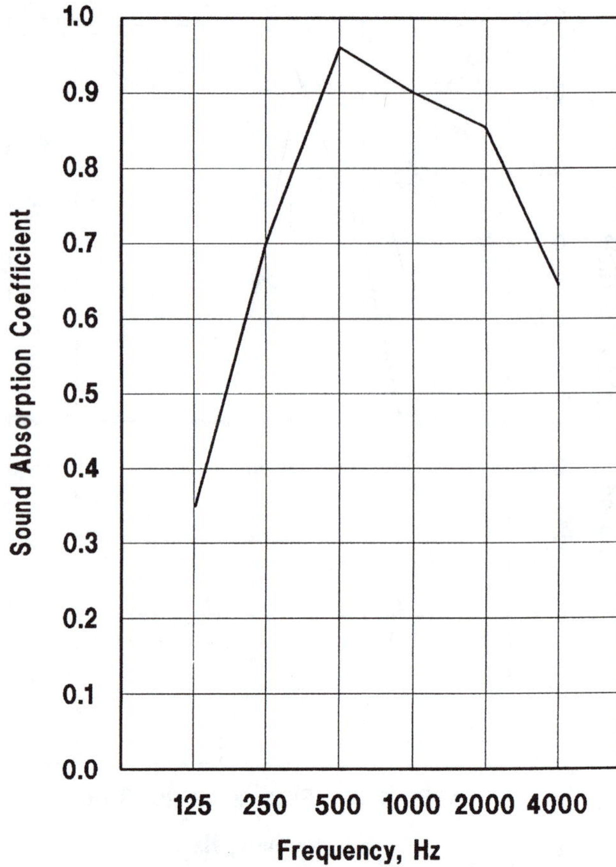


Perforated metal pans with absorbent pads
NRC 0.85

This product has an easily washable surface, but is seldom used today.

**Figure 2-16.** Perforated metal pans with absorbent pads (estimated absorption).

to 152 mm). If the vinyl covering is perforated, the thicker materials have excellent absorption at all frequencies. If the vinyl is not perforated, some high frequency absorption will be lost (which may be desirable in cases where there is excess high frequency absorption from other materials or the air). Densities range from less than 1 lb/cu ft (16 kg/cu m) to about 10 lbs/cu ft (160 kg/cu m). The ab sorption usually increases with thickness, and to a lesser extent with density. For a given thickness, mineral wool may be more dense and have a slightly higher absorption.

Applications for materials of this type include industrial shops and large "warehouse" type facilities. Blankets are also commonly used on top of suspended acoustical ceilings to increase the room-to-room ceiling attenuation (see Section 2.2). *Caution*: When blanket is used in this manner, care must be exercised not to cover light fixtures which require air circulation for cooling. See Fire Endurance Ratings Section 2.6 regarding the addition of acoustical blankets in a fire rated ceiling system.

Insulation boards may consist of glass, mineral or cellulose fibers. The glass fiber boards usually have higher absorption than the other two types. The lower density boards, about 3 lbs/cu ft (48.1 kg/cu m), are semirigid, while boards with densities in the 10 lbs/cu ft (160 kg/cu m) range are rigid enough to be used as formboards in roof systems. The insulation boards are used uncovered in mechanical rooms and shops for wall absorption, and for acoustical wall panels with fabric coverings (see Acoustical Wall Panels category).

Manufacturers: 5, 7, 13, 14, 17, 18, 23, 37, 38, 44, 49, 51, 52, 58, 65, 67 & 68.

*Solid Plastic Panels (Figure 2-18).* Panels are constructed of thin plastic and are intended to either fit in a grid system or attach to a hard surface. They are shaped to add diffusion to the reflected sound. Because they flex at

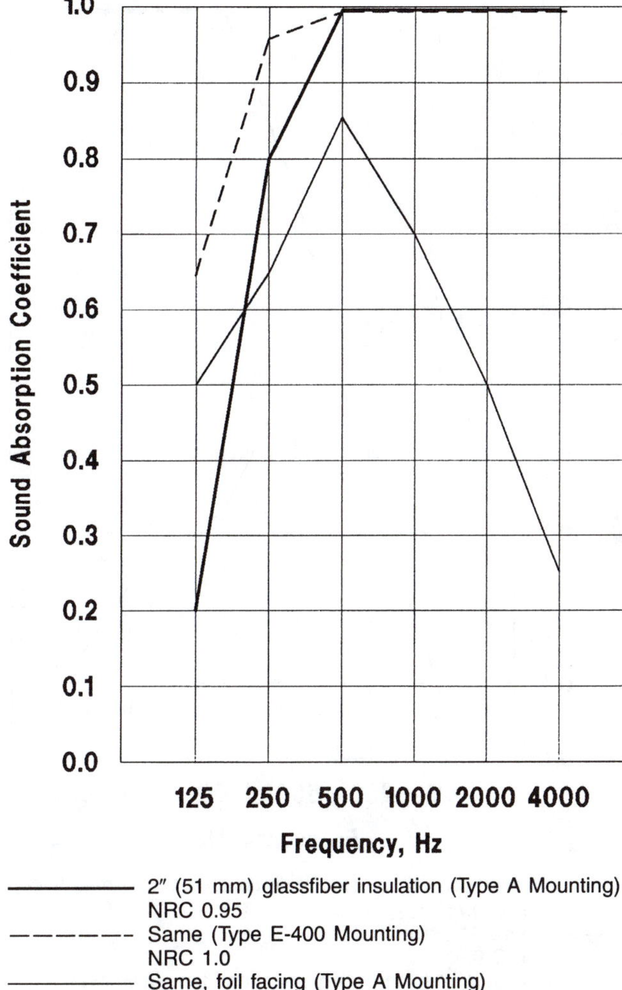


——————— 2″ (51 mm) glassfiber insulation (Type A Mounting) NRC 0.95

– – – – – – Same (Type E-400 Mounting) NRC 1.0

——————— Same, foil facing (Type A Mounting) NRC 0.65

These materials provide excellent broad-band absorption if the porous surface is exposed. The foil facing restricts high frequency absorption.

**Figure 2-17.** Fibrous blankets and boards.

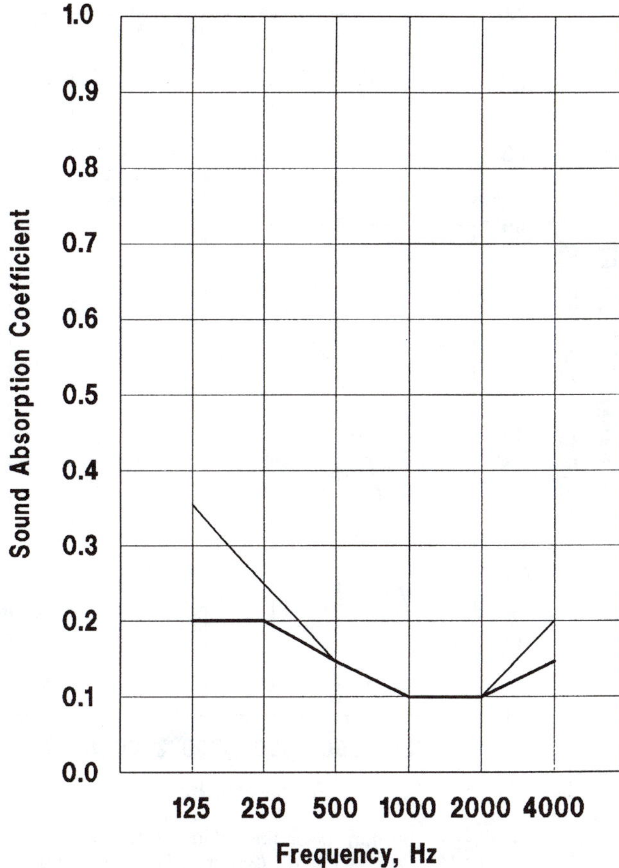


——————— Wenger Type I wall diffuser 4′-0″ × 4′-0″ (1.22 m × 1.22 m), Type B Mounting NRC 0.15

——————— Wenger Type II wall diffuser, Type B Mounting 4′-0″ × 4′-0″ (1.22 m × 2.44 m), Type B Mounting NRC 0.15

Solid panels that are flexible have low frequency absorption, and provide diffusion at higher frequencies.

**Figure 2-18.** Solid plastic panels.

low frequencies, considerable sound absorption may occur in this frequency range. Some types made of heavier plastic are intended for wall application (see Wall Systems, Section 2.5.2). Caution should be exercised in using panels of this type because of combustibility and accompanying development of toxic fumes.

Manufacturers: 68 & 81.

*Absorbent Ceiling Clouds and Baffles (Figure 2.19; see also Figure 2-29).* Panels of glass fiber insulation board suspended in a suitable frame creating a floating "cloud" can be a very effective acoustical treatment. Because of the free area around the cloud, the sound is able to reach the back surface as well, increasing the effective absorption. Clouds may be spaced between or below structural elements, or allow space between for lighting or air distribution fixtures. It is important that the suspension system incorporate seismic restraints.

The insulation board panels may be factory finished (see category on Glass Fiber Lay-in Panels), or covered with a suitable open-weave fabric. There are many installation and material possibilities with this type of acoustical treatment.

Manufacturers: 4, 5, 7, 10, 14, 15, 19, 23, 25, 32, 33, 37, 40, 47, 52, 60, 65, 66, 70, 75, 77, 80, 81 & 83.

### 2.5.2 Wall Systems

*Acoustical Wall Panels (Figure 2-20).* Within the past 10 or so years the importance of acoustical wall treatments has been recognized by the design profession, and now

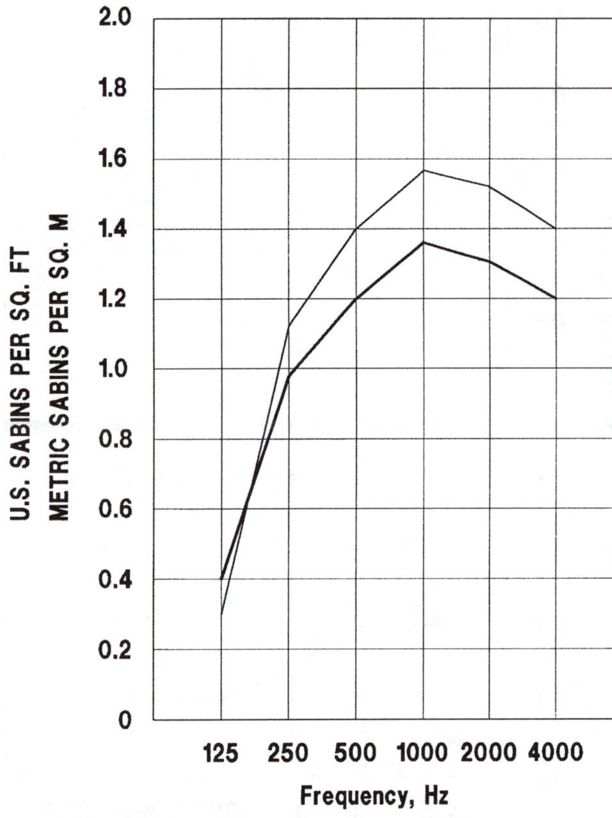


——— 1-1/2″ (38 mm) glass fiber cloud, 8′-0″ × 8′-0″ (2.44 m × 2.44 m), 10″ (254 mm) from hard surface.

——— 1-1/2″ (38 mm) glass fiber cloud, 4′-0″ × 4′-0″ (1.22 m × 1.22 m), 4 clouds spaced 24″ (610 mm) apart, 10″ (254 mm) from hard surface

Notice the sabins per sq ft (sq m) exceed 1.0 at the mid and high frequencies because of sound exposure on both sides of the cloud.

**Figure 2-19.** Absorbent ceiling clouds.

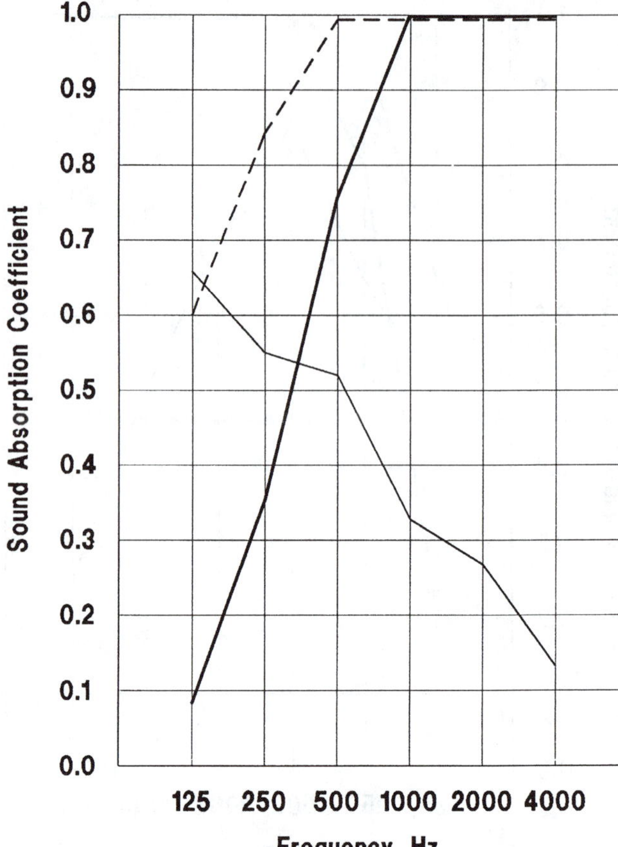


——— Acoustical wall panels, fabric covered 1″ (25 mm) thick NRC 0.75

– – – – Same 2″ (51 mm) thick NRC 0.95

——— Same 2″ (51 mm) thick, nonperforated vinyl cover NRC 0.40

Many types of acoustical wall panels are now on the market, some with fabric cover, some with perforated vinyl. Watch out for nonperforated covers.

**Figure 2-20.** Acoustical wall panels.

many such products are available. Typically such panels are composed of mineral or glass fiber boards, thickness 1″ to 2″ (25 mm to 51 mm), covered with a porous fabric or finely perforated vinyl. Several methods of mechanical attachment to walls are available, some allowing reinstallation of the panels at different locations. Typical applications include music rehearsal rooms, lecture halls, large conference rooms and open plan screens.

Manufacturers: 1, 4, 5, 6, 7, 9, 10, 13, 14, 15, 19, 21, 23, 25, 32, 33, 37, 40, 41, 44, 46, 47, 49, 52, 54, 58, 60, 65, 66, 68, 69, 71, 73, 74, 75, 78, 80, 81, 82 & 83.

*Reflecting/Diffusing Panels (Figure 2-21).* Panels of this type may be considered an acoustical material in that they reflect sound at various angles to increase sound diffusion and prevent detrimental reflections. They may be flat surfaced, splayed or convex. If the panels are to provide much diffusion, their dimensions must be related to the wavelength of the sound they are reflecting. Low frequencies will not be redirected by surfaces which are much smaller than the wavelength (see Chapter 1, Section 1.4.2, Figure 1-10). For further information on applications of reflecting panels to control room modes, flutter echoes and standing waves, see Sections 4.3, 4.5 and 5.2.3 in Chapters 4 and 5, respectively.

The panels may have significant low frequency absorption if they are not too thick (stiff), and if installed with an enclosed back air space. An absorbent filling the airspace will provide even more low frequency absorption. See Appendix A.5 for data showing how the resonant frequency can be varied with the thickness of the panel and depth of the air space. Hinged panels may be used on walls to change reflection patterns, and if alternated with absorptive panels, can be used to control reverberation by covering the absorptive areas.

Manufacturers: 4, 5, 7, 14, 15, 23, 33, 37, 40, 41, 47, 52, 60, 65, 66, 70, 73, 77, 80, 81 & 83.

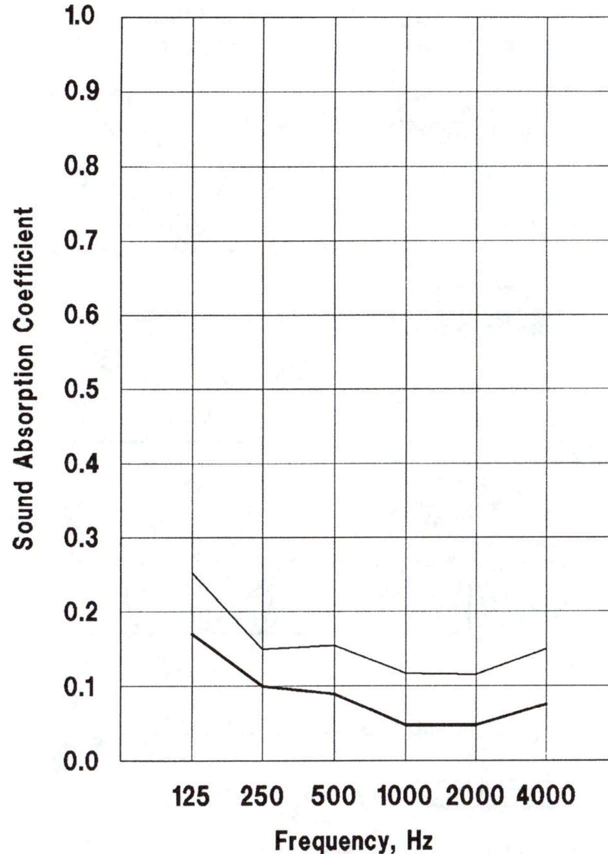


——— 1/4″ (6.4 mm) plywood panels, no absorbent in airspace NRC 0.10
——— Same, insulation in airspace NRC 0.15

There must be an airspace behind the panel for low frequency absorption, which can be enhanced by placing insulation in the airspace.

**Figure 2-21.** Reflecting/diffusing panels.

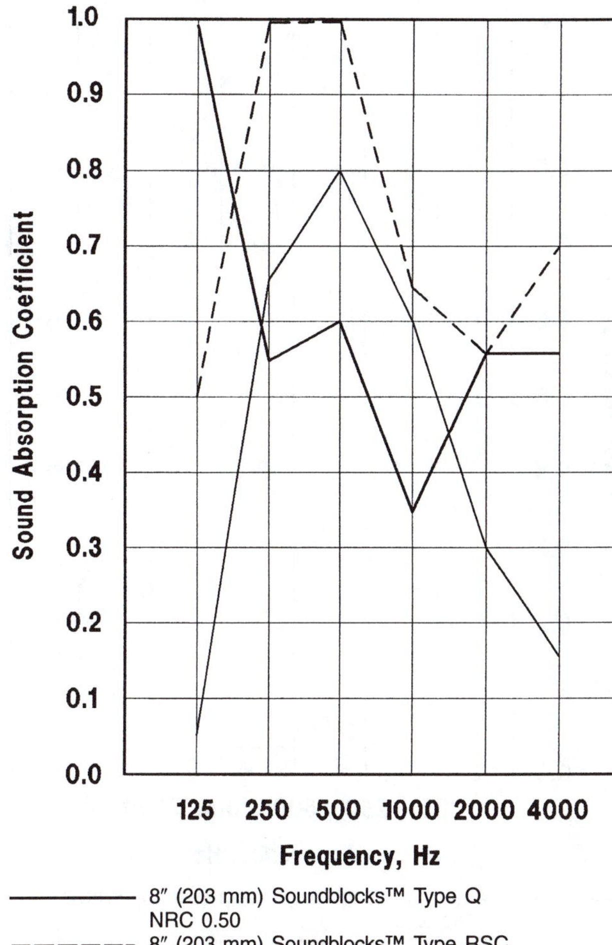


——— 8″ (203 mm) Soundblocks™ Type Q NRC 0.50
– – – – 8″ (203 mm) Soundblocks™ Type RSC NRC 0.80
——— 4″ (102 mm) perforated clay tile NRC 0.60

The absorption as a function of frequency varies substantially with the configuration of the slot opening, the design of the core space and perforation pattern. Many more patterns are available than are shown here.

**Figure 2-22.** Concrete masonry and perforated clay tile.

*Concrete Masonry and Perforated Clay Tile (Figures 2-22 and 2-23).* Hollow core concrete blocks may be manufactured with slots or holes in the face leading into the cavities. The cavities may have acoustical material or diaphragms to alter the absorptive characteristics, and are sometimes referred to as Helmholtz resonators. They may be designed to have peak absorption at a certain frequency. One application is a product called SOUNDBLOCKS™, consisting of hollow concrete blocks with slits in the face leading into the hollow core spaces. Clay tiles are also available with perforated faces and glass fiber in the cavities. Materials of this type are useful for surfaces subjected to impacts, and some types may be used out-of-doors.

Concrete blocks with certain types of light weight aggregate, such as pumice or expanded shale, may exhibit considerable porosity resulting in significant sound absorption. It is risky to rely on this type of absorption, however, because it is difficult to predict and may vary considerably from one batch to another. The porosity may also reduce the sound transmission loss of a wall made up of such block (see Section 3.3.3 in Chapter 3).

Manufacturers: 5, 49, 58 & 60.

*QRD Diffusers (Figure 2-24).* This is a special type of panel that has been developed by RPG™ Diffusor Systems based on a quadratic residue formula. The panels are constructed with a series of slots of various widths and depths which break up the sound waves, resulting in highly diffused reflections. Panels are usually constructed of wood, and are typically 4″ to 8″ (102 mm to 203 mm) in depth. These panels are particularly useful on the back walls of control rooms and adjacent walls of recording studios to provide diffusion without additional, unwanted absorption.

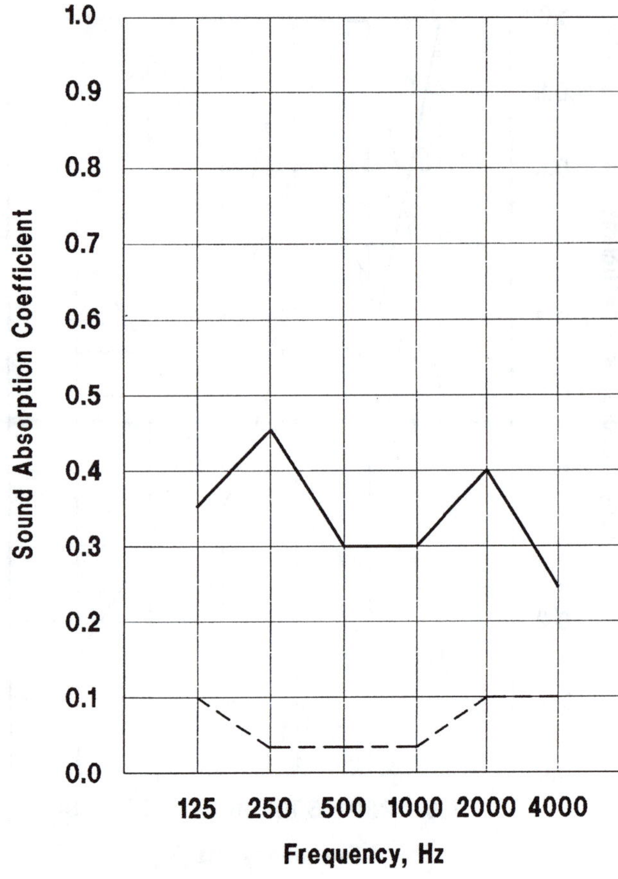


——————— 4″ (102 mm) lightweight aggregate, unpainted NRC 0.35

– – – – – – – Same, 2 coats paint NRC 0.05

A porous block can have considerable absorption. However, the absorption can almost be eliminated with a sealing coat of paint. The unpainted block may be a poor sound barrier, however, because sound is transmitted through the air passages.

**Figure 2-23.** Concrete masonry.

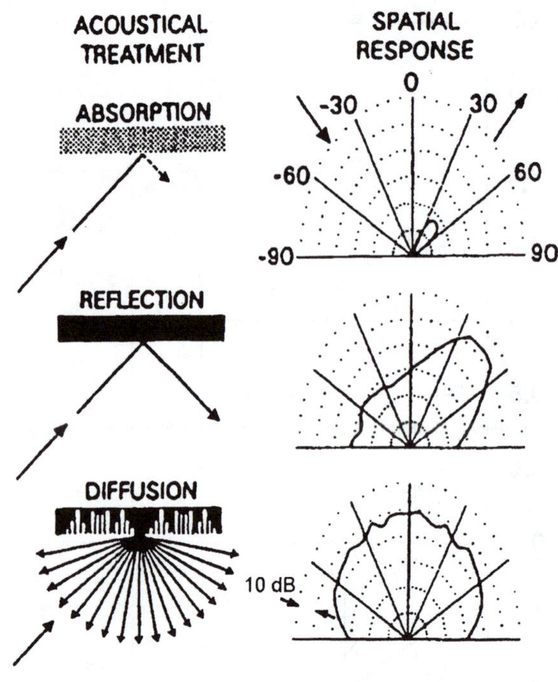


This graphic illustrates how sound is:

ATTENUATED by absorption

REDIRECTED by reflection

UNIFORMILY DISTRIBUTED by diffusion

(Courtesy of RPG Diffusor Systems, Inc.)

**Figure 2-24.** Patterns of reflected sound.

Manufacturers: 7, 15, 60, 73 & 81.

*Drapery (Figure 2-25).* Heavy velour drapes weighing at least 20 oz/sq yd (0.14 lb/sq ft, or 0.68 kg/sq m) and draped to at least 50% of its width are advantageous in spaces where adjustable reverberation is desired. Such spaces could include music rehearsal rooms and multiuse auditoriums. Absorption characteristics will depend on fabric weight, lining (if any), amount of draping and spacing from the wall. Disadvantages are that they require cleaning and re-fireproofing from time to time. They can be used to advantage to vary the acoustics in a space by drawing them back into pockets when higher reverberation is desired.

*Absorbent Blankets and Boards (Figure 2-26).* Materials of this type may be applied directly to wall (or ceiling) surfaces where a large amount of broad-band sound absorption is desired at minimum cost and where appearance is not an important consideration. Typical applications of this type could be in shops or mechanical rooms. Insulation thickness may be from 1″ to 4″ (25 mm to 102 mm) or more; thicker materials will provide better low frequency absorption.

Some acoustical blankets are backed with a solid, dense material such as a barium-loaded vinyl weighing 3/4 to 1 lb/sq ft (3.7 to 4.9 kg/sq m). They are used as hanging curtains to screen off noisy activities from adjacent occupied areas, yet may be removed if access is needed to either space (see Section 9.3.2).

Manufacturers: 5, 7, 13, 14, 17, 18, 23, 37, 38, 44, 49, 51, 52, 58, 65, 67 & 68.

*Acoustical Wedges (Figure 2-27).* Wedge-shaped acoustical materials are frequently used where very high absorption is required, such as in anechoic test chambers or control rooms where no reflected sound can be tolerated.

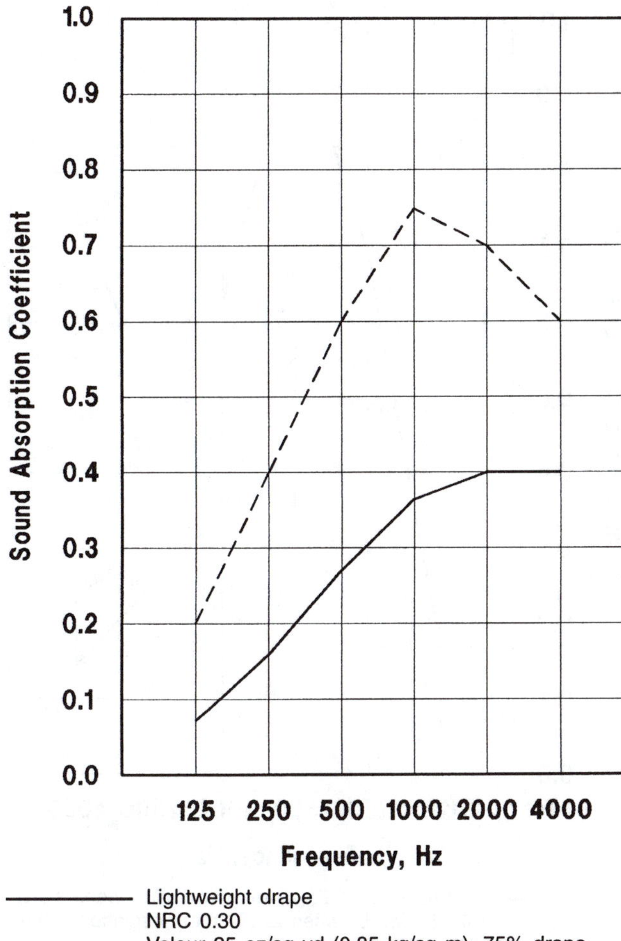


——— Lightweight drape
NRC 0.30
– – – – Velour 25 oz/sq yd (0.85 kg/sq m), 75% drape
NRC 0.60

Low frequency absorption can be enhanced by spacing the drape up to 1 ft (0.3 m) away from the wall. Full 100% drape (twice the fabric to cover the area) increases the absorption at all frequencies.

**Figure 2-25.** Drapery.

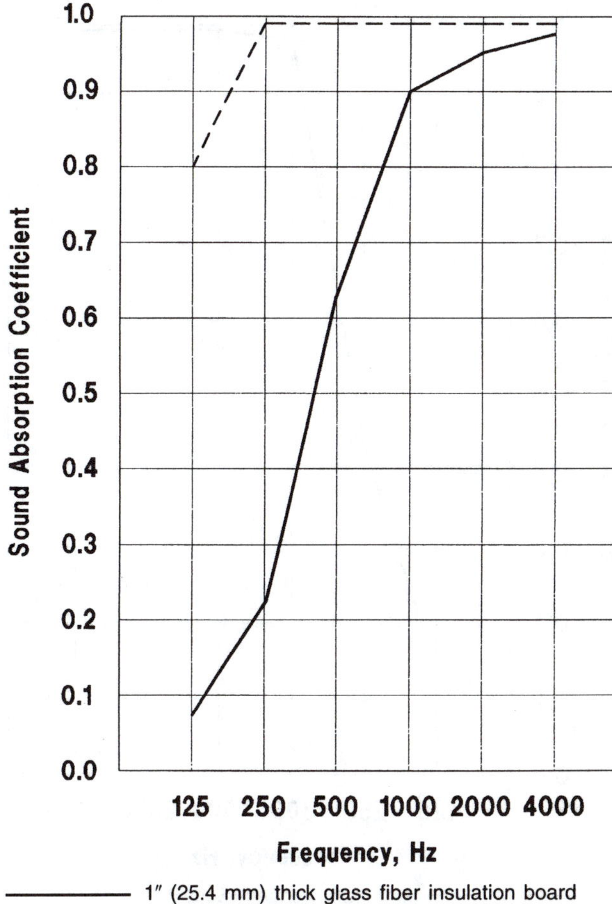


——— 1″ (25.4 mm) thick glass fiber insulation board mounting Type A
NRC 0.65
– – – – 4″ (102 mm) thick glass fiber insulation board mounting Type A
NRC 1.00

These materials provide high absorption at a low cost in applications where appearance is not important.

**Figure 2-26.** Absorbent blankets and boards.

They are usually made of solid glass fiber insulation with different depths for very high absorption of all frequencies above a cut-off frequency. These wedges may be about 6″ to 12″ (152 mm to 305 mm) square at the base and project out from 12″ to 48″ (305 mm to 1.22 m), depending upon the required performance. The longer the wedge, the lower the cut-off frequency.

Some wedge products are made of foamed plastic, which must be of the open-cell type to be effective acoustically. These wedges are typically 3″ to 4″ (76 mm to 102 mm) in depth and are often used in studios and control rooms because of their excellent mid and high frequency sound absorption, and because of their "high tech" appearance. Caution: The plastic used may be combustible and some give off noxious fumes when burned. For this reason, careful selection is essential. The manufacturer should supply copies of the laboratory test reports authenticating all combustibility and fume production tests under the appropriate ASTM procedures. The binder used in glass fiber products may also have to be certified for combustibility and smoke developed.

Manufacturers: 5, 7, 10, 15, 19, 33, 37, 47, 49 & 73.

*Spaced Wood Slats (Figure 2-28).* Wood slats with a space between them can provide a decorative surface which protects the absorbent blanket or insulation board behind. The sound absorption as a function of frequency will vary depending on the width and shape of the slat, spacing between slats and absorbent thickness. A slat configuration with good absorption across the standard frequency range consists of 3/4″ × 1 1/2″ (19 mm × 38 mm) wood slats with the narrow side exposed in a "bull-nose" shape. Slats that are wider than about 1″ (25 mm) will cause a loss of absorption at high frequencies, unless the spacing between slats is at least as wide as the slat. VENTWOOD™ is a product with a variety of slat panel

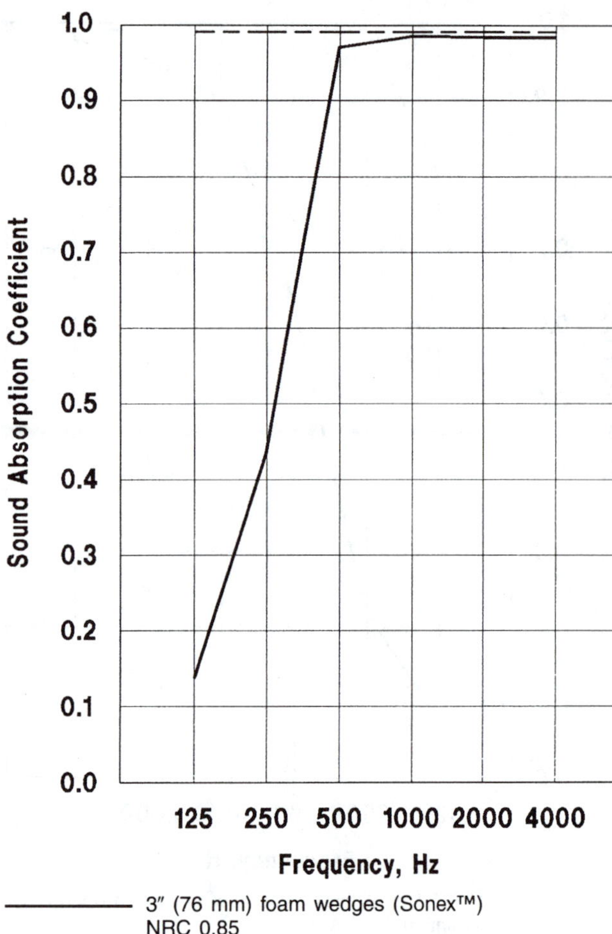


——————— 3″ (76 mm) foam wedges (Sonex™) NRC 0.85

– – – – – – 2′-0″ (610 mm) glass fiber wedges NRC 1.00

The longer wedges are used in special test room applications where a free field (no reflection) is required.

**Figure 2-27.** Acoustical wedges.

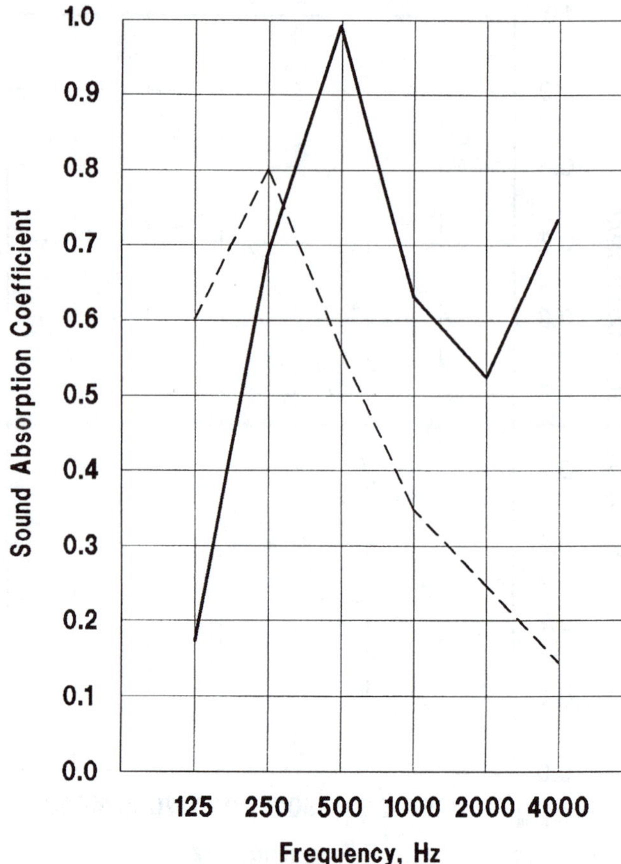


——————— "Bull nose" 1″ × 2″ (19 mm × 38 mm) wood slats 1/4″ (6 mm) between 2″ (51 mm) insulation behind NRC 0.70

– – – – – – 1″ × 4″ (19 mm × 89 mm) wood slats 1/2″ (13 mm) between 2″ (51 mm) insulation behind NRC 0.50

The wood slats protect the insulation for low wall applications, but will affect the absorption depending on the slat width and spacing between slats.

**Figure 2-28.** Spaced wood slats.

configurations. This material may also be used for ceiling treatments.

Manufacturers: 5 & 31.

### 2.5.3 Special Absorbers

*Suspended Absorbers (Figure 2-29).* A number of absorptive products are available for hanging from ceiling surfaces or structures. They may be in the form of glass fiberboard panels suspended either vertically or horizontally (clouds), or may take on other geometric forms such as cylinders. They may be wrapped in fabric or perforated or nonperforated vinyl. The nonperforated covering will result in some loss of high frequency absorption. There have been occasional problems of the vinyl covering developing cracks or slits with aging. An extended warranty may be good insurance.

Suspended absorbers are very efficient acoustically because sound strikes them from behind as well as in front. Sound absorption data are usually given as U.S. or metric sabins per unit rather than as coefficients. Typical applications are industrial shops, swimming pools or other large spaces where the suspended units would not be subject to damage. They are not recommended for gymnasiums

Manufacturers: 1, 4, 5, 7, 10, 13, 14, 15, 19, 23, 25, 33, 37, 40, 44, 47, 49, 52, 58, 60, 65, 66, 68, 80, 81 & 83.

*Duct Liners (Figure 2-30).* Air distribution ducts may be lined with glass fiber boards to control the transmission of noise in the ducts (see Chapter 7, Section 7.4.1). Duct liners are typically 1/2″ or 1″ (13 mm or 25 mm) thick, and the surface is usually coated to avoid erosion of the fibers from the air movement. For architectural applications, one type of duct liner is coated with a black neoprene, which is

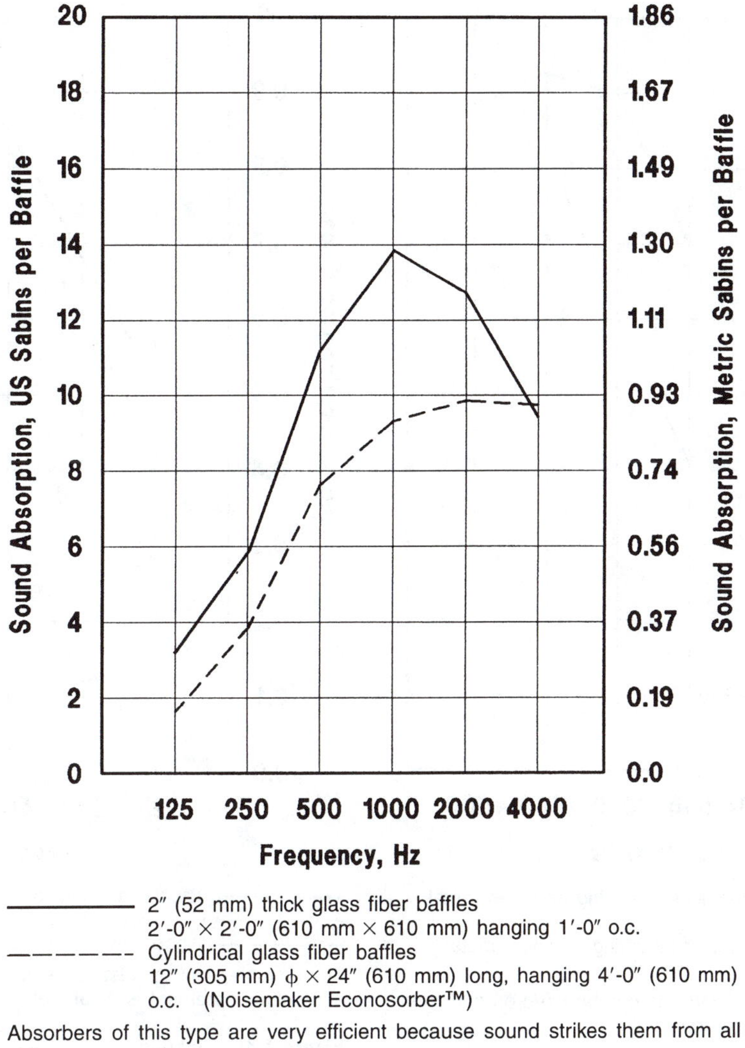


**Figure 2-29.** Suspended absorbers.

useful for surfaces that the designer may want to visually "disappear," such as behind spaced wood slats (see the earlier category on Spaced Wood Slats).

Manufacturers: 5, 33, 38, 49 & 51.

### 2.5.4 Building Materials and Other Elements

*Seating and Furniture (Figure 2-31).* The sound absorption of upholstered seats is an important consideration in determining the total acoustical absorption in many occupied spaces. The absorption is dependent on the type and thickness of the padding, the porosity of the cover material, and the spacing of the seats.

For large areas of seating, there are two methods for specifying the total sound absorption of the area. One is by multiplying the sabins per seat (U.S. or metric) by the number of seats. (A typical theater seat will occupy about 6.5 to 7.0 sq ft, or 0.6 to 0.65 sq m.) If the aisles are carpeted, this constitutes additional absorption. The other is by assigning coefficients to the total seating area, including aisle widths of up to 4 ft (1.22 m).

When the seats are occupied, most of the absorptive material is covered up, but the absorption of the person is added (see People, following). The absorption increment between unoccupied and occupied, therefore, is usually quite small unless the seat was not upholstered, such as a wood church pew. It is customary to specify seating absorption as either unoccupied and occupied.

The sound absorption of other types of furniture, such as sofas or beds, is usually specified as sabins per unit. Sound absorption data for some common surfaces or objects, such as stacks of books that may be found in a library or study, and many types of furniture, have never been measured and must be estimated based on absorption for similar materials.

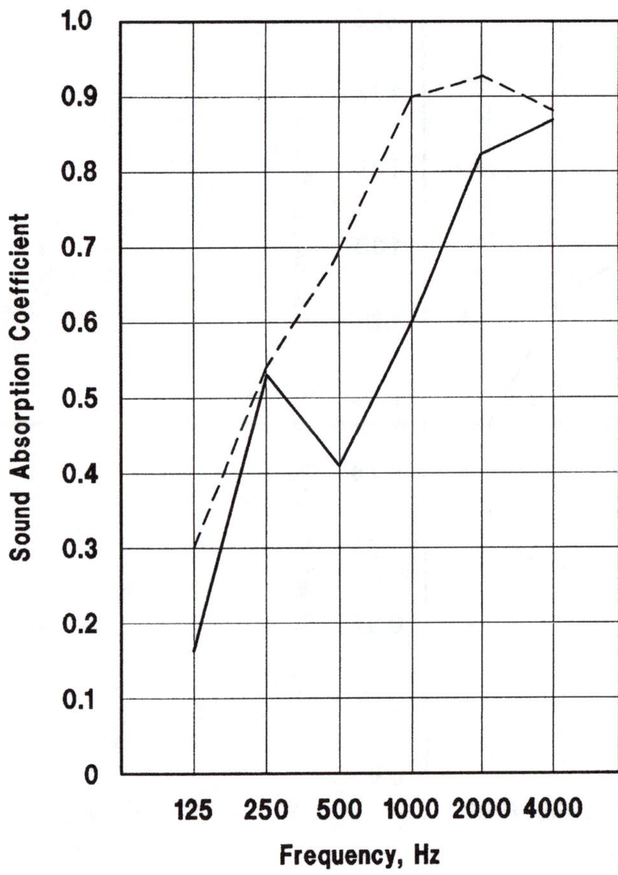


**Figure 2-30.** Duct liners.

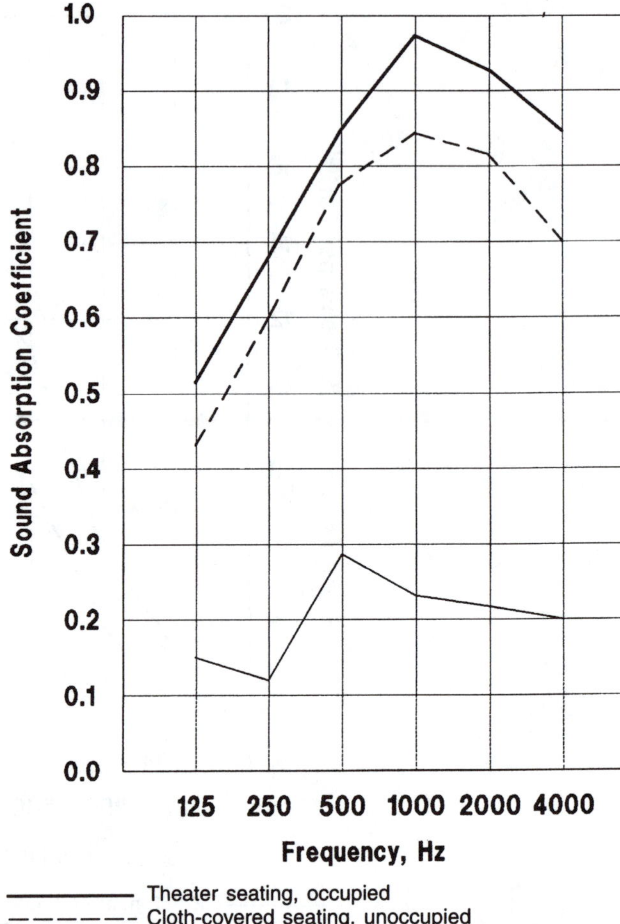


**Figure 2-31.** Seating.

*People (Figure 2-32).* The sound absorption of people depends on the type and quantity of clothing they wear. A fully clothed member of a theater audience will have several times as much absorption as someone dressed in a swimming suit. Because of the difficulty in determining the area of a person's clothing, people absorption is stated as sabins per person (U.S. or metric). Some reduction in absorption should be made for a person surrounded by other people in a crowded room as compared to standing alone or in a small group.

*Carpeting (Figure 2-33).* The sound absorption of floor carpeting is dependent on the thickness (including pad, if installed) and type of pile. Because it is a relatively thin material, it has very little absorption at low frequencies but can have considerable absorption at high frequencies. Carpet is sometimes used as an abuse-resistant covering over glass fiber insulation board applied to lower wall surfaces. Caution: For this application, the back of the carpet must not be sealed with latex or other impervious material. Also, it is a requirement that carpet installed on wall surfaces have a flame spread rating.

*Air Absorption (Table 2-1).* Sound waves are attenuated as they pass through air, whether out-of-doors or being reflected back and forth by room surfaces. The amount of absorption is dependent on the relative humidity and the frequency. In room acoustics, air absorption below 1000 Hz is not significant.

In order to have consistent acoustics in large halls, air conditioning with humidity control is essential. Pipe organs are particularly sensitive to humidity, and pianos must be retuned if subjected to humidity (or temperature) changes. The procedure for including air absorption in reverberation time calculations is presented in Chapter 5, Section 5.9. Air absorption is also important in the propa-

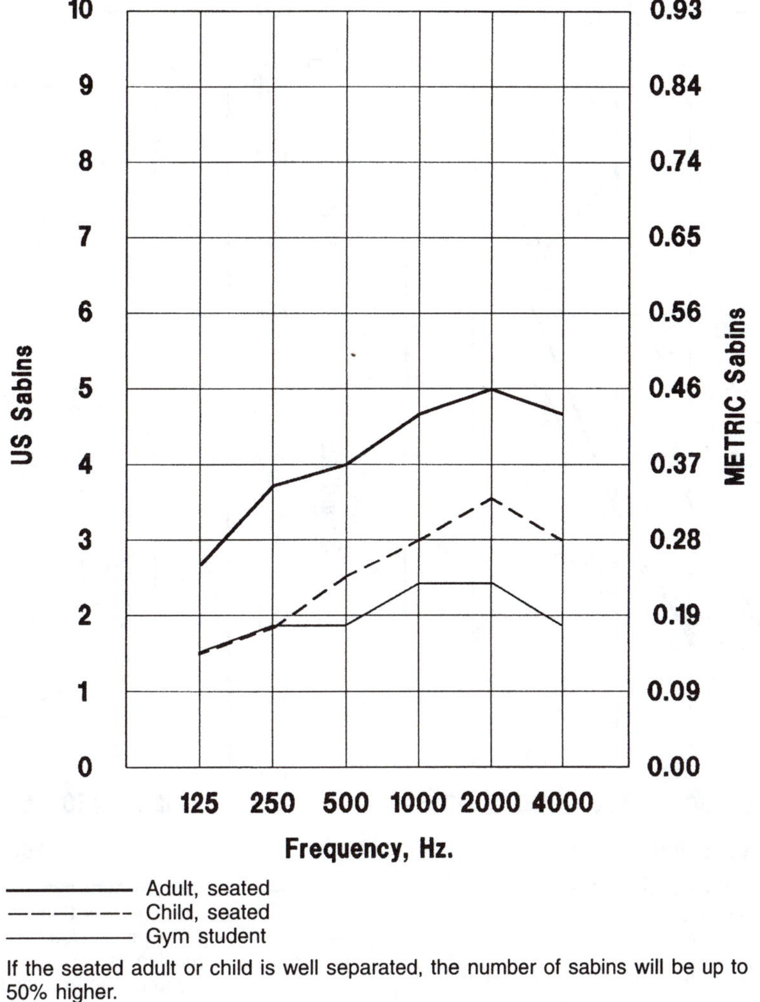


**Figure 2-32.** People.

gation of environmental noise, as discussed in Chapter 8, Section 8.3.1.

Air absorption data for enclosed spaces are presented in Table 2-1. The values are in U.S. sabins per 1000 cu ft, and in metric sabins per 30 cu m.

For example, in an auditorium with a cubic volume of 186,000 cu ft (5264 cu m), the absorption provided by the air at a frequency of 2000 Hz and a relative humidity of 40% would be 186 × 3.16 = 588 U.S. sabins (175 × 0.31 = 54.2 metric sabins).

*Gypsum Board, Wood Paneling and Wood Floors (Figure 2-34).* Building materials of this type are frequently used for wall and partition surfaces. Their sound absorption is mainly at the low frequencies due to flexing of the surface when subjected to the pressure fluctuations of impinging sound waves. The thinner the material, the more low frequency absorption. While the absorption coefficients are quite low compared to most other acoustical materials, their contribution to the total low frequency absorption in a room can be substantial because of their large areas.

*Glass, Large Panes (Figure 2-35).* Like wood paneling and gypsum board, sound absorption of glass is primarily at the low frequencies due to flexing. Low frequency absorption decreases as the glass thickness is increased. High frequency absorption is very low for all thicknesses. Indeed, glass (and some types of clear plastic) is a good sound reflector for the higher frequencies, and is occasionally used for that purpose to create "see-through" reflectors.

*Concrete and Stone (Figure 2-36).* Included in this category would be ceramic tile, terrazzo, vinyl floor coverings and water. Unless these materials exhibit some degree of porosity (see the Concrete Masonry category), they

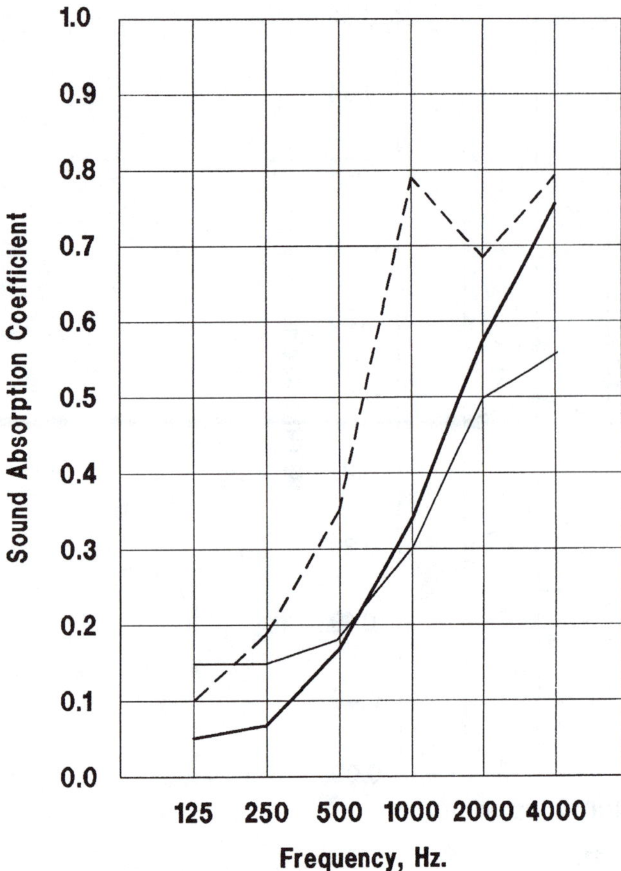


——— 20 oz/sq yd (0.68 kg/sq m) carpet, no pad
– – – – Same, with hair pad
——— Same, on wood riser (no pad)

The wood riser backing adds slightly to the low frequency absorption.

**Figure 2-33.** Carpeting.

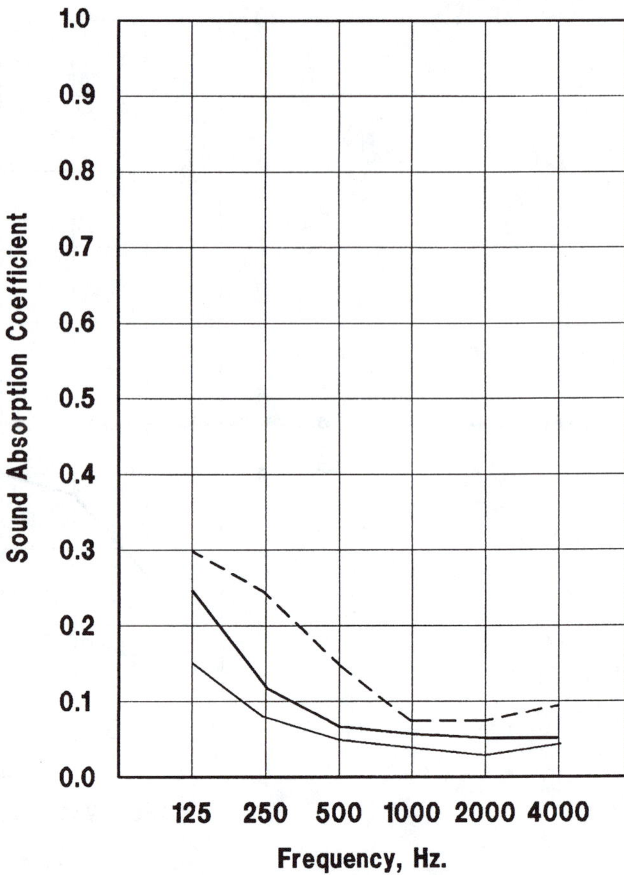


——— 5/8″ (16 mm) gypsum board on studs @ 16″ (406 mm) o.c. (For 2 layers, use wood flooring on joists)
– – – – 1/4″ (6 mm) wood paneling with air space and insulation behind
——— Wood flooring on joists

The absorption characteristics of thin wood paneling is highly dependent on the paneling thickness, depth of airspace behind and the presence of insulation. See Appendix A.5.

**Figure 2-34.** Gypsum board, wood paneling and wood floors.

have virtually no sound absorption. They are too massive and stiff to exhibit any low frequency flexural absorption. Sound absorption coefficients at all frequencies are typically 0.03 or less.

## 2.6 Fire Endurance Ratings

Suspended acoustical ceiling systems are rated for fire endurance, in minutes and/or hours. *Caution*: The ratings are based on the total system as tested, not for the acoustical material alone, or any individual part of the assembly. Changing the system in any way may have the effect of changing the fire rating, and it may not be accepted if significant changes to the system are made. For example, putting an acoustical blanket on top of the suspended ceiling

**TABLE 2-1.** Sound Absorption in Air in Enclosed Spaces

Air absorption in U.S. sabins for a volume of 1000 cu ft at 72° F. (Metric sabins are shown in parenthesis for a volume of 30 cu m at 22° C.)

| | Relative Humidity in Percent | | | |
|---|---|---|---|---|
| Freq., Hz | 15 | 20 | 40 | 60 |
| 1000 | 1.96 (0.19) | 1.64 (0.16) | 1.28 (0.13) | 1.10 (0.11) |
| 2000 | 6.62 (0.65) | 4.88 (0.48) | 3.16 (0.31) | 2.75 (0.27) |
| 4000 | 22.16 (2.18) | 16.47 (1.62) | 8.28 (0.81) | 6.83 (0.67) |
| 6000 | 43.10 (4.23) | 32.45 (3.19) | 16.04 (1.58) | 11.81 (1.16) |
| 8000 | 64.74 (6.36) | 54.19 (5.32) | 26.80 (2.63) | 18.63 (1.83) |

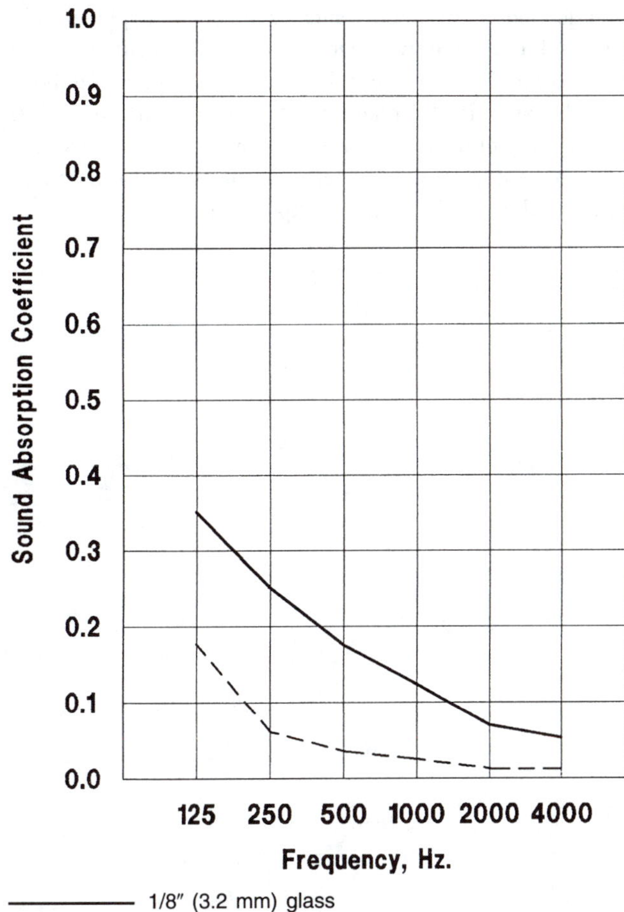


———— 1/8″ (3.2 mm) glass
– – – – 1/4″ (6.4 mm) glass

Large panes of glass exhibit considerable low frequency flexural absorption. Small panes in stiff frames will have much less absorption at the low frequencies.

**Figure 2-35.** Glass (large panes).

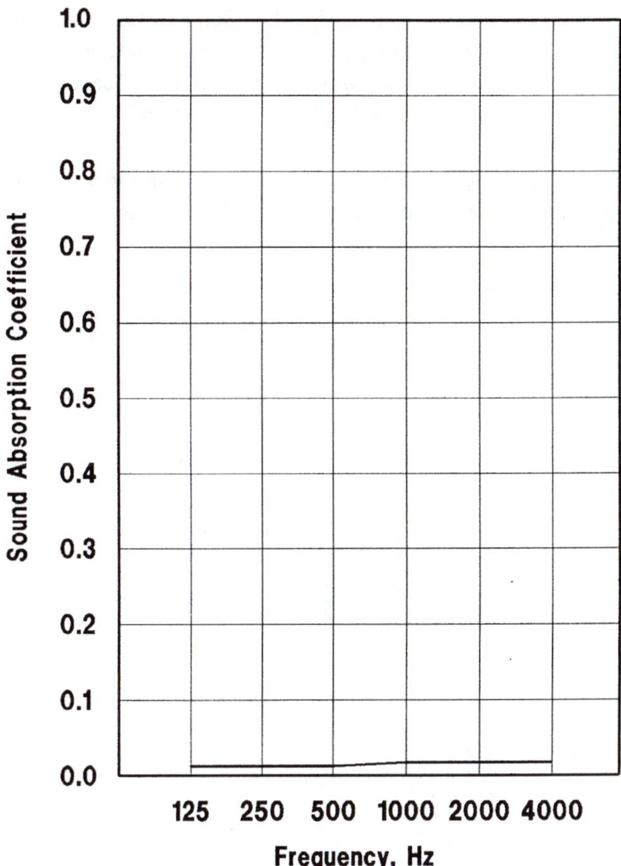


———— Concrete, stone, water

Most sound striking these massive, nonporous materials is reflected away.

**Figure 2-36.** Concrete, stone, water.

will retard the flow of heat, causing the metal suspension system to reach higher temperatures and, thus, premature failure. A tile or ceiling panel identified as "Fireguard" or a similar term does not mean that the material alone will provide a fire rating unless incorporated in a complete tested system (see Section 2.4 and Appendix A.9).

The fire ratings required for different building categories are specified by the applicable building codes, and are typically identified by a construction type designation and associated hour rating (1 through 4). Building departments should be consulted for the appropriate rating, as well as the insurance company where insurance is involved. Refer to the Underwriters Laboratory or Factory Mutual Research Corporation listings of tested assemblies.

### 2.6.1 Flame Spread

Acoustical materials are rated for "flame spread," which is the rate at which flames propagate along the surface of the material. Special types of intumescent paint have been developed which can be applied to nonconforming materials, to retard the spread of flames. See ASTM Designation: E84, "Test Method for Surface Burning Characteristics of Building Materials."

## *2.7 Light Reflectance*

Architects and lighting engineers are concerned about the light reflectance characteristics of suspended acoustical ceilings when designing for satisfactory illumination in rooms, particularly when indirect lighting is used. The reflectance is defined as the fraction of light reflected from the material surface. The test procedure for measuring light reflectance is specified in ASTM Designation: E1477, "Luminous Reflectance Factor by Using the Integrating Reflectometer."

## *2.8 Active Sound Cancellation*

Sound cancellation is a recent development that operates on the principle of wave cancellation. The sound generated by a noise source is sensed by a microphone placed in the propagation path. This signal is then amplified and fed into a loudspeaker, but exactly 180° out of phase with the original signal. The two waves combine by one cancelling the other, resulting in substantial attenuation.

At the present time active sound cancellation has limited application. The method is best suited to sound in well defined environments, such as air ducts, airplane cockpits and immediately adjacent to fixed machinery. The method requires considerable care in its application, however, because if the phase shift is much different from 180°, signal reinforcement rather than cancellation can occur. The system requires energy input and periodic maintenance, thus access to the equipment is required.

# Chapter 3

# Building Noise Control—Sound Transmission

Most buildings support activities such as speech, music, studying, work, or rest and relaxation, that can be adversely affected by noise or vibration. These activities can be impacted by each other, or by noisy or vibrating mechanical equipment, plumbing or machinery located elsewhere in the building. Noise from outside the building may also intrude and interfere with activities inside. This Chapter is primarily concerned with controlling the transmission of noise by the elements of building construction. Chapter 8 is devoted to the external noise environment, but the methods for controlling the intrusion of this noise into buildings are included in this Chapter. Also, architects and builders should be aware of building codes that regulate noise in buildings. See Section 3.1.1.

The control of noise in buildings can be subdivided into five parts:

1. Treating the source of noise;
2. Treating the space containing the source;
3. Providing insulation construction for airborne sound;
4. Providing isolating construction for impact sound;
5. Providing for isolation of noise and vibration from mechanical equipment.

*Treating the Source of Noise.* The control of noise emitted by a source depends on the characteristics of the source. If it is the result of some human activity, certainly the person producing the noise bears some responsibility for abatement, and many ordinances and regulations address this responsibility. Noise from equipment is basically the responsibility of the manufacturer, whether it is household appliances, office equipment, or mechanical equipment. It may be possible to reduce equipment noise by treating the source directly, but care must be exercised so as not to adversely affect the operation nor void the warranty. A further discussion of noise and vibration control for mechanical equipment is contained in Chapters 7 and 9.

*Treating the Space Containing the Source.* The addition of acoustical absorption to a room will reduce the sound level in the reverberant field (see Section 4.6.2). If it is desired to reduce direct sound radiation (sound that has not reached a room surface), it may be possible to insert a barrier in the transmission path, or locate the source in a partial enclosure. Again, the barrier or partial enclosure must not adversely affect operation of the machine. See Chapter 9.

*Providing Insulation Construction for Airborne Sound.* The airborne sound transmitting characteristics of building materials is an important part of architectural acoustics, and is treated extensively in Sections 3.2 and 3.3 of this Chapter.

*Providing Isolating Construction for Impact Sound.* Building construction behaves quite differently when subjected to direct impacts compared to airborne sound. Impact sound transmission is treated in Sections 3.5 and 3.6 of this Chapter.

*Providing for Isolation of Vibration from Mechanical Equipment.* Mechanical equipment in buildings is often placed over critical spaces, such as executive offices, performing areas or living units. Control of vibration (and airborne noise) is critical in such instances. See Chapter 7.

## 3.1 Methods for Rating Noise

Many methods for rating the impact that noise may have on different activities have been developed over the years. Indeed, the proliferation of noise descriptors has probably resulted in more confusion than it has helped to define the effects of noise. Noise descriptors are defined throughout the chapters of this book, but it may be useful to list the more important of them here, with reference to where more complete definitions can be found (most definitions are in Chapter 1).

*Decibels (dB).* The basic unit for specifying noise levels. See Section 1.2.1.

*Decibels-A (dBA).* Commonly used for rating environmental noise, both in buildings and out-of-doors. The unit appears in many noise regulations, and is the basic unit used in more complex rating schemes. See Section 1.6.

*Decibels-C (dBC).* Similar to dBA but with a virtually flat frequency weighting characteristic. See Section 1.6.

*Room Criteria Curves (RC).* A recent derivation of NC curves, but more restrictive at the low and high frequencies. For this reason, it is preferred to using NCs in many situations. See Section 1.7.

*Noise Criteria Curves (NC).* A family of curves defining maximum permissible octave band (or one-third octave band) sound levels. Commonly used for specifying background noise from HVAC systems and other sources. See Section 1.7.

*Speech Interference Level (SIL).* An average of mid-frequency octave band sound levels, once used extensively for assessing speech interference but now losing favor to dBA, NC and AI (see below). See Section 1.8.2.

*Articulation Index (AI).* A method for determining the effect that voice level, distance and background noise will have on speech intelligibility. See Section 1.8.2.

*Sones/Phons.* A method for calculating perceived loudness (sones) and loudness level (phons), See Section 1.5.1.

*Noys.* Similar to sones, but the attribute is noisiness rather than loudness. The difference between the two is not particularly significant. See Section 1.8.4.

*Perceived Noise Level (PNL, PNdB).* A single-number rating for noise based on a one-third octave band frequency analysis. See Chapter 8, Section 8.2.1.

*Energy Average Level ($L_{eq}$).* A type of average, usually stated in dBA, based on the total energy contained in a signal for a specified time period. See Section 1.3.2.

The noise descriptors listed below are used for rating community noise. Definitions are given in Section 8.2.1.

Day-night Sound Level (Ldn or DNL).

Community Noise Exposure Level (CNEL).

Single Event Level (SEL or SENEL).

Noise Exposure Forecast (NEF)

Composite Noise Rating (CNR).

Noise Level Percentile ($L_x$). (The x specifies the percentage of time the specified noise level is exceeded.)

### 3.1.1 Building Codes

Acoustical performance standards and guidelines for airborne and impact sound isolation of walls and floor-ceiling assemblies in multifamily dwellings and motel and hotel guest rooms are contained in numerous municipal building codes and ordinances, but most have their origin in the Uniform Building Code (UBC), Appendix Chapter 35. This Chapter establishes minimum performance requirements, but experience has shown that just meeting the standards will still result in numerous complaints of poor acoustical privacy.

The Federal Housing Administration (FHA) publishes guidelines that are more realistic in terms of occupant satisfaction and are recommended for most design applications.

The acoustical performance measures and test methods specified in these standards and guidelines are presented in the following Sections 3.2 and 3.5, and Appendix A.10. Applications to multifamily housing and hotels and motels are discussed in Chapter 5, Sections 5.4.2 and 5.4.3.

## 3.2 Airborne Sound Transmission Loss (TL)

Airborne sound transmission loss, abbreviated TL, refers to sound that is transmitted by the air until it encounters a building element. At this encounter, varying portions of the sound will be reflected back toward the source, absorbed within the element, or transmitted on through where it continues as airborne sound in the adjacent space, or "receiving room." This Section is concerned with that portion of the incident sound that is transmitted through the building element.

For the purposes of describing the airborne sound transmission loss characteristics of building materials, it is convenient to divide them into single panel and multiple panel constructions. The main physical properties of single panels that affect TL are mass, stiffness and damping. A lead sheet is a good example of a material where the TL is mass controlled, because it has virtually no stiffness and high internal damping. Such materials are sometimes referred to as a "limp mass." The Sound Transmission Class (STC—see Section 3.2.1) that may be expected for masonry walls based only on the mass of the wall is shown in Figure 3-1, showing curves based on a theoretical limp mass model, and another lower curve derived from actual TL measurements.

For most single panel materials, such as sheet steel, stiffness becomes a significant factor in determining TL. The mass elements are now coupled together, and do not vibrate independently of each other. Stiffness can either raise or lower the TL compared to what mass alone would

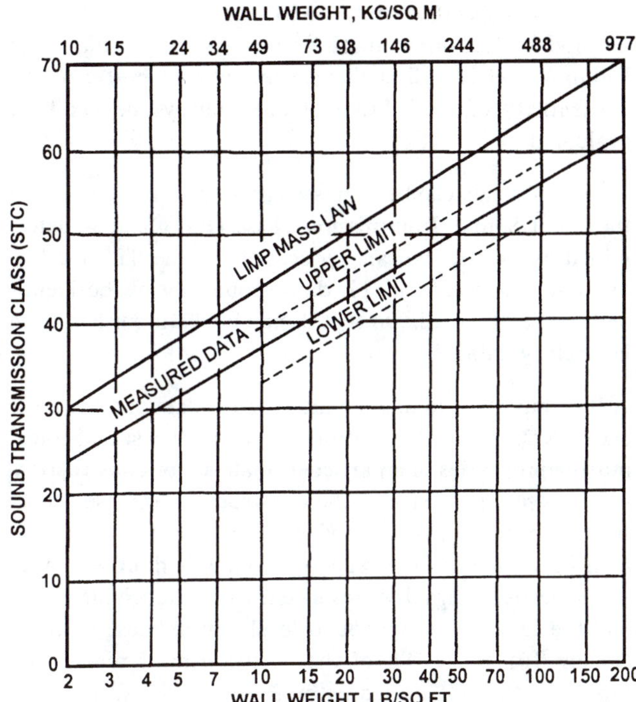


Data are mean square fit for 22 concrete masonry walls from 3 to 200 lb/sq ft (14.6 to 977 kg/sq m). Data based on sound transmission loss tests conducted by Riverbank Acoustical Laboratories.

**Figure 3-1.** STC ratings for single wythe concrete masonry walls.

account for, and the effects of stiffness on TL are usually frequency dependent.

Damping in a material affects the rate at which vibration of the material dies away. The sheet steel panel will ring when tapped because it has relatively low internal damping. If a material such as automotive undercoating is applied to the steel, the damping is increased and the ringing is diminished to a dull thud. Damping materials are often laminated to sheet material to improve TL. Figure 3-2 contains a generalized curve showing how the TL of a material is affected by mass, stiffness and damping as a function of frequency. This Figure will be referred to later in this Chapter with reference to the TL characteristics of specific materials.

In order to achieve high values of TL, it becomes inefficient to keep increasing the mass of a single panel. By subdividing the mass into two or more separate elements with an airspace between, a new property is introduced, that of isolation. The air itself does couple the two elements together to some extent, which can be minimized by increasing the separation, and by introducing an acoustical absorbent within the enclosed space.

The following sections of this Chapter contain a discussion of the methods by which TL is measured and re-

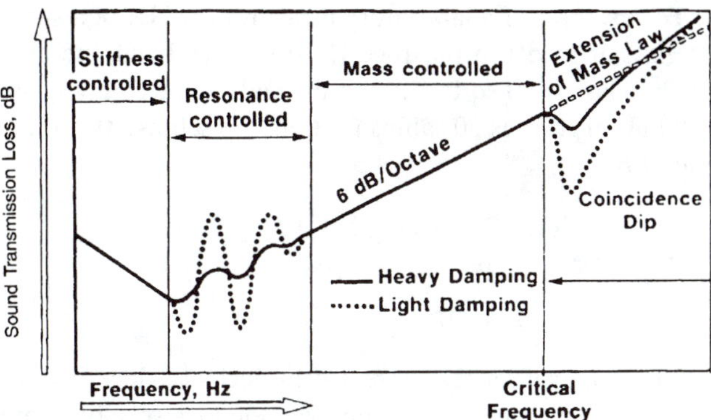


Notice the importance of damping in restricting the effects of resonances and the coincidence dip.

**Figure 3-2.** Factors affecting the sound transmission loss of a material.

ported, and TL performance data are presented for many types of building materials and constructions.

### 3.2.1 STC—FSTC—NIC

Airborne sound transmission loss is measured in the laboratory by the two-room method, where a sound source is located in one of the rooms and the test specimen is installed in an opening in the wall between the two rooms. The construction of the rooms is very massive, and the rooms are structurally isolated from each other, so that the only sound transmitted between rooms is through the test specimen. The laboratory test method is specified by ASTM Designation: E-90, "Test Method for Laboratory Measurement of Airborne Sound Transmission Loss of Building Partitions."

The source test signal consists of 18 contiguous one-third octave bands of "pink" noise (see Section 1.1.2) with center frequencies from 100 to 5000 Hz. The sound level is measured in both the source and receiving rooms (Ls and Lr, respectively), and the difference between them is the noise reduction (NR), expressed in decibels. Because the sound level measured in the receiving room depends not only on the sound transmitted through the test specimen, but on the area of the test specimen and the total absorption of the room as well (see Chapter 4, Section 4.6.3), the NR must be adjusted, or "normalized," to account for these factors. The result is the TL at each of the 18 frequency bands. The relationship between NR and TL is given by

$$\mathrm{TL} = \mathrm{Ls} - \mathrm{Lr} + 10\ \log\ \mathrm{S} - 10\ \log\ \mathrm{A},\ \mathrm{dB}$$
$$= NR + 10\ \log\ \mathrm{S} - 10\ \log\ A,\ \mathrm{dB}$$

where S = test specimen area, sq ft (sq m)

and A = total receiving room absorption, U.S. sabins (metric sabins).

For example, if a laboratory measures the NR of a sheet of glass at 35 dB in the 1000 Hz frequency band, the area of the glass is 20 sq ft (1.86 sq m) and the receiving room total absorption is 50 sabins (14.6 metric sabins), the TL is found by

$$\begin{aligned} TL &= 35 + 10 \log 20 - 10 \log 50, \text{ dB} \\ &= 35 + 13 - 17, \text{ dB} \\ &= 31 \text{ dB} \end{aligned}$$

TL is appropriately expressed in decibels, since it is, in effect, the *difference* between two sound levels, or more precisely the ratio of incident acoustic energy to transmitted energy. The TL is independent of the source room sound level; if the source level is increased by 10 dB, the receiving room level also increases by 10 dB, and the NR remains unchanged.

TL data can be expressed as a single-number rating called the Sound Transmission Class (STC), which is often used for specification purposes. The procedure for determining the STC is specified in ASTM Designation: E-413, "Classification for Rating Sound Insulation." It is a curve-fitting procedure following certain rules in which a set of contours are compared with the individual TL values and the appropriate contour selected. An explanation of the method for calculating STC and the associated table of contour values are presented in Appendix A.6.

A typical STC contour emphasizes the frequency range important for the understanding of speech (see Section 1.8.1). For this reason, an STC performance specification will not be appropriate for controlling the transmission of a sound which may contain strong low frequency components, such as music, mechanical noise, and some transportation and industrial noise sources. Where low frequency noise may be a problem, the TL data at those specific frequencies should be consulted.

The limitations of attempting to describe the TL characteristics of a material or construction by a single number should be understood. An STC rating of a certain value can represent the performance of different constructions that have widely different TL curves. Figure 3-3 shows TL curves for two very different constructions, but both curves are rated at STC-46. If low frequency TL was a concern, for example, the masonry wall would be the obvious choice, where at 100 Hz the difference in TL is just over 20 dB.

Sometimes there is a need to measure the TL of a construction in a field environment, in which case the rigid controls to eliminate sound transmission by transmission paths other than the test specimen (termed "flanking transmission") are often not present. There are a number of techniques for identifying and accounting for flanking transmission, so that a reasonably accurate measurement of TL can be made. The test method for field TL measurements is specified in ASTM Designation: E-336, "Test Method for Measurement of Airborne Sound Insulation in Buildings." It is a difficult measurement at best, and the authenticity of field TL data must always be carefully evaluated.

Field TL data can be expressed as a single number called the Field Sound Transmission Class (FSTC). It is determined by exactly the same procedure as the STC method discussed earlier. Field TL data should always be identified by the "F" prefix, so that it may be distinguished from laboratory data.

It is important to understand that TL data and the associated STC (or FSTC) ratings relate to the sound-transmitting properties of a particular material or construction, *independent* of the size of the construction and the acoustical properties of the receiving space. For example, if the area of a partition between two rooms is doubled, twice the acoustic energy is transmitted to the receiving room, and the receiving room sound level will increase by 3 dB (10 log $S_2/S_1$). The TL of the partition must remain the same, but the NR decreases by 3 dB. Similarly, if the receiving room acoustical absorption is doubled (−10 log $A_2/A_1$), the sound level in the room will be 3 dB less, and the NR will increase by 3 dB. Accounting for these variables is inherent in the TL equation given earlier, by the +10 log S and −10 log A terms.

In solving actual sound transmission problems, it is usually the NR between spaces that is ultimately of interest. This is the descriptor that tells the user what actual sound levels will be experienced in the receiving space. The NR data may be expressed as a single number in the same manner as the STC and FSTC. This single number is called the Noise Isolation Class (NIC), and is determined by the same ASTM Designation: E-413. Once again, STC and FSTC ratings are for individual building materials or constructions; NIC ratings are for the actual NR between sound source and receiver, and is often determined by sound transmitted through *several* building elements or sound leakage paths.

Because the sound transmitted between rooms often involves several building components, it is necessary to consider the TL of each separate component to calculate the NR. This technique is called "Composite TL," and is discussed in Section 3.4.

It is useful to have some appreciation of how NIC values relate to the degree of acoustical privacy. Remember that the method for determining NIC (like STC) emphasizes the speech frequencies; where strong low frequency components may be present, the TL at the lower frequencies should be considered.

NIC <30. Usually the result of uncontrolled sound leakage paths (see Section 3.8). Complete lack of speech privacy.

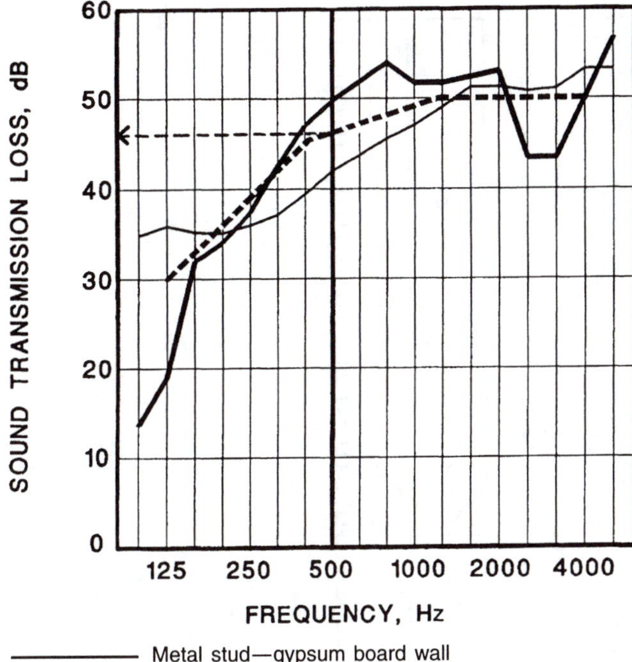


——— Metal stud—gypsum board wall
——— 8 inch concrete block wall
- - - - - STC 46 contour

The STC rating is determined by the TL value where the STC contour coincides with the 500 Hz line.

**Figure 3-3.** Two walls with STC 46 ratings.

NIC 30–35. Typical of the attenuation of many suspended ceilings with a ceiling-high partition between rooms. Speech is easily understood, though some words may be lost. Speech privacy is poor.

NIC 35–40. Typical for standard interior home construction with light weight frame walls and hollow core doors. Neighboring conversations are clearly audible and may be understood if listening conditions are good (i.e., low background noise).

NIC 40–45. Average conditions for good commercial construction, with well sealed walls and no common air ducts. Reasonably good speech privacy, unless a concerted listening effort is made.

NIC 45–50. Usually adequate for most private offices, but marginal for between units in multifamily housing. Some louder adjacent noise may be noticeable. Speech privacy is very good except for loud talking coupled with ideal listening conditions.

NIC 50–55. Confidential speech privacy assured for all but the most severe requirements. Heavy, multiple wall construction, carefully isolated air ducts and heavy, well sealed doors are required. Flanking transmission must be controlled.

NIC >55. Assurance that most noise and speech will be inaudible. Very adequate for between units of luxury multifamily housing.

## 3.3 Sound Transmission Loss Data

Sound transmission loss (TL) data are shown in Sections 3.3.1–3.3.9 for the most commonly used categories of materials or constructions, subdivided as follows:

| Section | Category |
|---|---|
| 3.3.1 | Single Panels |
| 3.3.2 | Wood and Metal Frame Partitions |
| 3.3.3 | Concrete and Masonry Walls |
| 3.3.4 | Glass Windows |
| 3.3.5 | Doors, Operable Partitions |
| 3.3.6 | Facades |
| 3.3.7 | Acoustical Curtains |
| 3.3.8 | Wood Frame Floors |
| 3.3.9 | Concrete Floors |

All TL data are presented as graphs as well as in numerical tables. The intent is to give the reader a quick visual impression of how the performance of a material or construction varies with frequency, or how one construction compares with another. Graphs achieve this purpose, while interpreting tabular data in making such evaluations is more difficult.

In most cases the TL data presented are based on laboratory measurements, and can be so identified by the STC rating designation. If the data are based on field measurements, the rating designation will be either FSTC or NIC. It is customary to make some allowance when using laboratory test data for applications in the field. Some references, such as Chapter 35 of the Uniform Building Code, suggest subtracting 5 dB from the laboratory values, and in the absence of special conditions which would suggest another value, 5 dB is probably a reasonably safe adjustment. Some categories of constructions require special attention during installation to achieve acceptable performance, and this information is presented in the discussions that accompany each category.

The TL data presented in the following graphs were derived mostly from laboratory test data (a list of acoustical testing laboratories is contained in Appendix A.10). For some materials the results of only one test are presented, while for others data from as many as four tests are averaged. Unfortunately, test data for presumably identical constructions from different laboratories (and sometimes even from the same laboratory) often display considerable variation. In such cases deriving a representative TL curve relies on the knowledge and experience of the interpreter. It could be argued that in such cases the data should be shown in ranges rather than specific data points, but that only places the burden of choosing what values to use on the reader.

The authors have carefully evaluated the available TL data in preparing the graphs that follow. The readers may encounter other data that will not agree with what is pre-

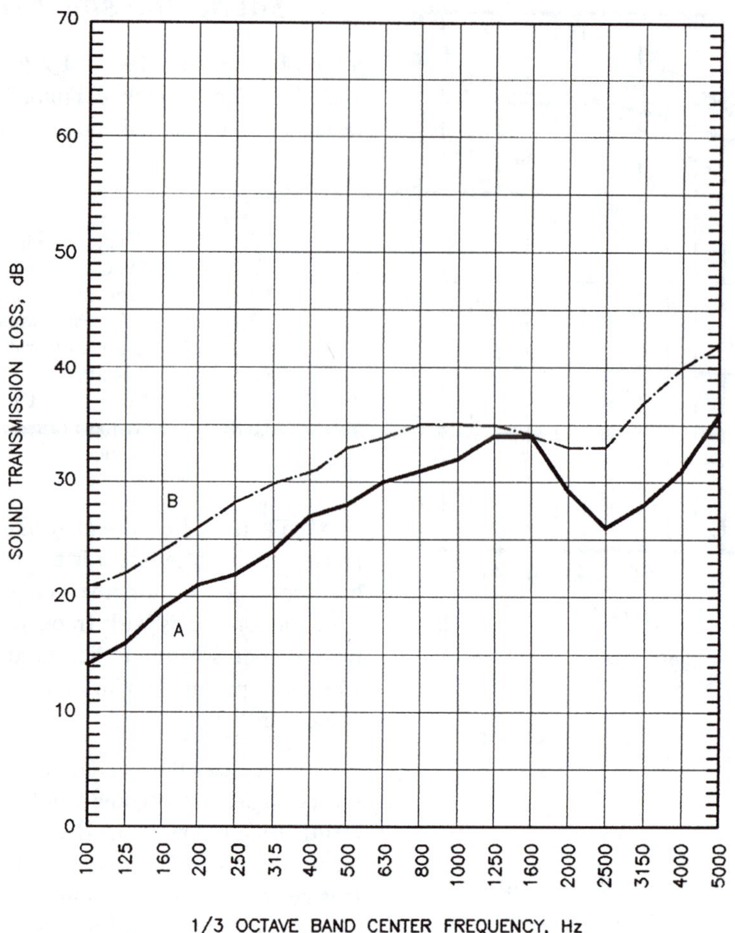


SOUND TRANSMISSION LOSS, dB

| CURVE | 100 | 125 | 160 | 200 | 250 | 315 | 400 | 500 | 630 | 800 | 1000 | 1250 | 1600 | 2000 | 2500 | 3150 | 4000 | 5000 | STC RATING |
|---|---|---|---|---|---|---|---|---|---|---|---|---|---|---|---|---|---|---|---|
| A | 14 | 16 | 19 | 21 | 22 | 24 | 27 | 28 | 30 | 31 | 32 | 34 | 34 | 29 | 26 | 28 | 31 | 36 | 30 |
| B | 21 | 22 | 24 | 26 | 28 | 30 | 31 | 33 | 34 | 35 | 35 | 35 | 34 | 33 | 33 | 37 | 40 | 42 | 34 |

———— A) one layer 5/8″ (16 mm) gypsum board, STC 30
—·—·—·—·· B) two layers 5/8″ (16 mm) gypsum board, STC 34

**Figure 3-4.** Sound transmission loss of gypsum board.

sented, and they should be advised to evaluate such data with care. Many tests are performed for product development purposes, and the constructions are not representative of typical field practice. This information may not be included in the test reports, or in the sales literature based on such reports.

### 3.3.1 Single Panels

*Figure 3-4.* Both of these curves of the TL of gypsum board by itself demonstrate a basic characteristic of the material that will be seen in many constructions using this material. The dip in the TL curve at frequencies around 2000 Hz is caused by the reinforcement of the flexural wave in the gypsum board by incident sound at a particular angle of incidence. This effect is termed the "coincidence dip," and the extent of the dip will depend on the physical properties of the material (many materials are subject to this phenomenon) such as stiffness and damping. In this figure, that dip is more pronounced for the single layer than for the double layer.

*Figure 3-5.* Notice that the plywood TL curve shows a coincidence dip at 2000 Hz, while the sheet steel curve

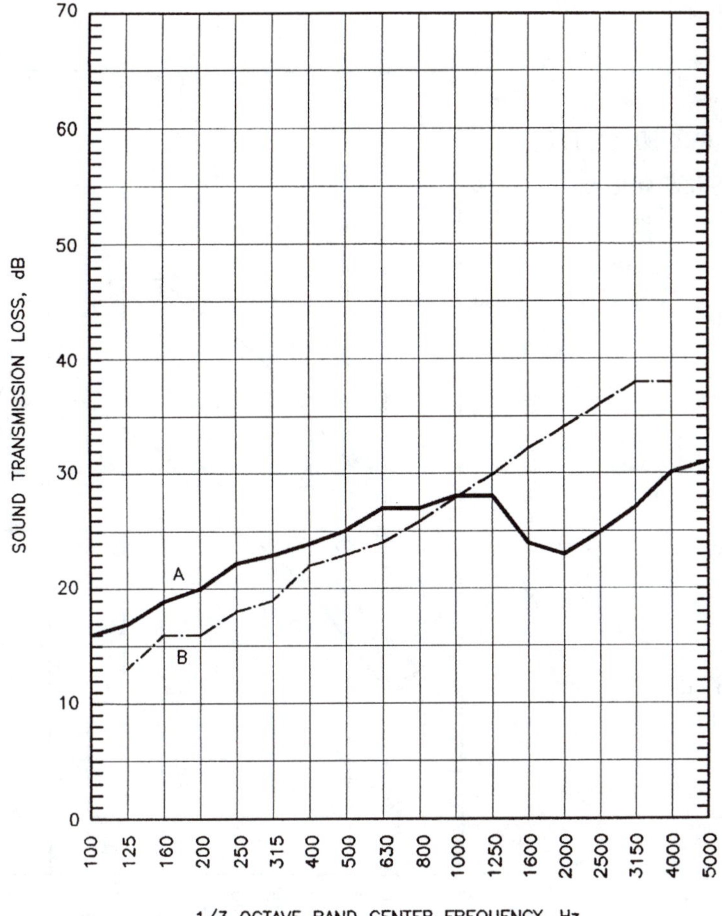


SOUND TRANSMISSION LOSS, dB

| CURVE | 100 | 125 | 160 | 200 | 250 | 315 | 400 | 500 | 630 | 800 | 1000 | 1250 | 1600 | 2000 | 2500 | 3150 | 4000 | 5000 | STC RATING |
|---|---|---|---|---|---|---|---|---|---|---|---|---|---|---|---|---|---|---|---|
| A | 16 | 17 | 19 | 20 | 22 | 23 | 24 | 25 | 27 | 27 | 28 | 28 | 24 | 23 | 25 | 27 | 30 | 31 | 26 |
| B | -- | 13 | 16 | 16 | 18 | 19 | 22 | 23 | 24 | 26 | 28 | 30 | 34 | 34 | 36 | 38 | 38 | -- | 27 |

——————— A) 3/4″ (19 mm) plywood
—·—·—·—·- B) 22 gauge sheet steel

**Figure 3-5.** Sound transmission loss of single panels.

does not. The greater stiffness of the plywood, however, provides slightly higher TL at the low and mid frequencies. A steel sheet of twice the weight (16 gauge) will have a TL curve approximately 5 dB higher.

### 3.3.2 Wood and Metal Frame Partitions

*Figure 3-6.* This figure shows the TL for three single wood stud partition constructions with a single layer of gypsum board on each side. While data for 5/8″ (16 mm) thick gypsum board is shown, similar data for 1/2″ (13 mm) gypsum board will be only slightly less. Notice the coincidence dip is present for all constructions.

The addition of insulation in the stud space with the gypsum board attached directly to the studs on both sides, shows only modest improvement, because the studs are the major transmission path between source and receiver sides of the partition. When the gypsum board is applied using resilient channel on one side, however, this transmission path is interrupted, the air space becomes a more important transmission path, and the presence of insulation in

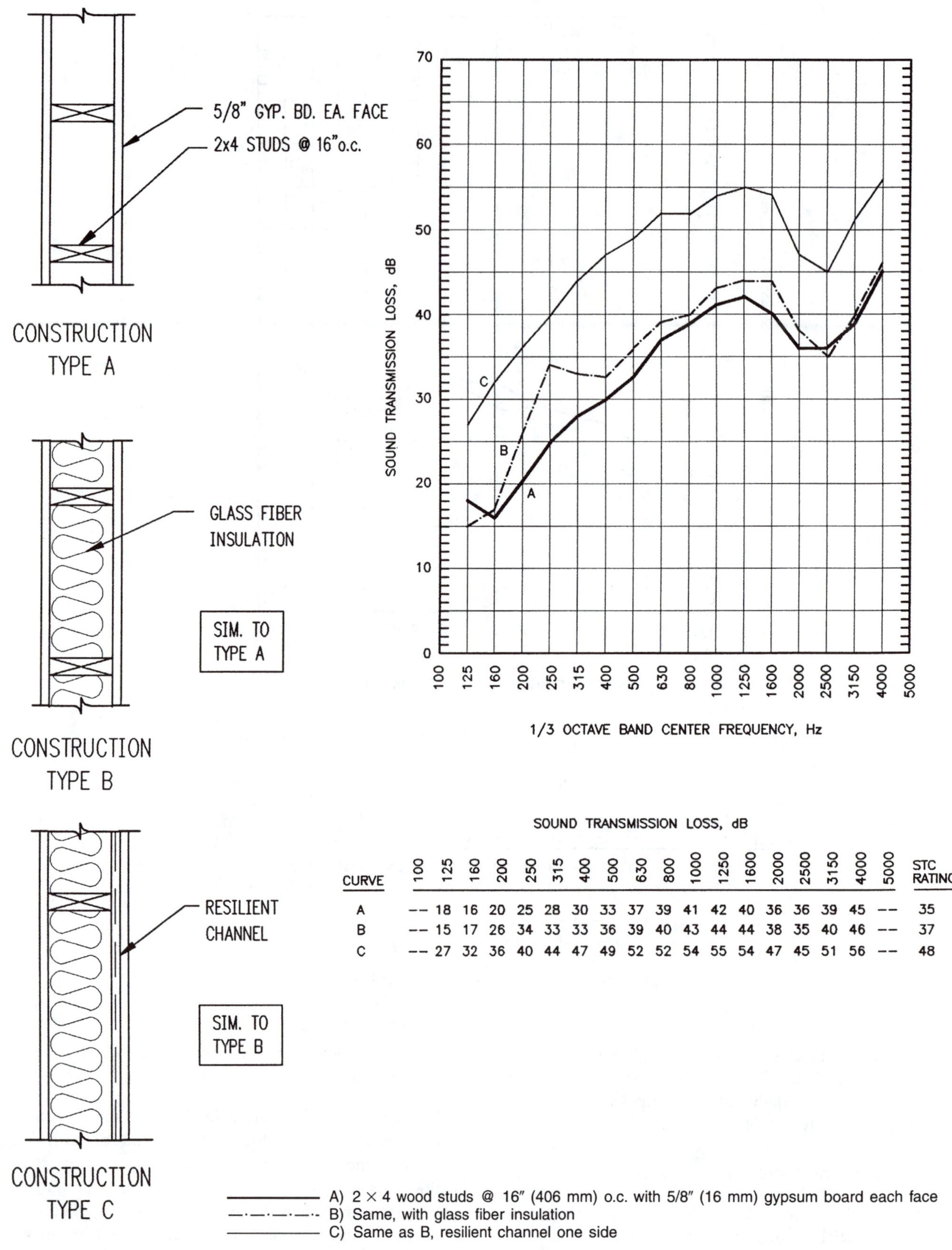


SOUND TRANSMISSION LOSS, dB

| CURVE | 100 | 125 | 160 | 200 | 250 | 315 | 400 | 500 | 630 | 800 | 1000 | 1250 | 1600 | 2000 | 2500 | 3150 | 4000 | 5000 | STC RATING |
|---|---|---|---|---|---|---|---|---|---|---|---|---|---|---|---|---|---|---|---|
| A | -- | 18 | 16 | 20 | 25 | 28 | 30 | 33 | 37 | 39 | 41 | 42 | 40 | 36 | 36 | 39 | 45 | -- | 35 |
| B | -- | 15 | 17 | 26 | 34 | 33 | 33 | 36 | 39 | 40 | 43 | 44 | 44 | 38 | 35 | 40 | 46 | -- | 37 |
| C | -- | 27 | 32 | 36 | 40 | 44 | 47 | 49 | 52 | 52 | 54 | 55 | 54 | 47 | 45 | 51 | 56 | -- | 48 |

**Figure 3-6.** Sound transmission loss of wood stud partitions.

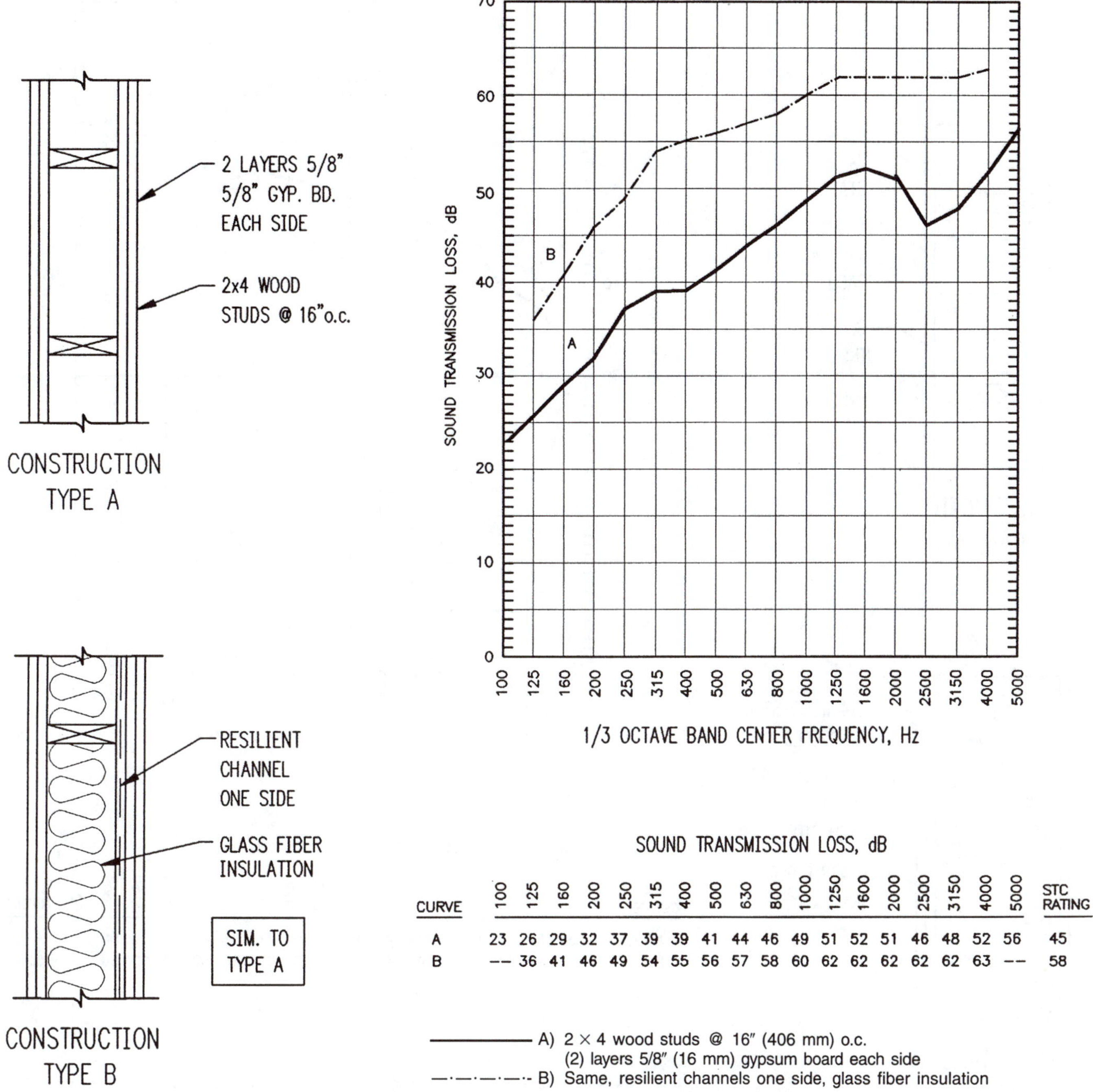


SOUND TRANSMISSION LOSS, dB

| CURVE | 100 | 125 | 160 | 200 | 250 | 315 | 400 | 500 | 630 | 800 | 1000 | 1250 | 1600 | 2000 | 2500 | 3150 | 4000 | 5000 | STC RATING |
|---|---|---|---|---|---|---|---|---|---|---|---|---|---|---|---|---|---|---|---|
| A | 23 | 26 | 29 | 32 | 37 | 39 | 39 | 41 | 44 | 46 | 49 | 51 | 52 | 51 | 46 | 48 | 52 | 56 | 45 |
| B | -- | 36 | 41 | 46 | 49 | 54 | 55 | 56 | 57 | 58 | 60 | 62 | 62 | 62 | 62 | 62 | 63 | -- | 58 |

**Figure 3-7.** Sound transmission loss of wood stud partitions.

the stud space is much more effective. While the use of resilient channels on one side causes a substantial improvement in TL compared to direct attachment, resilient channels on both sides (not shown) does not show twice the improvement.

If shelving or other fixtures are to be attached to the side with resilient channels, do not allow the attachment to contact the stud, as this will short circuit the isolation provided by the channel. Most fixtures can be attached directly to the gypsum board with appropriate devices (such as molly bolts).

*Figure 3-7.* Doubling the gypsum board to two layers on both sides of a single wood stud partition without insulation increases the STC rating from 35 (Figure 3-6) to 45. The use of resilient channels on one side and adding insulation increases the STC to 58, a substantial improvement.

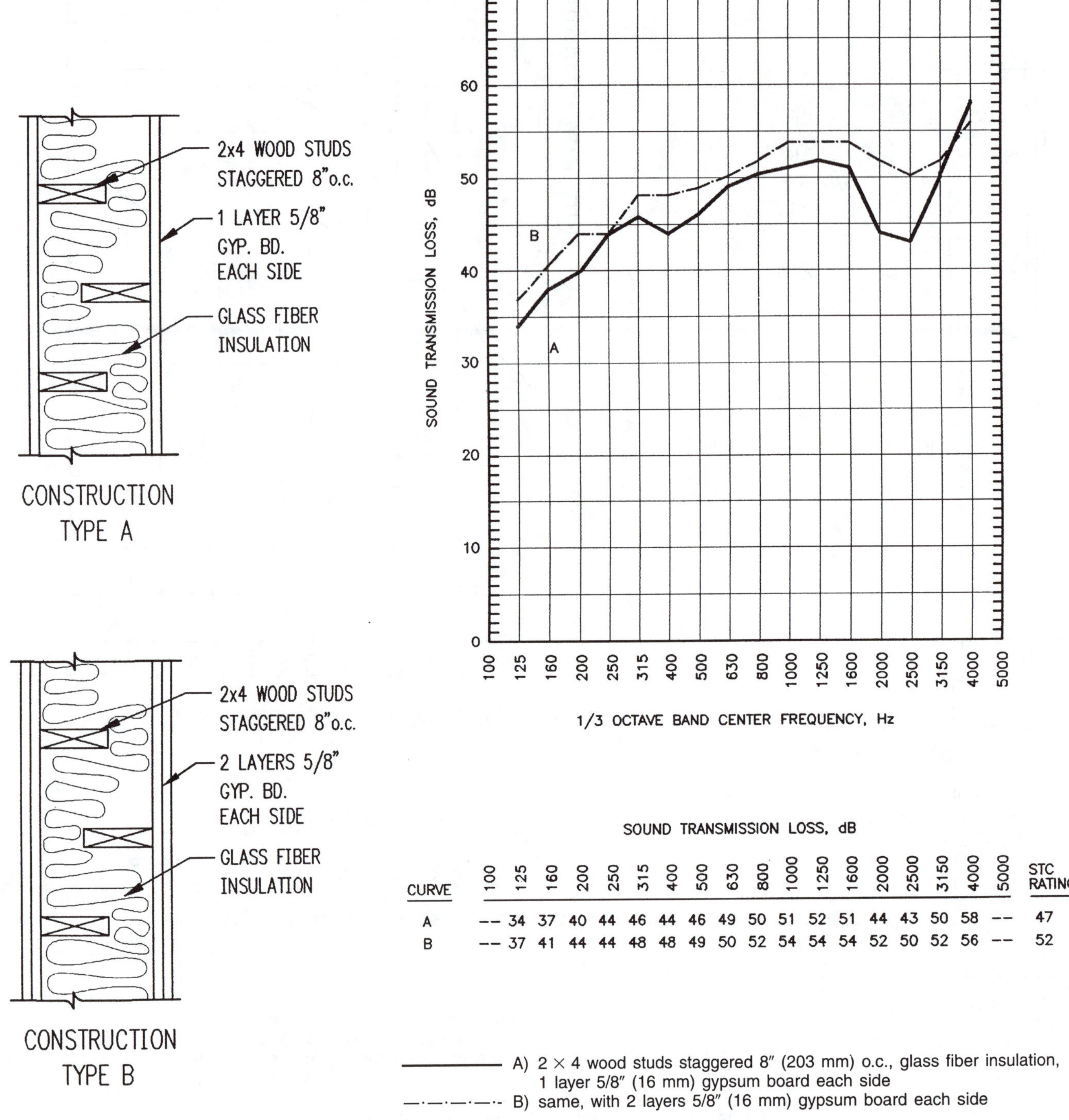


SOUND TRANSMISSION LOSS, dB

| CURVE | 100 | 125 | 160 | 200 | 250 | 315 | 400 | 500 | 630 | 800 | 1000 | 1250 | 1600 | 2000 | 2500 | 3150 | 4000 | 5000 | STC RATING |
|---|---|---|---|---|---|---|---|---|---|---|---|---|---|---|---|---|---|---|---|
| A | -- | 34 | 37 | 40 | 44 | 46 | 44 | 46 | 49 | 50 | 51 | 52 | 51 | 44 | 43 | 50 | 58 | -- | 47 |
| B | -- | 37 | 41 | 44 | 44 | 48 | 48 | 49 | 50 | 52 | 54 | 54 | 54 | 52 | 50 | 52 | 56 | -- | 52 |

**Figure 3-8.** Sound transmission loss of staggered wood stud partitions.

Notice the coincidence dip is not present in the data for the partition with the resilient channels.

*Figure 3-8.* Staggering the wood studs isolates one side of the partition from the other, but the common plates at the base and head of the partition constitute a transmission flanking path. Even with two layers of gypsum board on each side, the STC-52 value is still less than the single stud partition with two layers on each side and resilient channels (STC-58, see Figure 3-7). The coincidence dip is evident in the TL curve for both constructions, particularly for the single layer.

*Figure 3-9.* This final figure for wood stud partitions shows the TL for a double set of studs with separate plates top and bottom. Separate plates improves the isolation compared to staggered studs on single plates, or the use of resilient channels with single studs. Achieving ratings

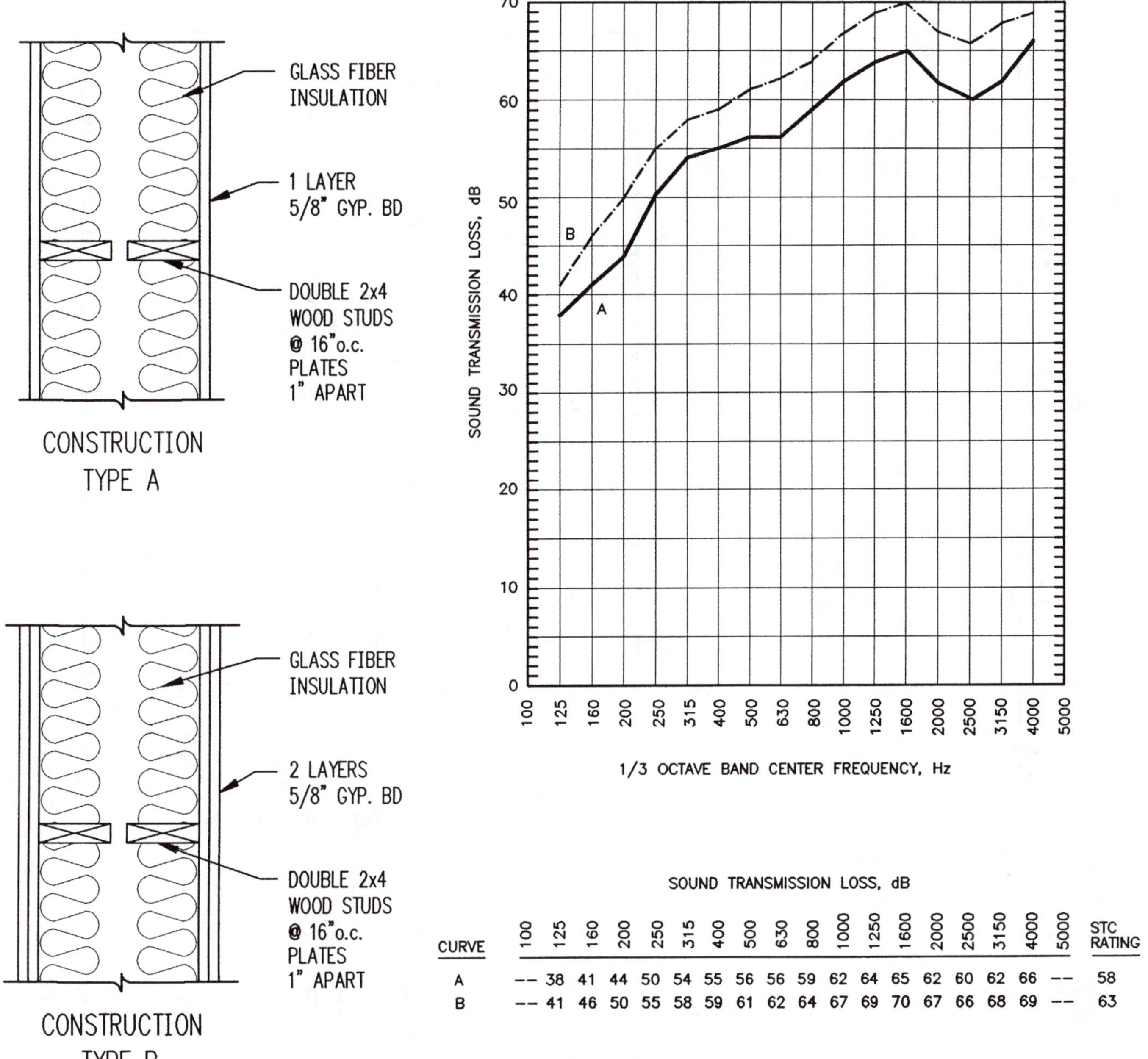


SOUND TRANSMISSION LOSS, dB

| CURVE | 100 | 125 | 160 | 200 | 250 | 315 | 400 | 500 | 630 | 800 | 1000 | 1250 | 1600 | 2000 | 2500 | 3150 | 4000 | 5000 | STC RATING |
|---|---|---|---|---|---|---|---|---|---|---|---|---|---|---|---|---|---|---|---|
| A | -- | 38 | 41 | 44 | 50 | 54 | 55 | 56 | 56 | 59 | 62 | 64 | 65 | 62 | 60 | 62 | 66 | -- | 58 |
| B | -- | 41 | 46 | 50 | 55 | 58 | 59 | 61 | 62 | 64 | 67 | 69 | 70 | 67 | 66 | 68 | 69 | -- | 63 |

———— A) 2 × 4 wood studs @ 16″ (406 mm) o.c. on separate plates 1″ apart with one layer of 5/8″ (16 mm) gypsum board each side
—·—·—·—·- B) Same, 2 layers 5/8″ (16 mm) gypsum board each side

**Figure 3-9.** Sound transmission loss of double wood stud partitions.

above the STC-63 for two layers of gypsum board on each side is about the upper practical limit for wood stud construction, though some additional TL could be achieved with more than two layers of gypsum board on one or both sides, or by increasing the separation between each side of the partition. The improvement by increasing the separation would be mostly at the lower frequencies. Do not add plywood or any other impervious material between the sets of studs. The internal separation between partition faces should be the maximum obtainable, and the addition of any such material in the middle of this space actually reduces the TL. If plywood is required for structural reasons, apply it to the outside face of the studs.

*Figure 3-10.* The purpose of this figure is to compare the TL of wood and metal single stud partitions. The remarkable improvement in TL with metal studs is caused by the much lower torsional (twisting) stiffness of metal studs compared to wood, resulting in less vibrational energy transmitted from one side to the other. It is important to

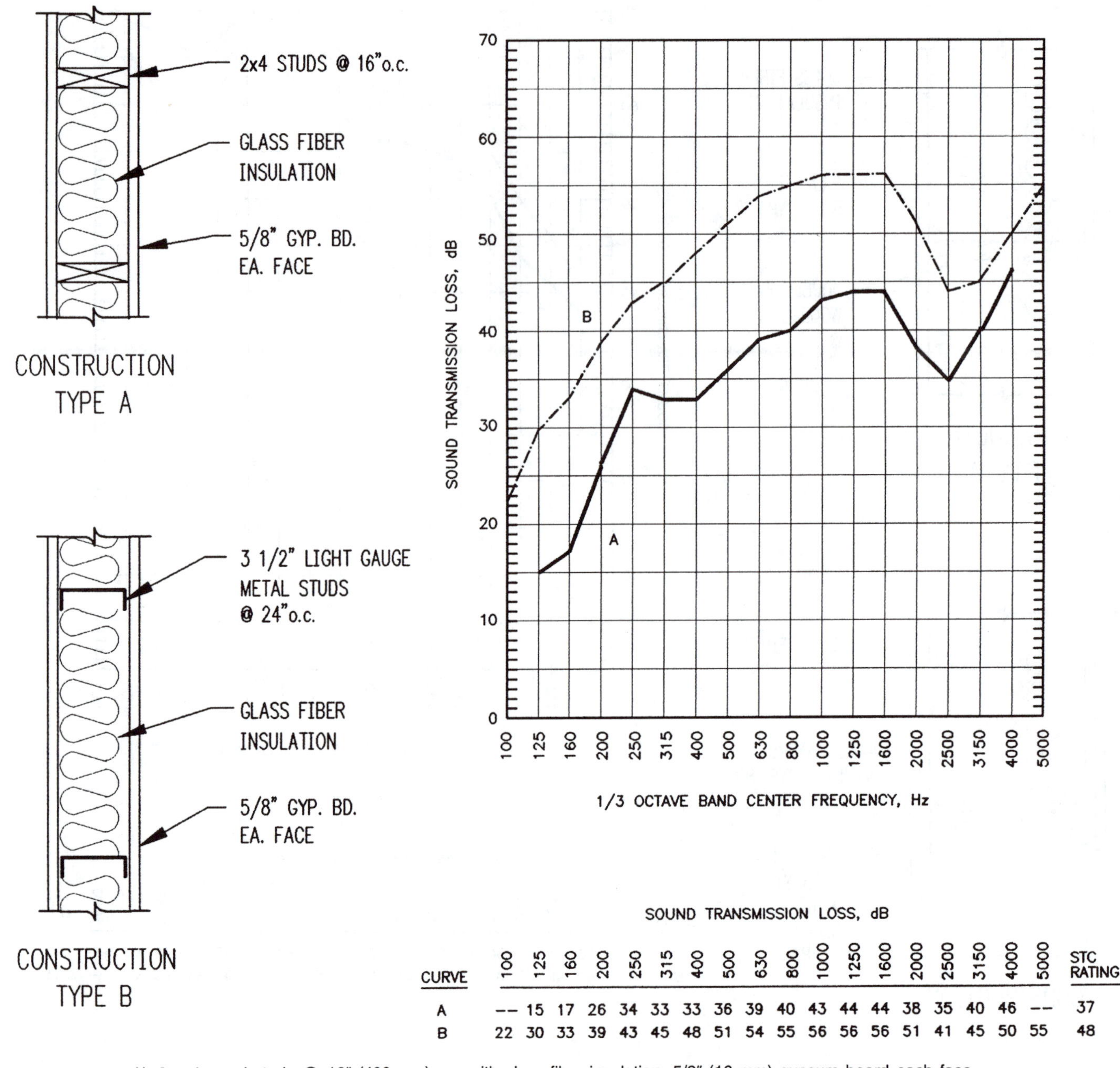


SOUND TRANSMISSION LOSS, dB

| CURVE | 100 | 125 | 160 | 200 | 250 | 315 | 400 | 500 | 630 | 800 | 1000 | 1250 | 1600 | 2000 | 2500 | 3150 | 4000 | 5000 | STC RATING |
|---|---|---|---|---|---|---|---|---|---|---|---|---|---|---|---|---|---|---|---|
| A | -- | 15 | 17 | 26 | 34 | 33 | 33 | 36 | 39 | 40 | 43 | 44 | 44 | 38 | 35 | 40 | 46 | -- | 37 |
| B | 22 | 30 | 33 | 39 | 43 | 45 | 48 | 51 | 54 | 55 | 56 | 56 | 56 | 51 | 41 | 45 | 50 | 55 | 48 |

———— A) 2 × 4 wood studs @ 16″ (406 mm) o.c. with glass fiber insulation, 5/8″ (16 mm) gypsum board each face

—·—·—·– B) 3 1/2 (89 mm) light gauge metal studs @ 24″ (610 mm) o.c. with glass fiber insulation, 5/8″ (16 mm) gypsum board each face

**Figure 3-10.** Sound transmission loss of wood & metal stud partitions (comparison).

note, however, that this applies *only* to light gauge metal studs. Heavy gauge studs or studs wider than 3 1/2″ (89 mm), usually required for tall partitions, will behave much like wood studs.

*Figure 3-11.* All metal studs for the partitions shown in this figure are light gauge, and all are single stud construction. The STC-61 rating for the partition with two layers of gypsum board on each side and resilient channels on one side would appear to perform about as well as the STC-63 rating obtained with the double wood stud partition and two layers of gypsum board on each side (Figure 3-9). Notice, however, the difference in TL at the low frequencies. The double wood stud partition has significantly better TL, and would be the better choice for reducing noise transmission in this frequency range. Also, the coin-

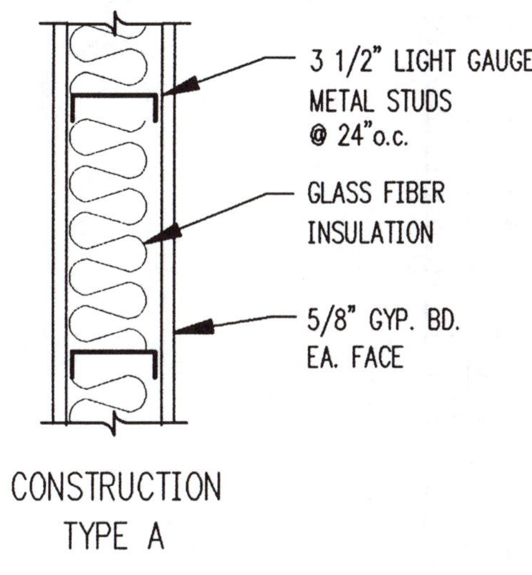


CONSTRUCTION TYPE A

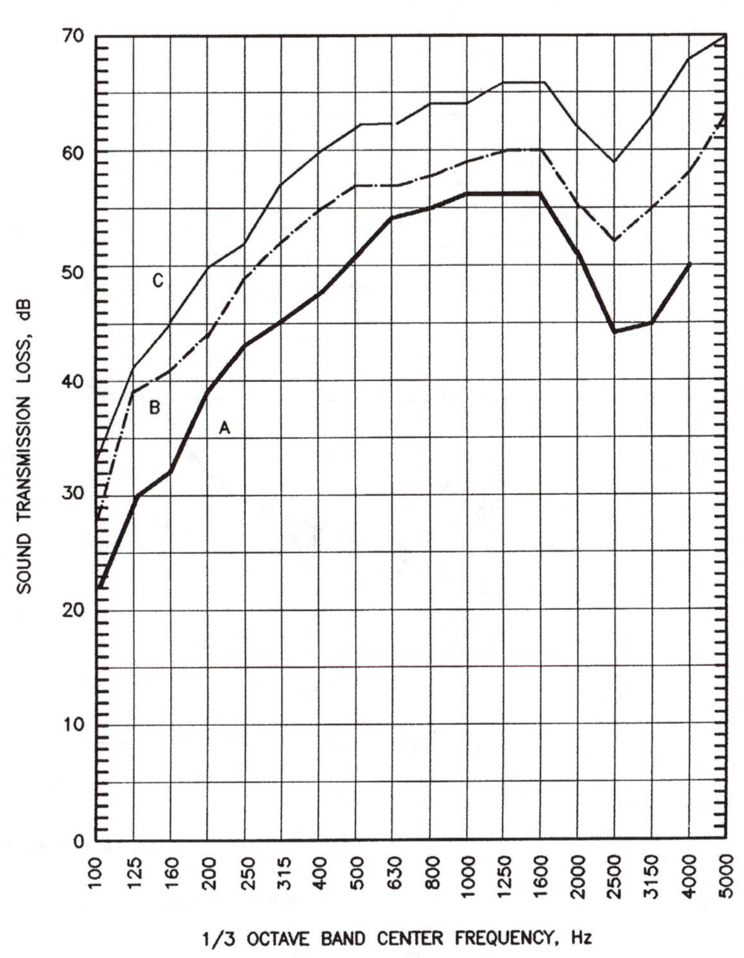


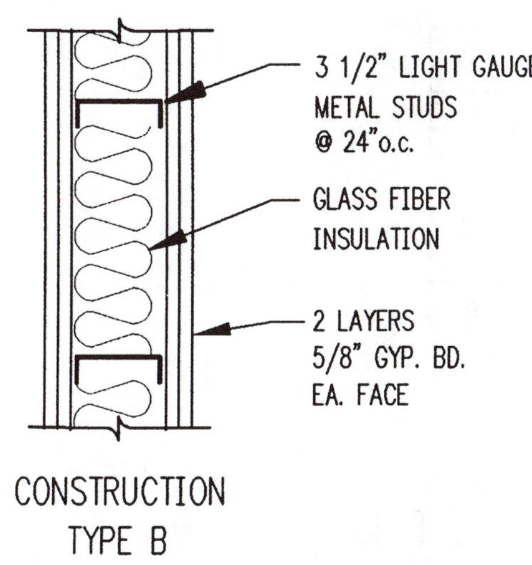


CONSTRUCTION TYPE B

SOUND TRANSMISSION LOSS, dB

| CURVE | 100 | 125 | 160 | 200 | 250 | 315 | 400 | 500 | 630 | 800 | 1000 | 1250 | 1600 | 2000 | 2500 | 3150 | 4000 | 5000 | STC RATING |
|---|---|---|---|---|---|---|---|---|---|---|---|---|---|---|---|---|---|---|---|
| A | 22 | 30 | 33 | 39 | 43 | 45 | 48 | 51 | 54 | 55 | 56 | 56 | 56 | 57 | 44 | 45 | 50 | -- | 48 |
| B | 28 | 39 | 41 | 44 | 49 | 52 | 55 | 57 | 57 | 58 | 59 | 60 | 60 | 55 | 52 | 55 | 58 | 63 | 56 |
| C | 33 | 41 | 45 | 50 | 52 | 57 | 60 | 62 | 62 | 64 | 64 | 66 | 66 | 62 | 59 | 63 | 68 | 70 | 61 |

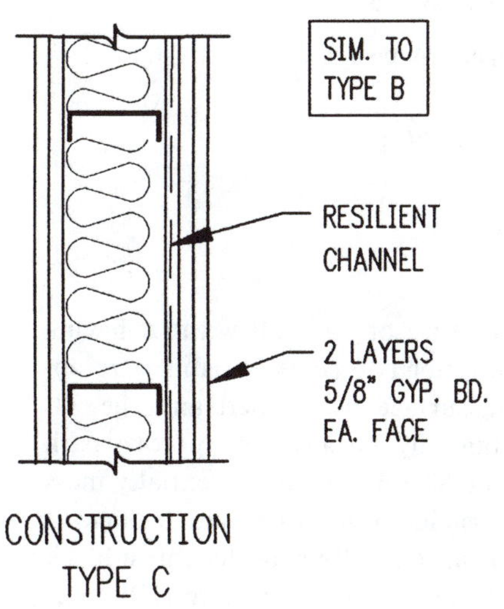


CONSTRUCTION TYPE C

A) 3 5/8″ light gauge metal studs 24″ o.c., glass fiber insulation
1 layer 5/8″ (16 mm) gypsum board each face
B) Same, 2 layers 5/8″ gypsum each face
C) Same as B, resilient channels added to studs 1 side

**Figure 3-11.** Sound transmission loss of metal stud partitions.

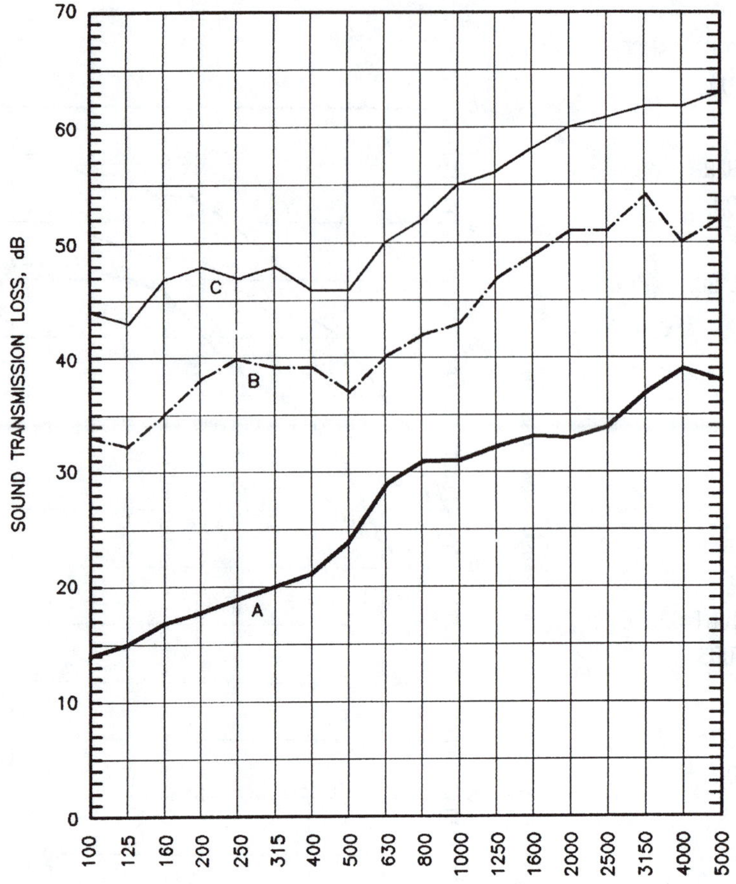


1/3 OCTAVE BAND CENTER FREQUENCY, Hz

SOUND TRANSMISSION LOSS, dB

| CURVE | 100 | 125 | 160 | 200 | 250 | 315 | 400 | 500 | 630 | 800 | 1000 | 1250 | 1600 | 2000 | 2500 | 3150 | 4000 | 5000 | STC RATING |
|---|---|---|---|---|---|---|---|---|---|---|---|---|---|---|---|---|---|---|---|
| A | 14 | 15 | 17 | 18 | 19 | 20 | 21 | 24 | 29 | 31 | 31 | 32 | 33 | 33 | 34 | 37 | 39 | 38 | 29 |
| B | 33 | 32 | 35 | 38 | 40 | 39 | 39 | 37 | 40 | 42 | 43 | 47 | 49 | 51 | 51 | 54 | 50 | 52 | 44 |
| C | 44 | 43 | 47 | 48 | 47 | 48 | 46 | 46 | 50 | 52 | 55 | 56 | 58 | 60 | 61 | 62 | 62 | 63 | 54 |

——— A) 4″ (102 mm) concrete block wall, expanded slag aggregate, 24 lb/sq ft (117 kg/sq m)

—·—·—·— B) Same, 2 coats paint each face, surfaces sealed.

——— C) 8″ concrete block wall, dense (sand & gravel) aggregate, 51 lb/sq ft (249 kg/sq m).

**Figure 3-12.** Sound transmission loss of concrete block walls.

cidence dip is more pronounced with metal studs than with wood studs.

### 3.3.3 Concrete and Masonry Walls

*Figure 3-12.* The type of aggregate used in the manufacture of concrete blocks affects both the weight of the blocks and their porosity, and these will in turn affect the TL performance. The expanded slag aggregate wall, for which TL data are shown in this Figure, was constructed from very porous block, permitting the sound to pass through the porous structure of the wall without having to move the mass of it. After two coats of paint were applied to each side, the surface was sealed and the TL values increased substantially because the porous path was no longer significant. The TL is now essentially mass controlled. Another possible source of sound leakage is through inadequate mortar joints. Be sure the entire block joint surfaces, both horizontal and vertical, are fully mortared.

It should be noted that the slag aggregate block was un-

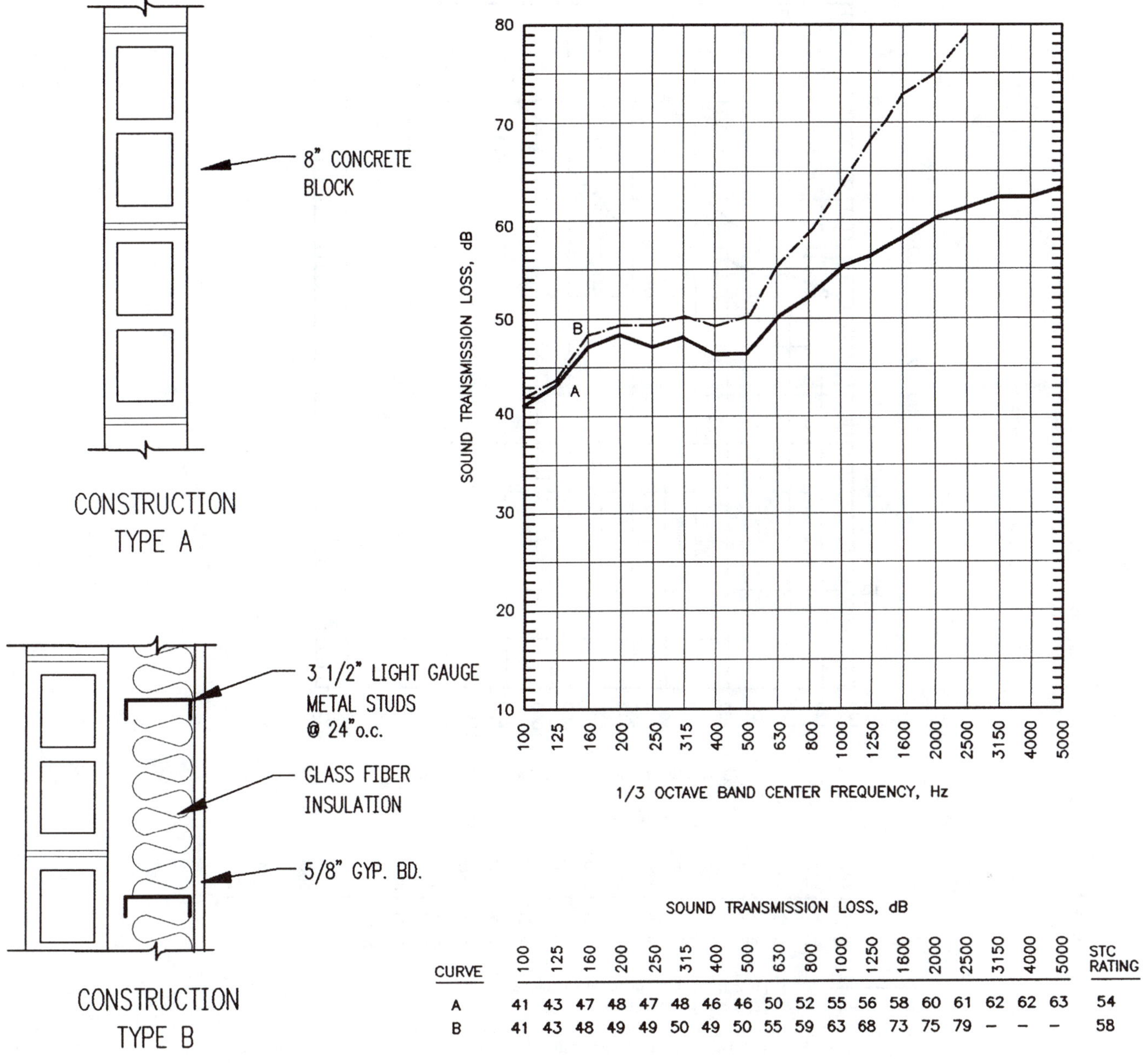


SOUND TRANSMISSION LOSS, dB

| CURVE | 100 | 125 | 160 | 200 | 250 | 315 | 400 | 500 | 630 | 800 | 1000 | 1250 | 1600 | 2000 | 2500 | 3150 | 4000 | 5000 | STC RATING |
|---|---|---|---|---|---|---|---|---|---|---|---|---|---|---|---|---|---|---|---|
| A | 41 | 43 | 47 | 48 | 47 | 48 | 46 | 46 | 50 | 52 | 55 | 56 | 58 | 60 | 61 | 62 | 62 | 63 | 54 |
| B | 41 | 43 | 48 | 49 | 49 | 50 | 49 | 50 | 55 | 59 | 63 | 68 | 73 | 75 | 79 | – | – | – | 58 |

——— A) 8″ (203 mm) dense aggregate concrete block 51 lb/sq ft (249 kg/sq m)

—·—·—·—· B) Same, with 3 1/2″ (89 mm) studs (metal or wood) spaced 1″ (25 mm) from CMU, no contact, insulation, 5/8″ (16 mm) gypsum board

**Figure 3-13.** Sound transmission loss of CMU/frame construction.

usually porous, and that most block types will not exhibit such an increase in TL by sealing the surface. It is always good practice, however, to seal the surface of concrete block walls if maximum TL is desired.

The TL of the unpainted dense aggregate concrete block wall (not shown) increased by only 1 or 2 dB by painting.

*Figure 3-13.* If a separate frame wall is constructed adjacent to but not in contact with a wall built of concrete masonry units (CMU) (or concrete wall—see Figure 3-14), the improvement in TL is very little at the low frequencies compared to the masonry wall by itself. As the frequency increases, however, the TL also increases and will result in TL values above 80 dB at the high frequencies. This TL characteristic is the result of the wavelength of the sound compared to the separation of the two walls. If the separation between the interior surfaces is more than one wavelength (which it is at the high frequencies), and absorption is

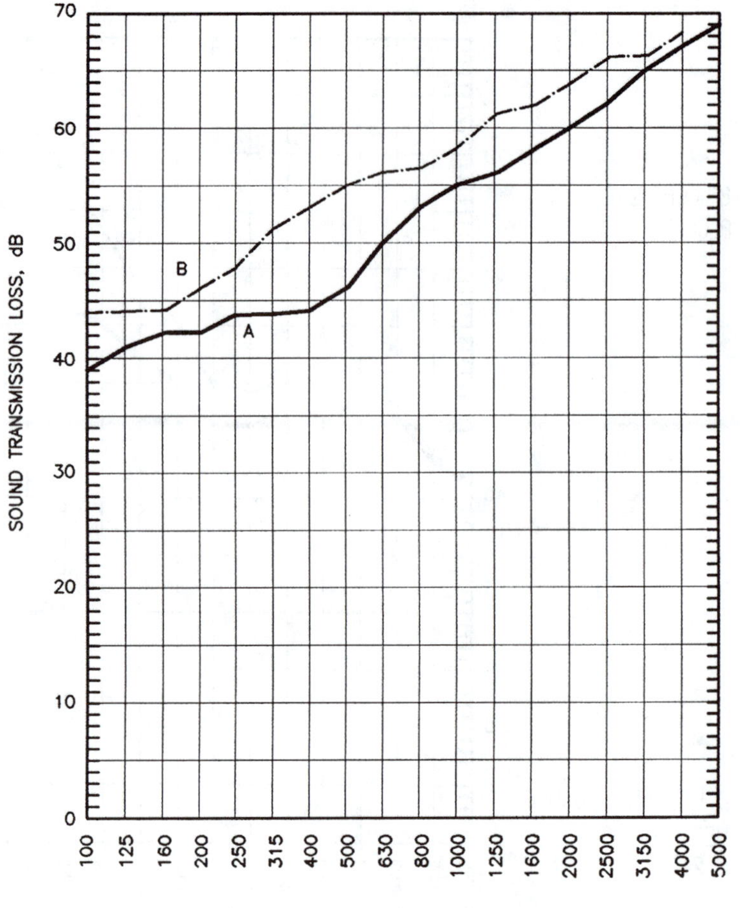


SOUND TRANSMISSION LOSS, dB

| CURVE | 100 | 125 | 160 | 200 | 250 | 315 | 400 | 500 | 630 | 800 | 1000 | 1250 | 1600 | 2000 | 2500 | 3150 | 4000 | 5000 | STC RATING |
|---|---|---|---|---|---|---|---|---|---|---|---|---|---|---|---|---|---|---|---|
| A | 39 | 41 | 42 | 42 | 44 | 44 | 44 | 46 | 50 | 53 | 55 | 56 | 58 | 60 | 62 | 65 | 67 | 69 | 53 |
| B | 44 | 44 | 44 | 46 | 48 | 51 | 53 | 55 | 56 | 56 | 58 | 61 | 62 | 64 | 66 | 66 | 68 | -- | 58 |

———————— A) 4″ (102 mm) thick concrete wall/slab
—·—·—·—·- B) 8″ (203 mm) concrete wall/slab

**Figure 3-14.** Sound transmission loss of concrete walls.

present in the airspace, the TLs of the two walls may almost be added without penalizing the TL of the second wall.

Another important consideration for preserving the TL of high performance constructions is structural flanking transmission. This subject is treated further in the discussion for Figure 3-27, and Section 3.8.

*Figure 3-14.* The TL performance of concrete walls is similar to that of nonporous concrete masonry with comparable weights, as can be seen by comparing the curves in this Figure with the top curve in Figure 3-12, representing the TL for a dense aggregate concrete block wall. Wall constructions in this category have considerably higher TL values at the low frequencies compared to frame partition constructions with similar STC ratings (see Figure 3-11, for example). The TL curves shown here apply also to concrete floors.

### 3.3.4 Glass, Windows

*Figure 3-15.* The TL performance of glass sheets, like gypsum board, exhibits a coincidence dip at the higher frequencies. The frequency range at which the dip occurs is dependent on the thickness of the glass, as can be seen by comparing the TL curves for the 1/8″ (3.2 mm) and 1/4″ (6.4 mm) glass. The dip for the 1/8″ glass appears to be centered at 5000 Hz, while the dip for the 1/4″ glass is cen-

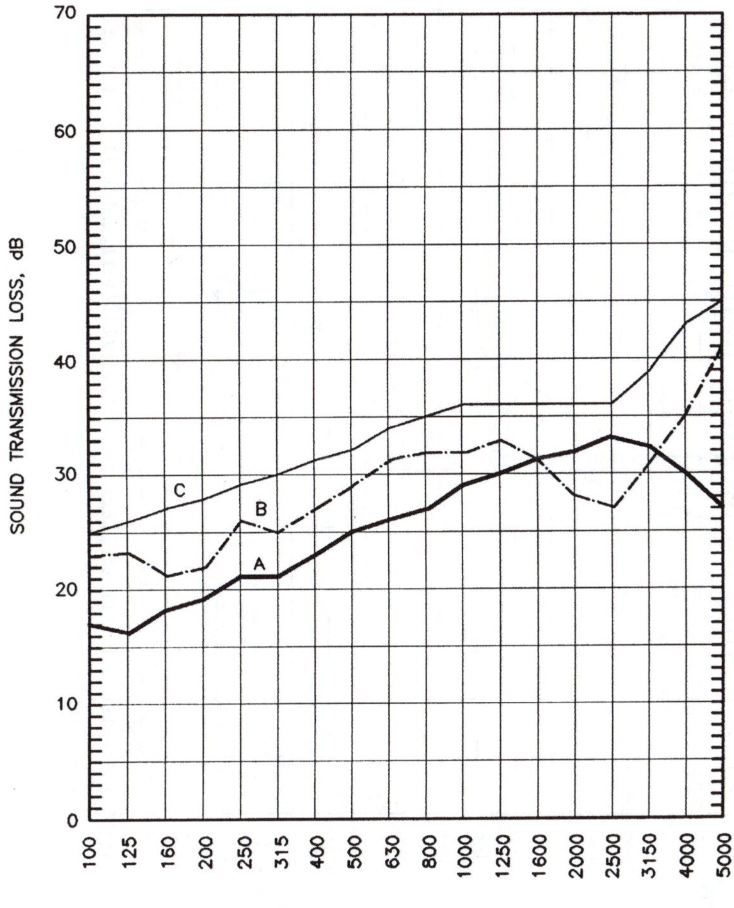


SOUND TRANSMISSION LOSS, dB

| CURVE | 100 | 125 | 160 | 200 | 250 | 315 | 400 | 500 | 630 | 800 | 1000 | 1250 | 1600 | 2000 | 2500 | 3150 | 4000 | 5000 | STC RATING |
|---|---|---|---|---|---|---|---|---|---|---|---|---|---|---|---|---|---|---|---|
| A | 17 | 16 | 18 | 19 | 21 | 21 | 23 | 25 | 26 | 27 | 29 | 30 | 31 | 32 | 33 | 32 | 30 | 27 | 28 |
| B | 23 | 23 | 21 | 22 | 26 | 25 | 27 | 29 | 31 | 32 | 32 | 33 | 31 | 28 | 27 | 31 | 35 | 41 | 30 |
| C | 25 | 26 | 27 | 28 | 29 | 30 | 31 | 32 | 34 | 35 | 36 | 36 | 36 | 36 | 36 | 39 | 43 | 45 | 35 |

———— A) 1/8″ (3.2 mm) glass
—·—·—·—·- B) 1/4″ (6.4 mm) glass
———— C) 1/4″ (6.4 mm) laminated glass

**Figure 3-15.** Sound transmission loss of glass.

tered one octave lower at 2500 Hz. Laminated glass, which is constructed with a thin layer of plastic between glass sheets, provides much more internal damping resulting in elimination of the coincidence dip.

Because much of the vibrational energy contained in glass excited by impinging sound is in the form of flexural waves, it is useful to install glass in resilient gaskets made of neoprene or similar "lossy" material. Such gaskets will absorb energy by removing it from the flexural wave, thus reducing the amount of sound radiated from the receiver side of the glass.

*Figure 3-16.* This figure shows TL curves for three thicknesses of laminated glass. The internal damping of the laminated construction reduces resonances that are found in regular glass, resulting in TL curves that are fairly smooth.

*Figure 3-17.* The TL of insulated glass is of concern because most energy codes require insulated units to reduce heat transfer for both hot and cold climates. It is interesting to compare the TL for 1/2″ (13 mm) laminated glass (Figure 3-16) with that for the insulated glass unit shown in this figure, consisting of 1/4″ (6.4 mm) laminated glass

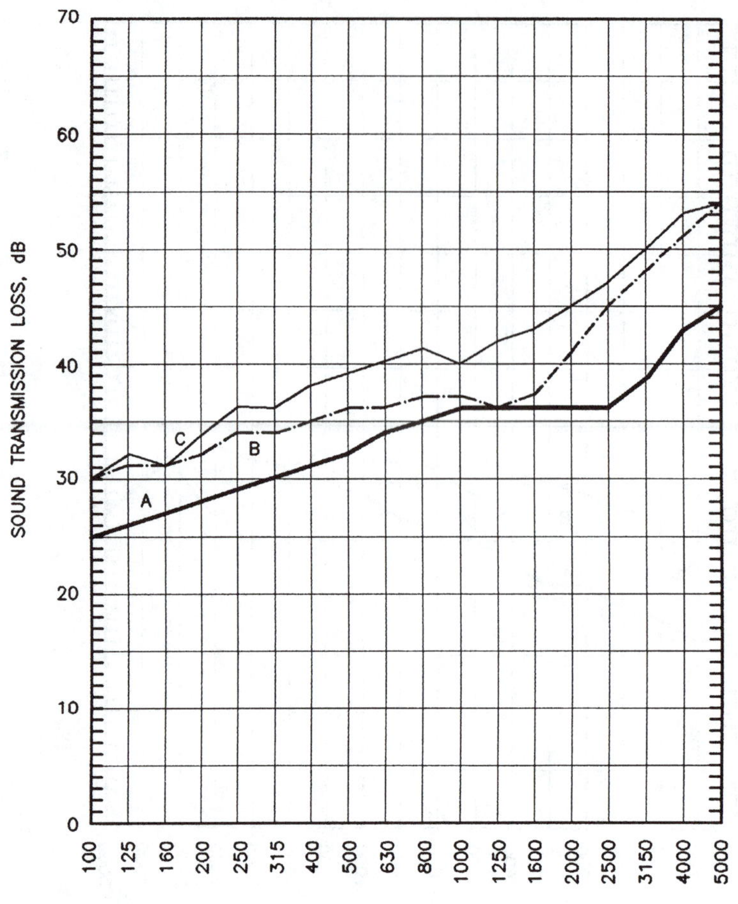


SOUND TRANSMISSION LOSS, dB

| CURVE | 100 | 125 | 160 | 200 | 250 | 315 | 400 | 500 | 630 | 800 | 1000 | 1250 | 1600 | 2000 | 2500 | 3150 | 4000 | 5000 | STC RATING |
|---|---|---|---|---|---|---|---|---|---|---|---|---|---|---|---|---|---|---|---|
| A | 25 | 26 | 27 | 28 | 29 | 30 | 31 | 32 | 34 | 35 | 36 | 36 | 36 | 36 | 36 | 39 | 43 | 45 | 35 |
| B | 30 | 31 | 31 | 32 | 34 | 34 | 35 | 36 | 36 | 37 | 37 | 36 | 37 | 41 | 45 | 48 | 51 | 54 | 38 |
| C | 30 | 32 | 31 | 34 | 36 | 36 | 38 | 39 | 40 | 41 | 40 | 42 | 43 | 45 | 47 | 50 | 53 | 54 | 42 |

———— A) 1/4″ (6.4 mm) laminated glass
—·—·—·—· B) 1/2″ (13 mm) laminated glass
———— C) 3/4″ (19 mm) laminated glass

**Figure 3-16.** Sound transmission loss of laminated glass

—1/2″ (13 mm) airspace—1/4″ regular glass. The total thickness of glass is the same for both, but the STC ratings are within one, and the 1/2″ glass has higher TL values at the low frequencies. All insulated units show dips and peaks in the TL curves resulting from resonances in the enclosed airspace, and also show what are probably pronounced coincidence dips at the higher frequencies.

*Figure 3-18.* The curves in this figure demonstrate that the TL increases as the separation between the glass increases. The effects of resonances in the enclosed airspace and coincidence dips are still present. The TL of double glazed units can be increased by lining the head, jambs and sill with an acoustical absorbent. There is also some evidence that installing the two panes of glass so they are not parallel helps to minimize the effects of resonances in the airspace. There does not appear to be any particular advantage in triple glazing. It is advantageous, however, to maintain maximum obtainable separation between the sheets of glass in double glazed units, and to set the glass in resilient gaskets.

### 3.3.5 Doors, Operable Partitions

*Figure 3-19.* The TL of a solid core door without acoustical seals is determined almost entirely by sound leakage, particularly at the bottom if excessive clearance is allowed

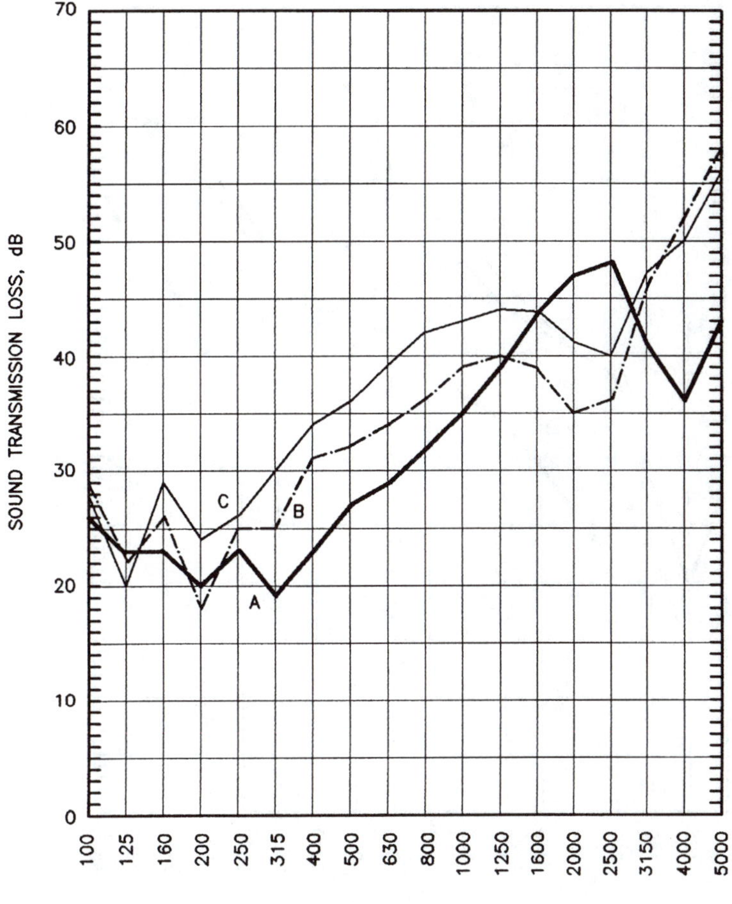


SOUND TRANSMISSION LOSS, dB

| CURVE | 100 | 125 | 160 | 200 | 250 | 315 | 400 | 500 | 630 | 800 | 1000 | 1250 | 1600 | 2000 | 2500 | 3150 | 4000 | 5000 | STC RATING |
|---|---|---|---|---|---|---|---|---|---|---|---|---|---|---|---|---|---|---|---|
| A | 26 | 23 | 23 | 20 | 23 | 19 | 23 | 27 | 29 | 32 | 35 | 39 | 44 | 47 | 48 | 41 | 36 | 43 | 31 |
| B | 29 | 22 | 26 | 18 | 25 | 25 | 31 | 32 | 34 | 36 | 39 | 40 | 39 | 35 | 36 | 46 | 52 | 58 | 35 |
| C | 28 | 20 | 29 | 24 | 26 | 30 | 34 | 36 | 39 | 42 | 43 | 44 | 44 | 41 | 40 | 47 | 50 | 56 | 39 |

———— A) Insulated glass, 1/8″ (3.2 mm) glass—3/8″ (9.5 mm) airspace—1/8″ (3.2 mm) glass

—·—·—·— B) Insulated glass, 1/4″ (6.4 mm) glass—1/2″ (13 mm) airspace—1/4″ (6.4 mm) glass

———— C) Insulated glass, 1/4″ (6.4 mm) laminated glass—1/2″ (13 mm) airspace—1/4″ (6.4 mm) glass

**Figure 3-17.** Sound transmission loss of insulated glass.

for air transfer. By equipping the door with adjustable seals all around, including a threshold closure at the bottom, the STC can be increased from 27 for the condition without seals to 33 with seals. The improvement is greater at the higher frequencies, which are important for speech intelligibility. Substantial further improvement can be achieved with a specially designed, factory-rated door system, usually including the door panel, frame, seals and all hardware.

There are three general type of seals for sound retardant door installations: (1), compression, consisting of soft sponge neoprene corrugations; (2), fin-type seals, which have one or several flexible fins to effect the seal; and (3), magnetic seals. The compression type provide a good seal if they are properly adjusted and maintained. However, since maintenance is seldom done, magnetic seals (for metal doors—see Figure 3-20) have the advantage of not requiring readjustment after installation. The fin-type seals usually are not as effective as the other two types, but can be improved by installing a second fin seal.

A potential problem for seals requiring adjustment occurs if a door closure is used. The closure may not exert

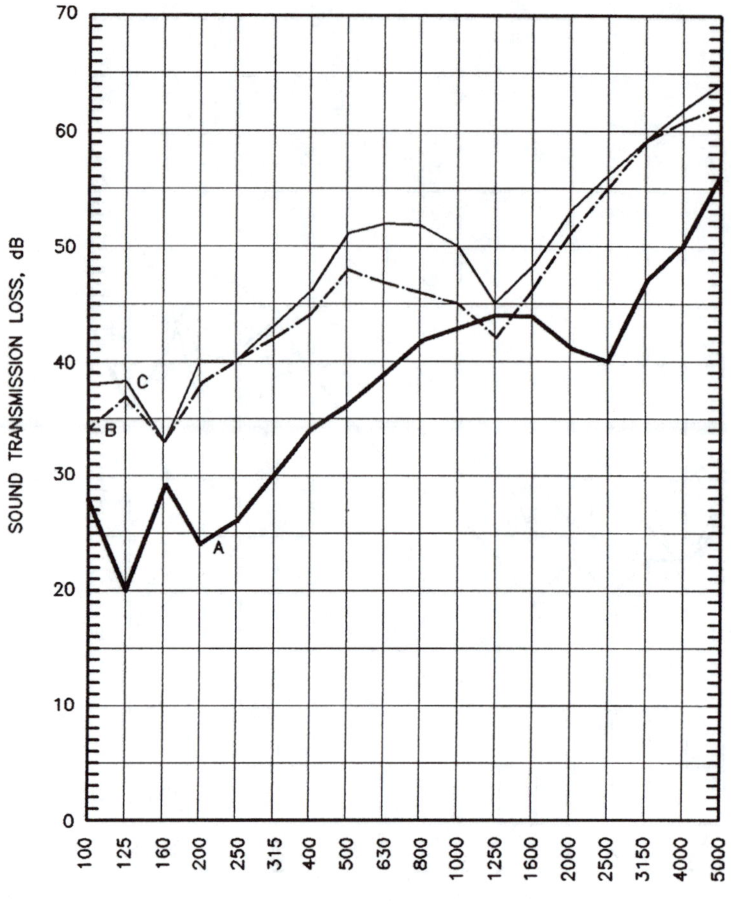


SOUND TRANSMISSION LOSS, dB

| CURVE | 100 | 125 | 160 | 200 | 250 | 315 | 400 | 500 | 630 | 800 | 1000 | 1250 | 1600 | 2000 | 2500 | 3150 | 4000 | 5000 | STC RATING |
|---|---|---|---|---|---|---|---|---|---|---|---|---|---|---|---|---|---|---|---|
| A | 28 | 20 | 29 | 24 | 26 | 30 | 34 | 36 | 39 | 42 | 43 | 44 | 44 | 41 | 40 | 47 | 50 | 56 | 39 |
| B | 34 | 37 | 33 | 38 | 40 | 42 | 44 | 48 | 47 | 46 | 45 | 42 | 46 | 51 | 55 | 59 | 61 | 62 | 46 |
| C | 38 | 38 | 33 | 40 | 40 | 43 | 46 | 51 | 52 | 52 | 50 | 45 | 48 | 53 | 56 | 59 | 62 | 64 | 49 |

———— A) Insulated glass, 1/4″ (6.4 mm) laminated glass—1/2″ (13 mm) airspace—1/4″ (6.4 mm) glass

—·—·—·—· B) Insulated glass, 1/2″ (13 mm) laminated glass—2″ (51 mm) airspace—3/8″ (9.5 mm) glass

———— C) Insulated glass, 1/2″ (13 mm) laminated glass—4″ (102 mm) airspace—3/8″ (9.5 mm) glass

**Figure 3-18.** Sound transmission loss of insulated-laminated glass.

enough force to latch the door to make the seals effective. The bottom seal on many door types may be compression, sealing against a raised threshold, or a drop-type which lowers as the door is closed with a plunger bearing against the door frame. These may have one or two drop bars which form a seal. Some of these do not seal over the top of the bar and are not completely effective. Yet another type of very effective bottom seal is achieved with a cam-type hinge that allows the door to drop slightly as it closes.

If head and jamb door seals are to be applied to a frame not supplied by the manufacturer of the door, be sure the door frame is flush and does not have integral stops. If the acoustical stops are applied on top of stops in the frame, the height and width of the opening is reduced, and the handle may require an extended offset for ease of operation. Be sure to caulk between the rough opening and the door frame.

Threshold closures operating on carpet do not provide good seals. Raised metal or wood thresholds work well acoustically, but are discouraged because of accessibility

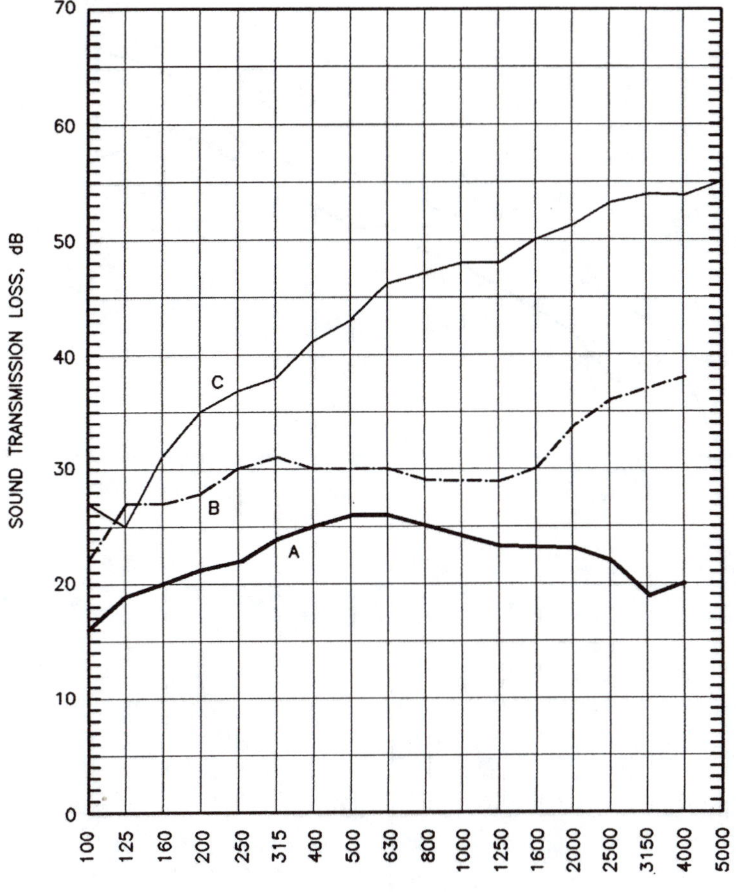


SOUND TRANSMISSION LOSS, dB

| CURVE | 100 | 125 | 160 | 200 | 250 | 315 | 400 | 500 | 630 | 800 | 1000 | 1250 | 1600 | 2000 | 2500 | 3150 | 4000 | 5000 | STC RATING |
|---|---|---|---|---|---|---|---|---|---|---|---|---|---|---|---|---|---|---|---|
| A | 16 | 19 | 20 | 21 | 22 | 24 | 25 | 26 | 26 | 25 | 24 | 23 | 23 | 23 | 22 | 19 | 20 | -- | 22 |
| B | 22 | 27 | 27 | 28 | 30 | 31 | 30 | 30 | 30 | 29 | 29 | 29 | 30 | 34 | 36 | 37 | 38 | -- | 33 |
| C | 27 | 25 | 31 | 35 | 37 | 38 | 41 | 43 | 46 | 47 | 48 | 48 | 50 | 51 | 53 | 54 | 54 | 55 | 46 |

A) 1 3/4″ (44 mm) solid core wood door, 5 lb/sq ft (24.4 kg/sq m), no seals
B) Same, with seals
C) 1 3/4″ (44 mm) sound-rated door, 7 lb/sq ft (34.2 kg/sq m), average of 3 tests

**Figure 3-19.** Sound transmission loss of wood doors.

requirements. Flush or slightly raised thresholds can overcome this difficulty.

There are many applications where more sound isolation is required than can be achieved by a single door. In such cases two doors installed in tandem will provide additional attenuation (see Figure 3-20). The amount of additional attenuation depends critically on frequency and the separation between the doors. If the separation is only a few inches, such as between interconnecting hotel rooms, the second door will add less than one-half its STC rating to that of the first door. By the time the separation becomes several feet or more, where vestibules may separate the doors, the STC values are virtually additive. The separation must be at least one wavelength for this condition to apply. In any case, the attenuation will increase if acoustical absorption is present in the enclosed space between the doors.

It is important that the person installing sound-rated doors understand the importance of careful adjustment of the seals after the door is hung. In critical applications, an in-place noise reduction test will ensure that the installation has been properly done.

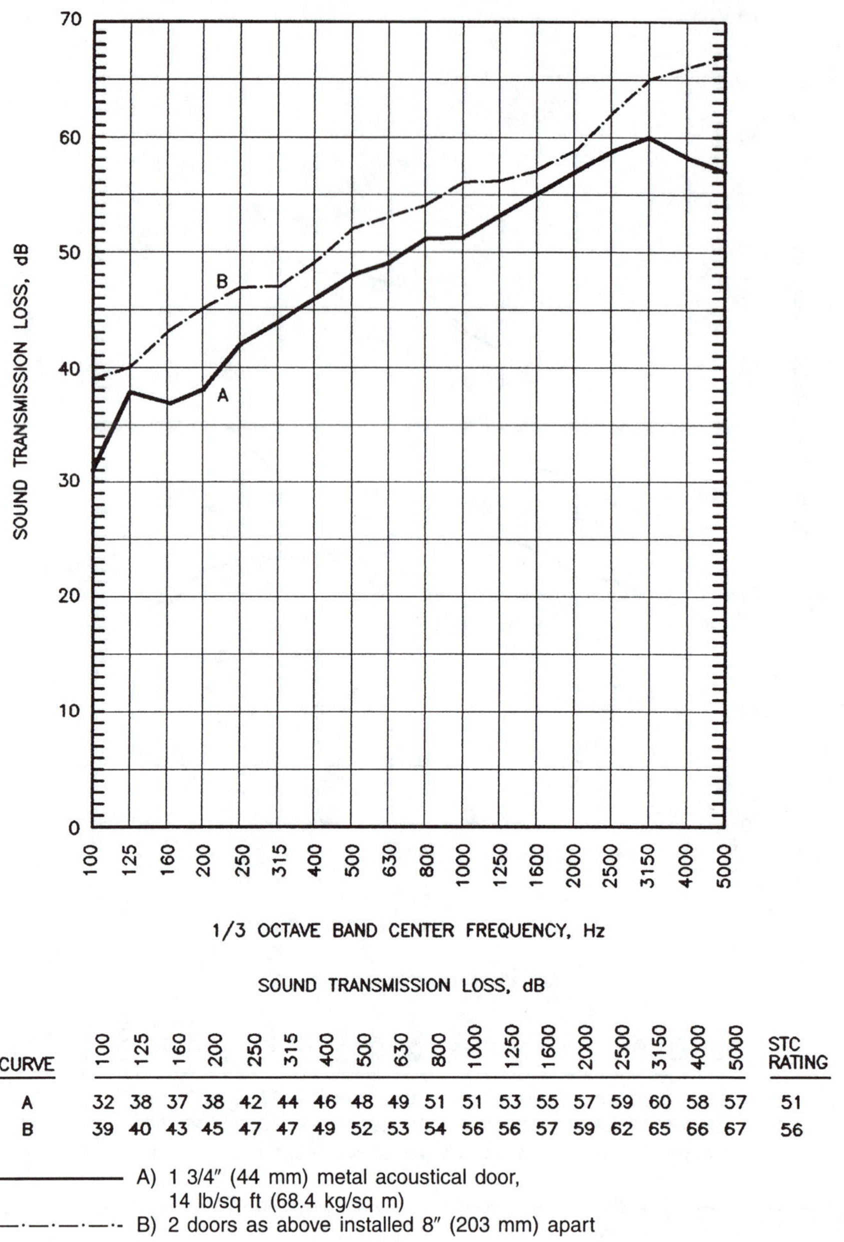


SOUND TRANSMISSION LOSS, dB

| CURVE | 100 | 125 | 160 | 200 | 250 | 315 | 400 | 500 | 630 | 800 | 1000 | 1250 | 1600 | 2000 | 2500 | 3150 | 4000 | 5000 | STC RATING |
|---|---|---|---|---|---|---|---|---|---|---|---|---|---|---|---|---|---|---|---|
| A | 32 | 38 | 37 | 38 | 42 | 44 | 46 | 48 | 49 | 51 | 51 | 53 | 55 | 57 | 59 | 60 | 58 | 57 | 51 |
| B | 39 | 40 | 43 | 45 | 47 | 47 | 49 | 52 | 53 | 54 | 56 | 56 | 57 | 59 | 62 | 65 | 66 | 67 | 56 |

———— A) 1 3/4″ (44 mm) metal acoustical door, 14 lb/sq ft (68.4 kg/sq m)

—·—·—·— B) 2 doors as above installed 8″ (203 mm) apart

**Figure 3-20.** Sound transmission loss of metal doors.

*Figure 3-20.* Metal sound-rated doors are available with higher STC values than are presently available for wood doors, though the metal doors are usually much heavier. If even higher TL values are required than can be obtained with a single door, two doors installed in tandem will provide some added performance. See the discussion for Figure 3-19. Also, higher TL values than shown in this Figure can be obtained with thicker and heavier industrial-type doors, with heavy duty seals and hardware. Usually metal sound-rated doors are provided complete with the frame, seals and all hardware. The manufacturer's recommendations for installation and adjustment should be carefully followed. It is important to fill hollow metal door frames with grout, or pack tightly with insulation.

*Figure 3-21.* Sliding glass doors need to be equipped with tight seals for maximum acoustical effectiveness. In this Figure the insulated glass panel for the door represented by the lower curve would have an STC rating of about 39 if completely sealed (see Figure 3-17). Sound leakage is accounting for the lower rating of STC-29. Referring to the upper curve, the installation of a second door parallel to the triple glazed door (separation unknown), achieved an STC rating of 42.

*Figure 3-22.* Extra caution is required interpreting TL data for operable partitions because the acoustical performance is so dependent on the quality of installation, particularly the effectiveness of the seals. Most manufac-

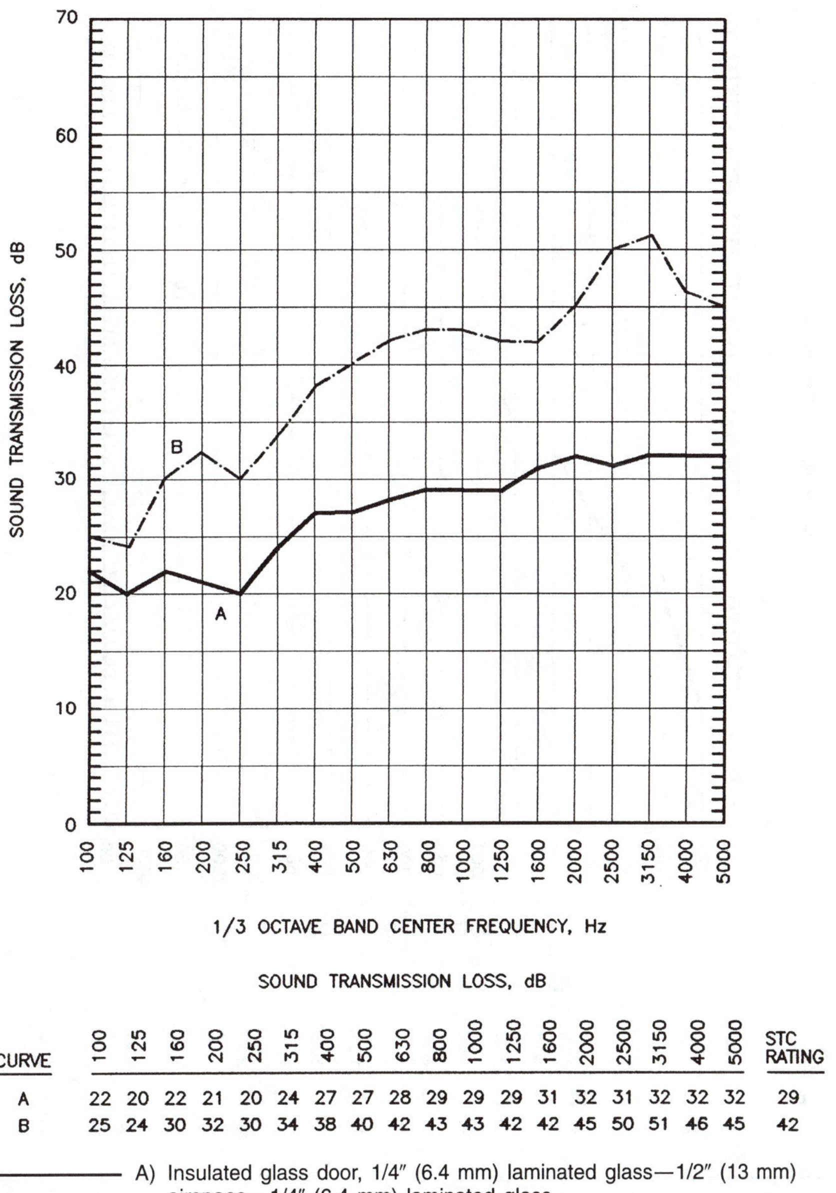


SOUND TRANSMISSION LOSS, dB

| CURVE | 100 | 125 | 160 | 200 | 250 | 315 | 400 | 500 | 630 | 800 | 1000 | 1250 | 1600 | 2000 | 2500 | 3150 | 4000 | 5000 | STC RATING |
|---|---|---|---|---|---|---|---|---|---|---|---|---|---|---|---|---|---|---|---|
| A | 22 | 20 | 22 | 21 | 20 | 24 | 27 | 27 | 28 | 29 | 29 | 29 | 31 | 32 | 31 | 32 | 32 | 32 | 29 |
| B | 25 | 24 | 30 | 32 | 30 | 34 | 38 | 40 | 42 | 43 | 43 | 42 | 42 | 45 | 50 | 51 | 46 | 45 | 42 |

———— A) Insulated glass door, 1/4″ (6.4 mm) laminated glass—1/2″ (13 mm) airspace—1/4″ (6.4 mm) laminated glass

—·—·—·—·· B) Triple glazed door, 3 sheets of 1/8″ (3.2 mm) glass with additional 1/4″ (6.4 mm) laminated glass door (glass spacings unknown)

**Figure 3-21.** Sound transmission loss of sliding glass doors.

turers have laboratory STC ratings for operable partitions in the low 50s, but *field* performance can be up to 15 points less. The noise reduction data shown in this Figure are for field tests performed on partitions representing about the highest quality of installation that can be expected in the field. The test results are given in terms of noise reduction (NR), recalling that the single-number rating for NR data is the Noise Isolation Class (NIC, see Section 3.2.1). Curves are shown for a single partition, then for two partitions of the same model separated by 19″ (483 mm). Useful suggestions for the proper installation of operable partitions are contained in ASTM Designation: E 557, "Architectural Application and Installation of Operable Partitions."

An in-situ acoustical performance specification should *always* be required for operable partition installations if sound isolation is important. The testing should be performed by a qualified acoustical consultant, following ASTM Designation: E-336, "Measurement of Airborne Sound Transmission Loss in Buildings." An FSTC or NIC requirement of about eight points below the laboratory STC rating is typical. Consult the manufacturer for current product TL/STC ratings.

There is potentially a serious problem that is often overlooked when the operable partition is supported by the overhead roof structure. The roof deflection caused by snow loads can make the partition inoperable. The deflec-

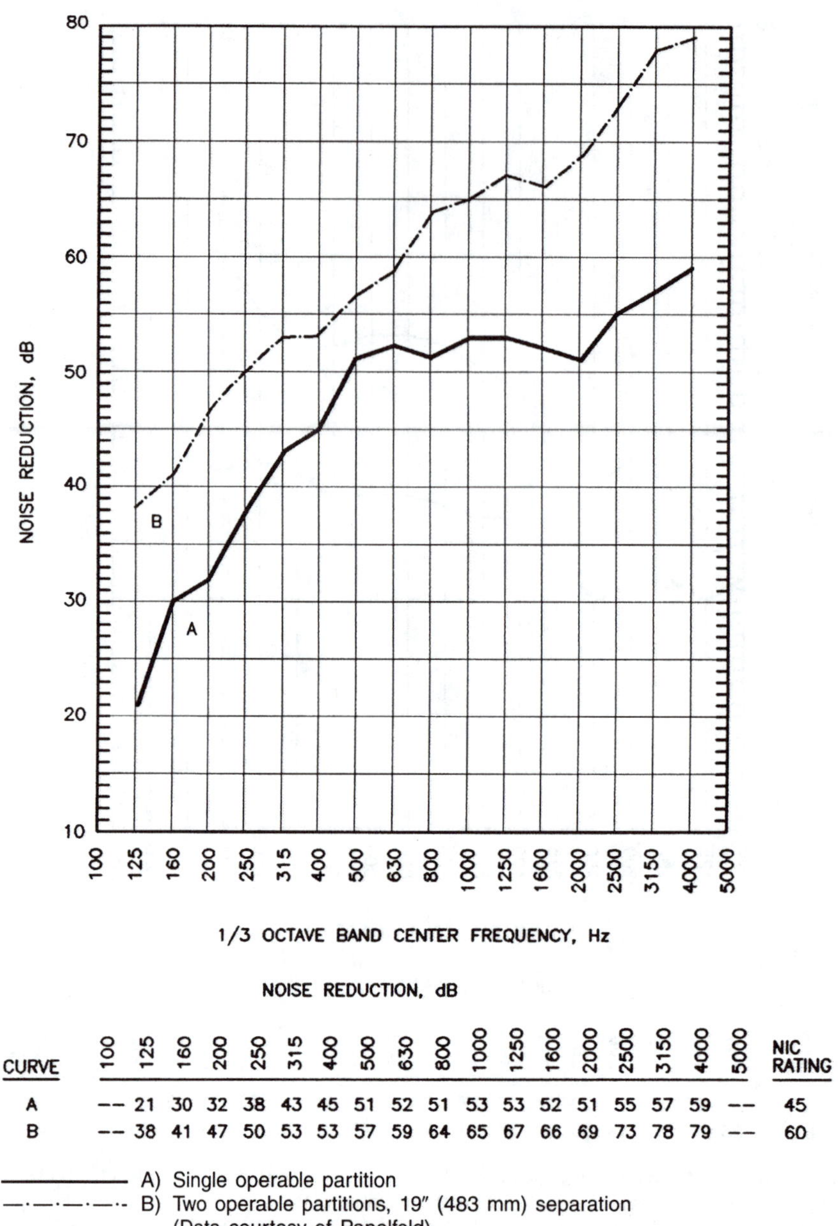


NOISE REDUCTION, dB

| CURVE | 100 | 125 | 160 | 200 | 250 | 315 | 400 | 500 | 630 | 800 | 1000 | 1250 | 1600 | 2000 | 2500 | 3150 | 4000 | 5000 | NIC RATING |
|---|---|---|---|---|---|---|---|---|---|---|---|---|---|---|---|---|---|---|---|
| A | -- | 21 | 30 | 32 | 38 | 43 | 45 | 51 | 52 | 51 | 53 | 53 | 52 | 51 | 55 | 57 | 59 | -- | 45 |
| B | -- | 38 | 41 | 47 | 50 | 53 | 53 | 57 | 59 | 64 | 65 | 67 | 66 | 69 | 73 | 78 | 79 | -- | 60 |

——— A) Single operable partition
—·—·—·—·- B) Two operable partitions, 19″ (483 mm) separation
(Data courtesy of Panelfold)

**Figure 3-22.** Noise reduction of operable partitions.

tion of the structure carrying these walls must not be morc than what thc partition can accommodate. A separate structural support system from that of the roof system is recommended.

### 3.3.6 Facades

*Figure 3-23.* This figure shows the noise reduction of typical exterior residential construction in cold climates (i.e., insulated walls and windows) for windows open and closed. The primary noise transmission path is the windows, even when they are closed. If there were no windows (not shown), the NIC would probably be in the range 35–40.

Determining the noise reduction (NR) of an exterior construction consisting of a walls, doors and/or windows involves a composite TL calculation, the procedure for which is given in Section 3.4. In essence, the acoustic energy transmitted through each building component is calculated, then all are summed together to arrive at the overall noise reduction. This is done for each frequency band of interest.

There are many variables that can affect the resultant noise level inside a building, such as the spectrum of the noise source, the angle at which the sound strikes the building, the size and acoustical absorption in the receiving room, and the types of construction. The complexity of measuring the noise reduction of a building facade is

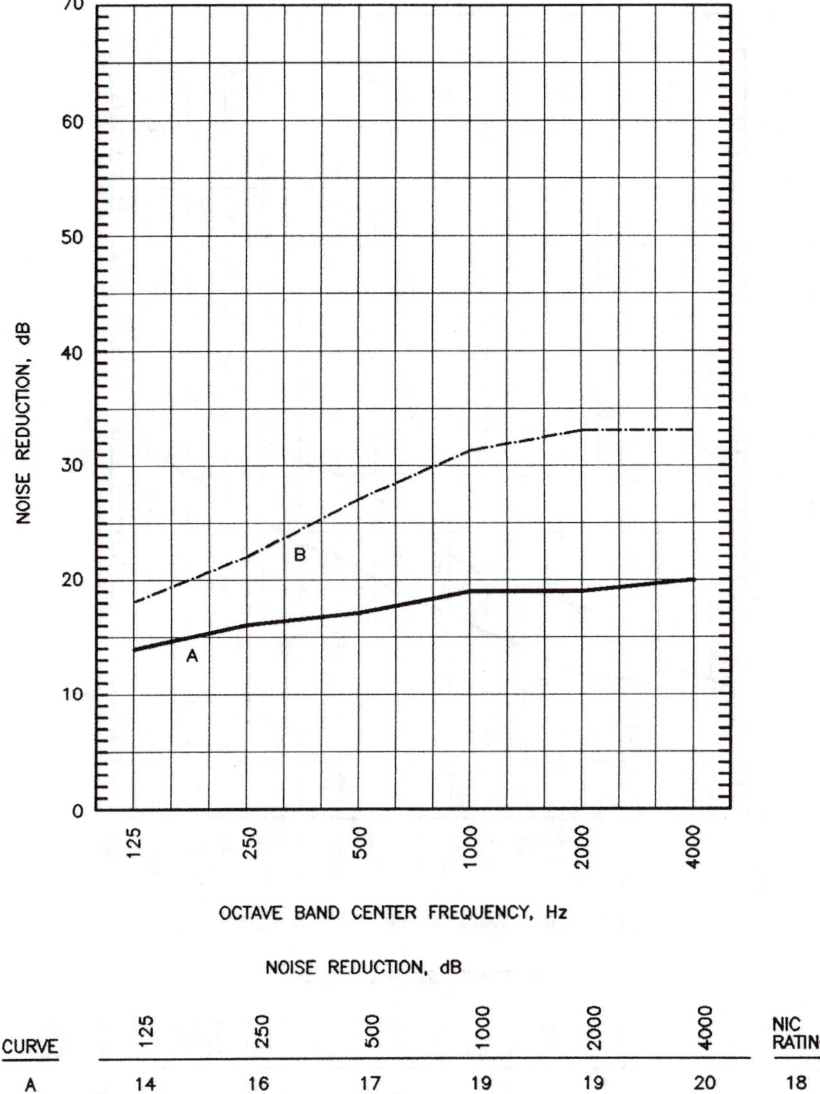


NOISE REDUCTION, dB

| CURVE | 125 | 250 | 500 | 1000 | 2000 | 4000 | NIC RATING |
|---|---|---|---|---|---|---|---|
| A | 14 | 16 | 17 | 19 | 19 | 20 | 18 |
| B | 18 | 22 | 27 | 31 | 33 | 33 | 30 |

———— A) Windows open
—·—·—·— B) Windows closed

**Figure 3-23.** Noise reduction through typical residential exterior construction (insulated).

evidenced by referring to the specified methodology in ASTM Designation: E 966, "Standard Guide for Field Measurement of Airborne Sound Insulation of Building Facades and Facade Elements." Another ASTM document defining the procedure for determining a single-number rating for the NR data (analogous to FSTC and NIC) is given in ASTM Designation: E 1332, "Standard Classification for Determination of Outdoor-Indoor Transmission Class (OITC)."

In situations that are not critical, the procedure for determining building facade NR may be simplified by following the composite TL calculation procedure. For uses where low interior background noise is critical, the services of an acoustical consultant should be employed.

### 3.3.7 Acoustical Curtains

*Figure 3-24.* The spreading of noise from machinery in an open shop can be minimized by surrounding the machine with a flexible curtain. These curtains usually consist of a barium-loaded vinyl and weigh from 0.5 to 1.5 lbs/sq ft (2.44 to 7.32 kg/sq m). Some are treated with an acoustical absorbent on one or both sides to reduce noise build-up within the enclosure. TL data are shown for 3 representative curtains in this Figure. The stitched blanket curtain, which consists of an absorbent on both faces, shows a significant TL increase in the mid to high frequency range, compared to the nonabsorptive curtains. If the ceiling above the enclosure is reflective, treating it with an absorbent will further reduce noise escaping into the shop.

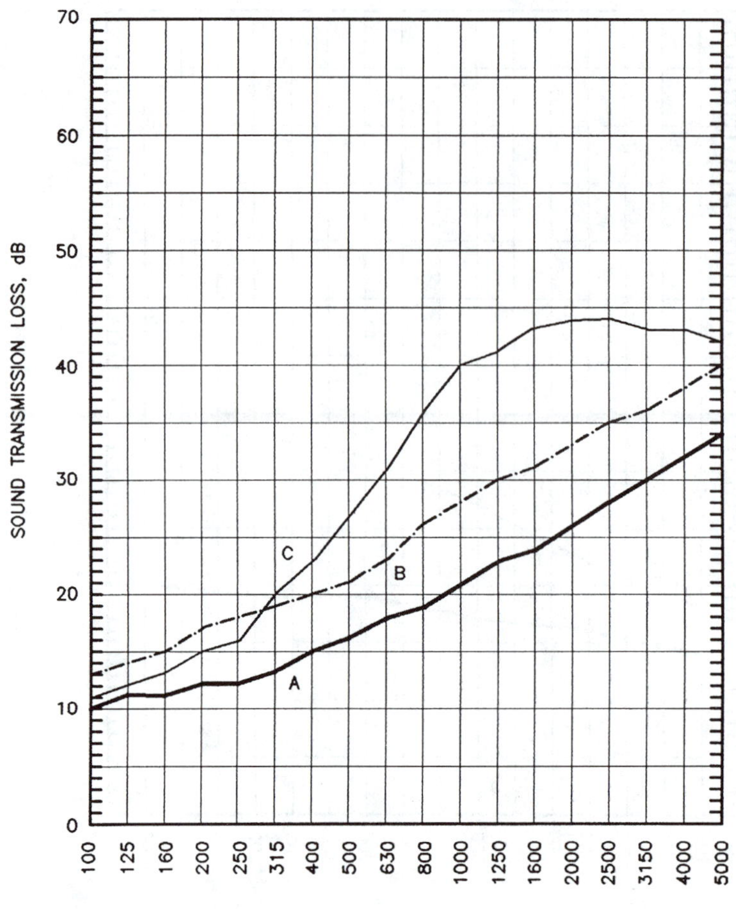


SOUND TRANSMISSION LOSS, dB

| CURVE | 100 | 125 | 160 | 200 | 250 | 315 | 400 | 500 | 630 | 800 | 1000 | 1250 | 1600 | 2000 | 2500 | 3150 | 4000 | 5000 | STC RATING |
|---|---|---|---|---|---|---|---|---|---|---|---|---|---|---|---|---|---|---|---|
| A | 10 | 11 | 11 | 12 | 12 | 13 | 15 | 16 | 18 | 19 | 21 | 23 | 24 | 26 | 28 | 30 | 32 | 34 | 21 |
| B | 13 | 14 | 15 | 17 | 18 | 19 | 20 | 21 | 23 | 26 | 28 | 30 | 31 | 33 | 35 | 36 | 38 | 40 | 27 |
| C | 11 | 12 | 13 | 15 | 16 | 20 | 23 | 27 | 31 | 36 | 40 | 41 | 43 | 44 | 44 | 43 | 43 | 42 | 29 |

A) Barium-loaded vinyl, 0.5 lb/sq ft (2.4 kg/sq m)
B) Same, 1 lb/sq ft (4.9 kg/sq m)
C) Stitched blanket, 1.5 lb/sq ft (7.3 kg/sq m)

**Figure 3-24.** Sound transmission loss of industrial curtains.

Be sure to investigate the flammability characteristics of the product if used around welding areas.

### 3.3.8 Wood Frame Floors

*Figure 3-25.* Wood frame structure for floor-ceiling systems is a common practice in low rise apartments and condominiums (usually four floors or less), and noise from units overhead is often a main source of complaints. This is unfortunate because the construction methods that will perform satisfactorily are well publicized in many sources available to design architects and developers. Some additional cost is involved, but surveys have been conducted where renters or purchasers were asked if they would pay a few hundred dollars extra for acoustical privacy, and almost without exception the answer has been yes.

This figure shows the noise reduction of three wood frame floor-ceiling systems specially constructed in a condominium project specifically for acoustical testing purposes. The source and receiving rooms were identical in size, so a direct comparison of the NR curves is appropriate.

Construction DA, with the ceiling gypsum board applied directly to the joists, received an NIC rating of only 38. This performance is almost certain to generate complaints should this construction be used in multifamily

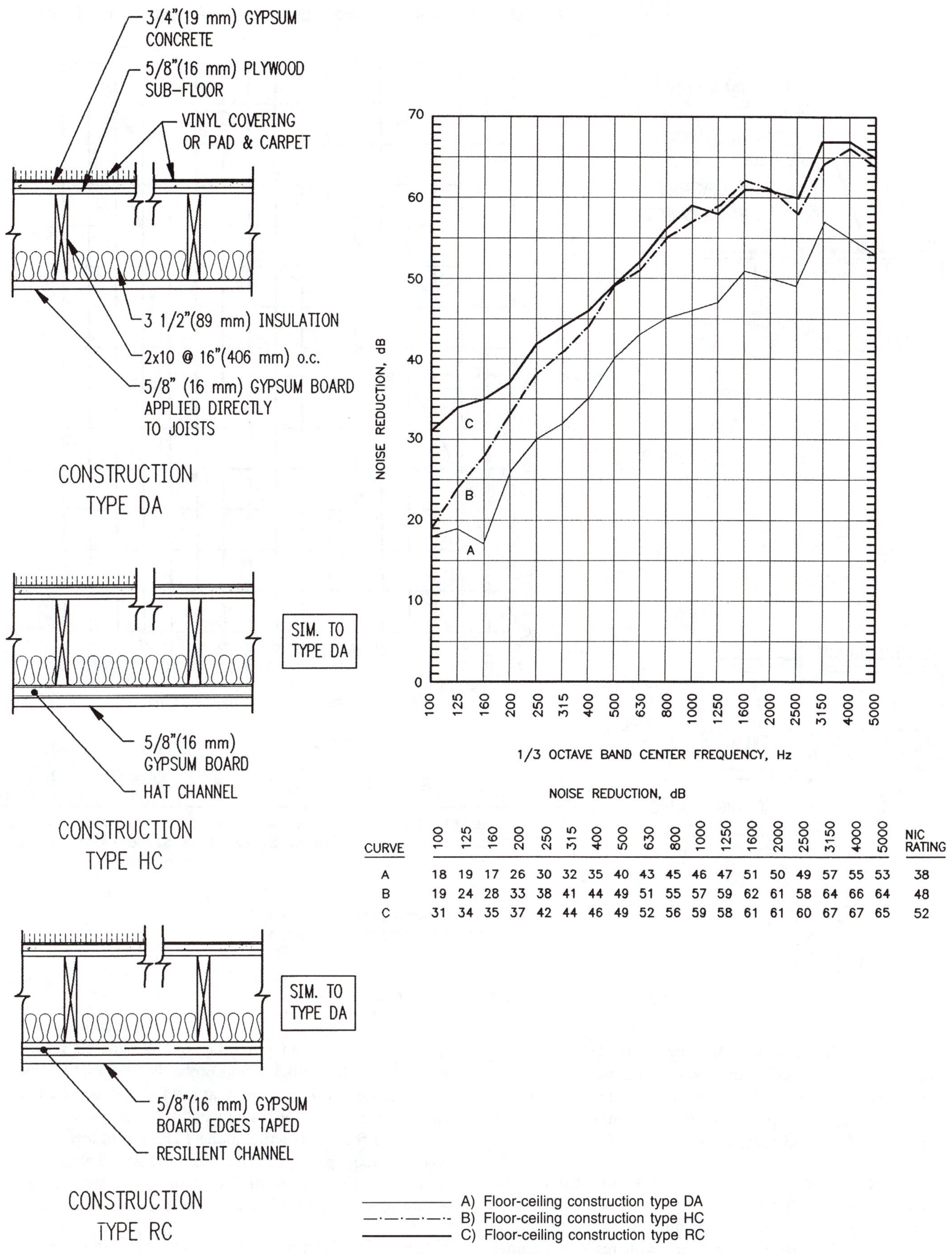


NOISE REDUCTION, dB

| CURVE | 100 | 125 | 160 | 200 | 250 | 315 | 400 | 500 | 630 | 800 | 1000 | 1250 | 1600 | 2000 | 2500 | 3150 | 4000 | 5000 | NIC RATING |
|---|---|---|---|---|---|---|---|---|---|---|---|---|---|---|---|---|---|---|---|
| A | 18 | 19 | 17 | 26 | 30 | 32 | 35 | 40 | 43 | 45 | 46 | 47 | 51 | 50 | 49 | 57 | 55 | 53 | 38 |
| B | 19 | 24 | 28 | 33 | 38 | 41 | 44 | 49 | 51 | 55 | 57 | 59 | 62 | 61 | 58 | 64 | 66 | 64 | 48 |
| C | 31 | 34 | 35 | 37 | 42 | 44 | 46 | 49 | 52 | 56 | 59 | 58 | 61 | 61 | 60 | 67 | 67 | 65 | 52 |

**Figure 3-25.** Noise reduction of wood frame construction.

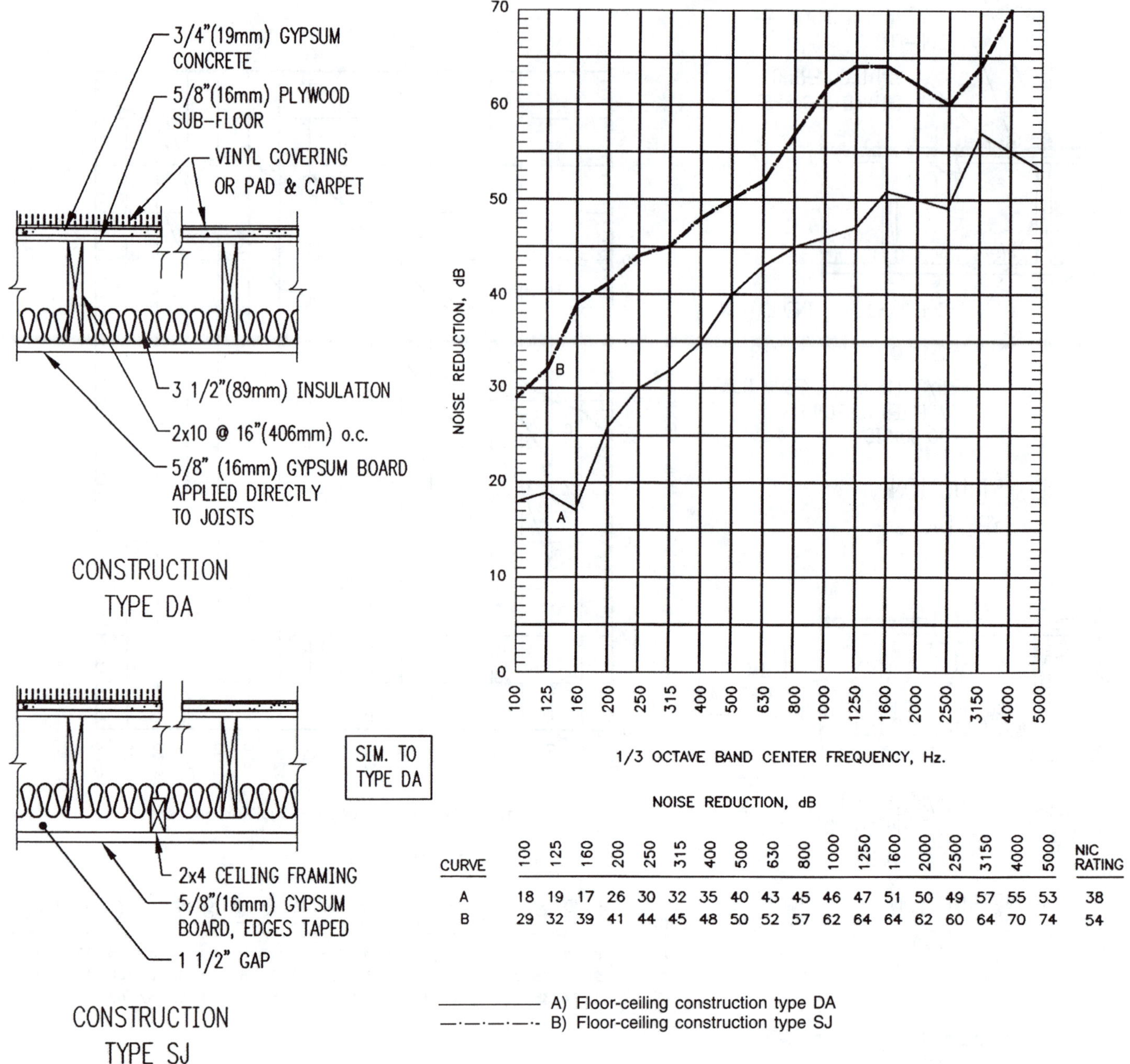


NOISE REDUCTION, dB

| CURVE | 100 | 125 | 160 | 200 | 250 | 315 | 400 | 500 | 630 | 800 | 1000 | 1250 | 1600 | 2000 | 2500 | 3150 | 4000 | 5000 | NIC RATING |
|---|---|---|---|---|---|---|---|---|---|---|---|---|---|---|---|---|---|---|---|
| A | 18 | 19 | 17 | 26 | 30 | 32 | 35 | 40 | 43 | 45 | 46 | 47 | 51 | 50 | 49 | 57 | 55 | 53 | 38 |
| B | 29 | 32 | 39 | 41 | 44 | 45 | 48 | 50 | 52 | 57 | 62 | 64 | 64 | 62 | 60 | 64 | 70 | 74 | 54 |

**Figure 3-26.** Noise reduction of wood frame construction.

residences. The Uniform Building Code (UBC, see Appendix 9.2) has a minimum field performance requirement of FSTC 45 (comparable to NIC for these tests), and many consultants believe the minimum performance should be at least 50 to avoid complaints.

Attaching the gypsum board ceiling using hat channel (construction HC) provides a significant improvement at most frequencies, and achieves an NIC rating of 48. Finally, attaching the gypsum board with resilient channel (construction RC) raises the NIC rating to a reasonably acceptable 52. The placement of insulation in the joist space is important in obtaining the designated ratings. Without it, NIC ratings would probably be 5 to 10 points less. It should also be noted that a floor surface of either vinyl or carpet does not significantly affect the noise reduction of the construction alone. (The sound level in the receiving room would be affected by the added absorption of the carpet.) However, the situation is completely different for impact sound, as explained in Sections 3.5 and 3.6.

*Figure 3-26.* The noise reduction curve shown in this figure for floor-ceiling construction SJ is the fourth construction in the series of tests described for the previous Figure.

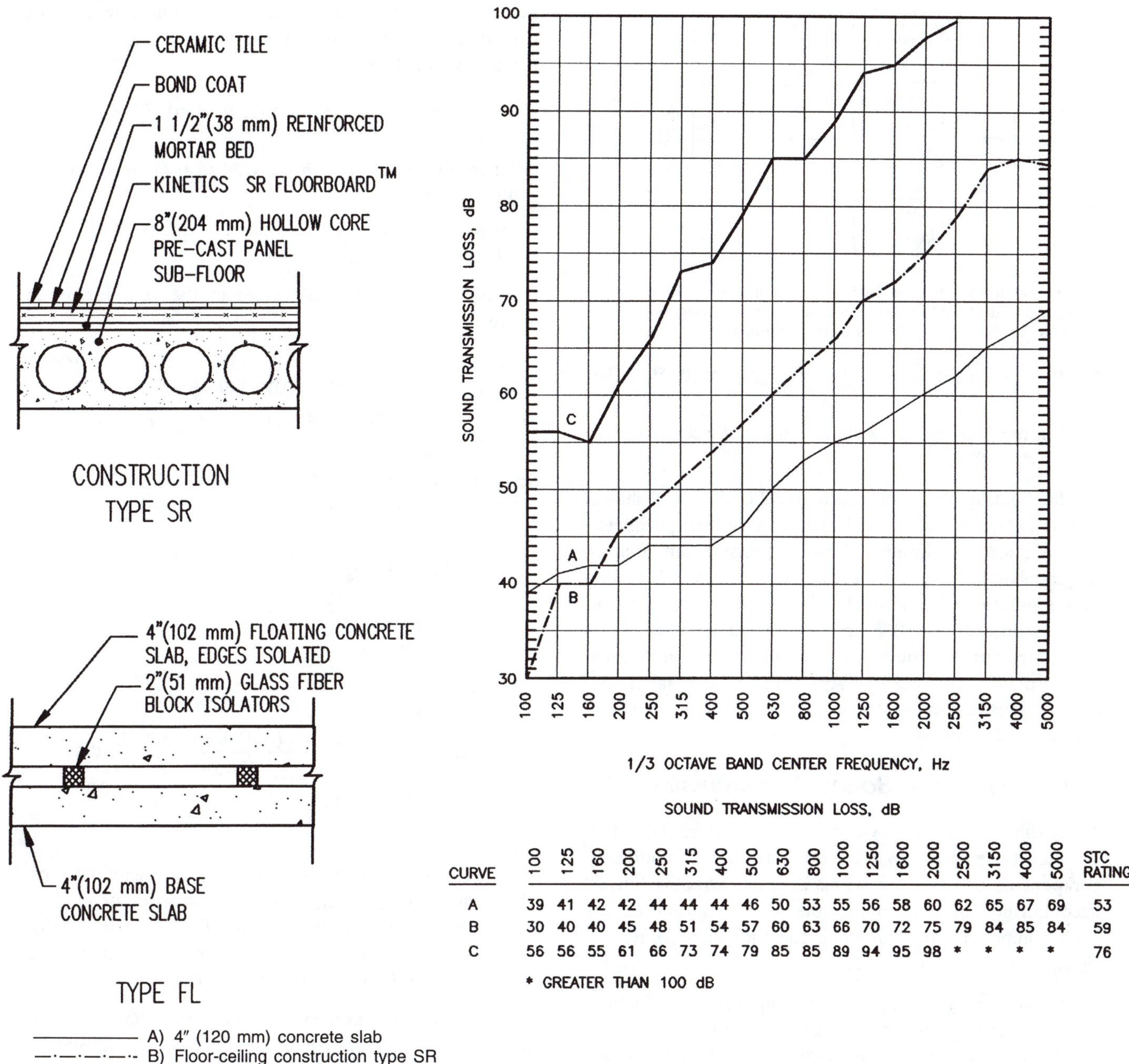


SOUND TRANSMISSION LOSS, dB

| CURVE | 100 | 125 | 160 | 200 | 250 | 315 | 400 | 500 | 630 | 800 | 1000 | 1250 | 1600 | 2000 | 2500 | 3150 | 4000 | 5000 | STC RATING |
|---|---|---|---|---|---|---|---|---|---|---|---|---|---|---|---|---|---|---|---|
| A | 39 | 41 | 42 | 42 | 44 | 44 | 44 | 46 | 50 | 53 | 55 | 56 | 58 | 60 | 62 | 65 | 67 | 69 | 53 |
| B | 30 | 40 | 40 | 45 | 48 | 51 | 54 | 57 | 60 | 63 | 66 | 70 | 72 | 75 | 79 | 84 | 85 | 84 | 59 |
| C | 56 | 56 | 55 | 61 | 66 | 73 | 74 | 79 | 85 | 85 | 89 | 94 | 95 | 98 | * | * | * | * | 76 |

* GREATER THAN 100 dB

A) 4″ (120 mm) concrete slab
B) Floor-ceiling construction type SR
C) Floor-ceiling construction type FL

**Figure 3-27.** Sound transmission loss of concrete floors.

In this case the gypsum board ceiling was attached to separate wood joists, providing an even greater degree of isolation than achieved by the resilient channel. The curve for construction DA is shown for comparison.

### 3.3.9 Concrete Floors

*Figure 3-27.* Concrete construction is frequently found in up-scale multifamily residences, particularly high-rise towers. The added mass and stiffness of concrete compared to wood frame construction is evidenced by the better performance at the low frequencies. The underlayment for a hard surface topping as represented by construction SR shows substantial improvement compared to bare concrete except for the low frequencies.

Extraordinarily high TL performance can be achieved by a "floating" concrete floor over a base slab, as represented by construction FL. There are three basic types of isolators supporting the floating slab, consisting of neoprene, steel springs and glass fiber blocks. Separation between the two slabs is usually in the range 1″ to 3″ (25 mm to 76 mm). This type of construction is used only where very high TL performance is required. It is sometimes

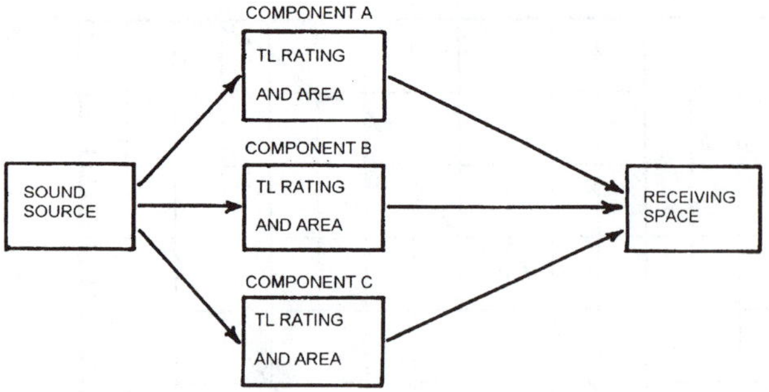


The sound transmitted by each component is determined separately, then summed to obtain the level in the receiving space. The process is followed for each frequency of interest.

**Figure 3-28.** Composite sound transmission between two spaces.

used in mechanical rooms above critical offices or conference rooms.

The floating floor for which the TL data are shown in this Figure was carefully isolated around the perimeter of the floor to minimize flanking transmission. Flanking transmission becomes a major concern for TL values above about 60 dB, and is very difficult to avoid when TL values approach 100 dB. Extremely careful detailing of the entire structure is required. The manufacturers of floating floor isolators are listed in Appendix A.12, numbers 37 and 43. It is also advisable to employ the services of an acoustical consultant.

## 3.4 Composite Sound Transmission

Many times it is necessary to calculate the noise reduction between two spaces separated by more than one building component, such as a wall with a door and window. This is accomplished by determining the sound transmitted by each component and summing the individual contributions, as illustrated in Figure 3-28.

The sound transmitted by each component is a function of the TL of the component and its area. It is necessary to convert the TL values of each component to a transmission coefficient, symbolized by the Greek letter tau ($\tau$), by taking the antilog of the TL divided by 10, multiplying by the area, then summing the products. This sum is then changed back to a TL value representing the composite sound transmission loss, TLc, for that particular combination of components. This procedure is represented by the following equation (to be determined for each frequency of interest):

$$\text{TLc} = -10 \log \left[ \frac{S_1\tau_1 + S_2\tau_2 + \ldots S_n\tau_n}{S} \right], \text{ dB}$$

where $\tau_n = -10^{-\text{TLn}/10}$

$\text{TL}_n$ = Sound transmission loss of the component n, dB

$S_n$ = Area of component n, sq ft (sq m)

For example, assume the construction separating 2 rooms consists of the following components (TL values for a frequency of 500 Hz):

Concrete block wall, 150 sq ft (13.9 sq m), $\text{TL}_{500}$ = 46 dB.

Solid core wood door with seals, 20 sq ft (1.86 sq m), $\text{TL}_{500}$ = 30 dB.

Glass window, 1/4″ (6.4 mm) thick, 15 sq ft (1.39 sq m), $\text{TL}_{500}$ = 25 dB.

Inserting these numerical values in the equation given above:

$S\tau$ for the concrete block wall $= 150(2.51 \times 10^{-5}) = 0.0038$ sq ft $[13.9(2.51 \times 10^{-5}) = 0.000349$ sq m$]$

$S\tau$ for the door $= 20 \times 0.001 = 0.02$ sq ft $(1.86 \times 0.001 = 0.00186$ sq m$)$

$S\tau$ for the window $= 15 \times 0.0316 = 0.0474$ sq ft $(1.39 \times 0.0316 = 0.0044$ sq m$)$

$$\text{TLc} = -10 \log \left( \frac{0.0038 + 0.02 + 0.0474}{185} \right) \text{dB}$$
$$= -10 \log 0.000385 \text{ dB}$$
$$= 34 \text{ dB}$$

In metric units

$$\text{TLc} = -10 \log \left( \frac{0.000349 + 0.00186 + 0.0044}{17.15} \right) \text{dB}$$
$$= -10 \log 0.000385 \text{ dB}$$
$$= 34 \text{ dB}$$

Notice that the $S\tau$ values have the dimension of area. This number does, in fact, represent the fraction of a square foot (or square meter) having zero TL. The window, for example, transmits as much acoustic energy as an opening with an area of 0.0474 sq ft (0.0044 sq m).

It is further evident from this calculation that the TL of the concrete block wall is considerably compromised by the sound transmitted through the door and, in particular, through the window. The relative amount of sound energy transmitted by each component can be determined by comparing the $S\tau$ values for each component. The window transmits 12.5 times the acoustic energy as that transmitted by the wall (0.0474/0.0038), and 2.4 times that transmitted by the door (0.0474/0.02). The procedure for determining the actual noise reduction between source and receiving spaces is explained in Section 3.2.

## 3.5 Impact Sound Insulation

The response of floor-ceiling structures to impacts is quite different from the response to airborne sound. Impacts impart vibrational energy directly to the structure, usually at a single point at a time, while airborne sound strikes the floor in a reasonably uniform manner over the entire surface. Impacts typically come from footfalls, but other impact-type sources are (to name a few) bouncing balls, scraping furniture, exercise machines, printing presses, appliances such as clothes washers or dryers, and even hi-fi loudspeaker enclosures in direct contact with the floor.

### 3.5.1 IIC—FIIC

Impact sound insulation is measured in both the laboratory and field using a standard tapping machine which is placed on the test floor surface. The machine consists of five hammers with specified characteristics, including weight, distance dropped and the rate at which they drop. The machine is located at a minimum of three positions on the test surface, since joist construction will respond differently to the hammer impacts depending on whether the machine is directly over a joist or between joists.

While the tapping machine is operating, the sound pressure level is measured in the room below in contiguous one-third octave bands with center frequencies from 100 to 3150 Hz. (There is considerable interest developing at this time in extending the low frequency range to 63 Hz because of problems being experienced by many multifamily residents in buildings with lightweight, wood frame floors.) No measurements are made in the room where the machine is operating. The levels for the three (or more) machine positions are averaged to obtain a single sound level for each frequency band. These levels are normalized to a standard room absorption, and are termed "Normalized Impact Sound Pressure Level," abbreviated in this handbook as NISPL. The data are normalized to facilitate comparison with the results of tests done in other rooms, since the sound levels measured in the receiving room are affected by room absorption as well as sound transmitted through the floor/ceiling assembly. The test procedure is specified in ASTM Designation: E 492, "Laboratory Measurement of Impact Sound Transmission Through Floor-Ceiling Assemblies Using the Tapping Machine."

The NISPL curve can be expressed as a single-number in much the same manner that TL data can be expressed as the single-number STC (see Section 3.2.1), except the procedure is the inverse of the STC method because compliance with the appropriate contour is determined by the *excess* noise rather than the deficiencies. For laboratory impact data, the Impact Insulation Class (IIC) is determined according to a curve fitting procedure defined in ASTM Designation: E 989, "Standard Classification for Determination of Impact Insulation Class (IIC)." The calculation procedure is explained in Appendix A.6.

The ability of the standard tapping machine test to properly rank-order different floor-ceiling assemblies in terms of judgments of acceptability of the radiated noise has been debated for many years by the manufacturers of floor covering materials, and within the acoustical consulting profession. Some people believe that the machine is too harsh on hard floor surfaces, or that it overrates carpeted surfaces. It is true that the relatively light weight hammers do not energize a floor structure in the same way that footfalls do. A number of different test methods and impact devices have been studied, but none seem to provide a better rank-ordering than the standard machine. Additional discussion of the test method is contained in the text associated with impact noise data presented on the following pages.

The procedure for measuring impact sound insulation in the field is very similar to that used for laboratory measurements. The significant difference in test results between laboratory and field data is due to the boundary conditions of the test floor. For laboratory tests, the floor assembly is isolated from the surrounding structure and flanking transmission through the surrounding structure is minimal. In the field, the test floor is an integral part of the building structure (except for properly installed floating floors), and flanking transmission can be quite significant.

The method for field tests is contained in ASTM Designation: E 1007, "Standard Test Method for Field Measurement of Tapping Machine Impact Sound Transmission Through Floor-Ceiling Assemblies and Associated Support Structures." The method is essentially the same as the laboratory method. The single-number rating for field data is the Field Impact Insulation Class (FIIC), determined in exactly the same manner as the IIC.

IIC and FIIC numerical ratings are intended to represent the same degree of acceptance for impact noise as that represented by the STC and FSTC numerical ratings for airborne noise. That is, if an STC 50 rating for a particular floor-ceiling construction is judged to be marginally acceptable for a multifamily residence, an IIC rating of 50 will be judged equally acceptable.

Before the IIC rating method was introduced many years ago, a similar single-number rating was in use, called the Impact Noise Rating (INR). The same curve fitting procedure was employed to determine the INR, except the method resulted in a number approximately 51 points less than the IIC. That is,

$$\text{IIC} \sim \text{INR} + 51$$

Should the reader encounter INR data, it may be converted to an IIC rating by adding 51. Bear in mind, however, that such data are probably more than 20 years old, and may not represent current construction materials or practices.

## 3.6 Impact Insulation Data

There is much less impact sound insulation data available on floor-ceiling assemblies than there is for airborne sound transmission loss, and there is less consistency in the data that are available. Nevertheless, different constructions and, in particular, floor coverings have clearly unique impact performance characteristics which the authors have attempted to demonstrate in the figures that follow.

### 3.6.1 Frame Construction

*Figure 3-29.* The impact sound transmission data shown in this figure were obtained from the same series of wood frame floor-ceiling constructions for which airborne sound transmission data were presented in Figures 3-25 and 3-26. The FIIC rating for the type DA construction is 65, and would be considered acceptable for multifamily use if a rating of 50 is considered the minimum acceptable. The reader will recall that the airborne NIC rating for this construction was an unacceptable 38 (see Figure 3-25).

The high FIIC rating for the DA construction demonstrates the effectiveness of a carpet and pad surface using the standard tapping machine test method. It was mentioned earlier, however, that the tapping machine does not energize a floor system in the same way as a footfall. Wood frame floor construction is prone to produce low frequency "thumping" sounds when subjected to footfalls, even when the floor is carpeted. Creaking is another common problem for floors of this construction. Neither thumping nor creaking is likely to be experienced using the tapping machine, thus the apparently "safe" rating of FIIC 65 may be misleading. Methods for avoiding these problems are discussed in Chapter 5, Section 5.4.2.

Attachment of the ceiling gypsum board using resilient channels (RC construction) reduces the impact sound level significantly in the low to mid frequency range, and even more reduction is achieved with separate ceiling joists (construction SJ). As for construction DA, the FIIC ratings above 65 indicate acceptable impact performance, but the low frequency thumping problem from footfalls may be present. Of the three constructions, the SJ construction will provide the best insurance against thumping.

*Figure 3-30.* Impact sound transmission data are shown in this figure for the same three constructions represented in the previous figure, but with a sheet vinyl surface instead of pad and carpet. FIIC ratings drop substantially, and none meet the minimum requirement of FIIC 50. However, the NISPL shows significant relative improvement in the progression from the DA to RC to SJ constructions.

FIIC ratings in the range 50–55 for wood frame construction with hard surfaces can be obtained with underlayments of a resilient material such as 1/4″ (6.4 mm) sponge neoprene. Also, there are several products produced specifically for this purpose available from manufacturer numbers 37 and 43 listed in Appendix A.12.

### 3.6.2 Concrete Construction

*Figure 3-31.* Concrete is a very good conductor of vibrational energy, resulting in poor impact sound insulation performance with a bare surface. By applying foam-backed vinyl sheeting to the surface, the NISPL at the mid and high frequencies drops by as much as 45 dB. With a carpet and pad surface, the reduction in NISPL compared to bare concrete occurs across the entire frequency range, showing again that carpet is an excellent treatment for reduction of impact noise. Furthermore, concrete construction is not likely to have the low frequency thumping problem from footfalls as does wood frame construction.

Excellent impact insulation performance can be achieved with a "floating" concrete slab on a base slab. The isolators supporting the floating slab are the same type as those described for the airborne TL test of the same construction (Figure 3-27). It is very important that the floating slab be isolated from the building structure around the entire perimeter of the slab. In this Figure, there is some evidence of flanking in the FL curve at frequencies above 630 Hz. Any connections between the two slabs, such as floor drains, must be installed in a manner that does not short circuit the isolation. Instructions provided by the isolator manufacturer must be carefully observed, as flanking transmission can seriously degrade the acoustical performance of the system.

## *3.7 Noise-Induced Vibration*

It is possible for high sound levels to cause noticeable vibration of large architectural surfaces, such as floors, walls, ceilings and windows. Frequencies below 100 Hz are most likely to cause vibration, because these are the frequencies where the surfaces vibrate most easily. The vibration may be unpleasant to experience, or it may cause annoying rattling of light fixtures, objects on shelves, etc.

The sources of low frequency noise causing vibration problems could be aircraft, surface vehicles, and some industrial operations such as earth moving equipment, natural gas pumping stations and batch asphalt plants. Vibration could also be caused by mechanical equipment located within or adjacent to the building. Large, low RPM fans and reciprocating compressors can have large unbalanced forces which, if improperly isolated, can cause vibration which is transmitted throughout the building. Refer to Chapter 7, Figure 7-6, for the approximate sensitivity and reaction of people to feelable vibration.

## *3.8 Noise Control Construction Techniques*

The acoustical performance of any building material or construction can be compromised if the installation or construction is improperly or carelessly done. The purpose of this Section is to outline procedures that if followed

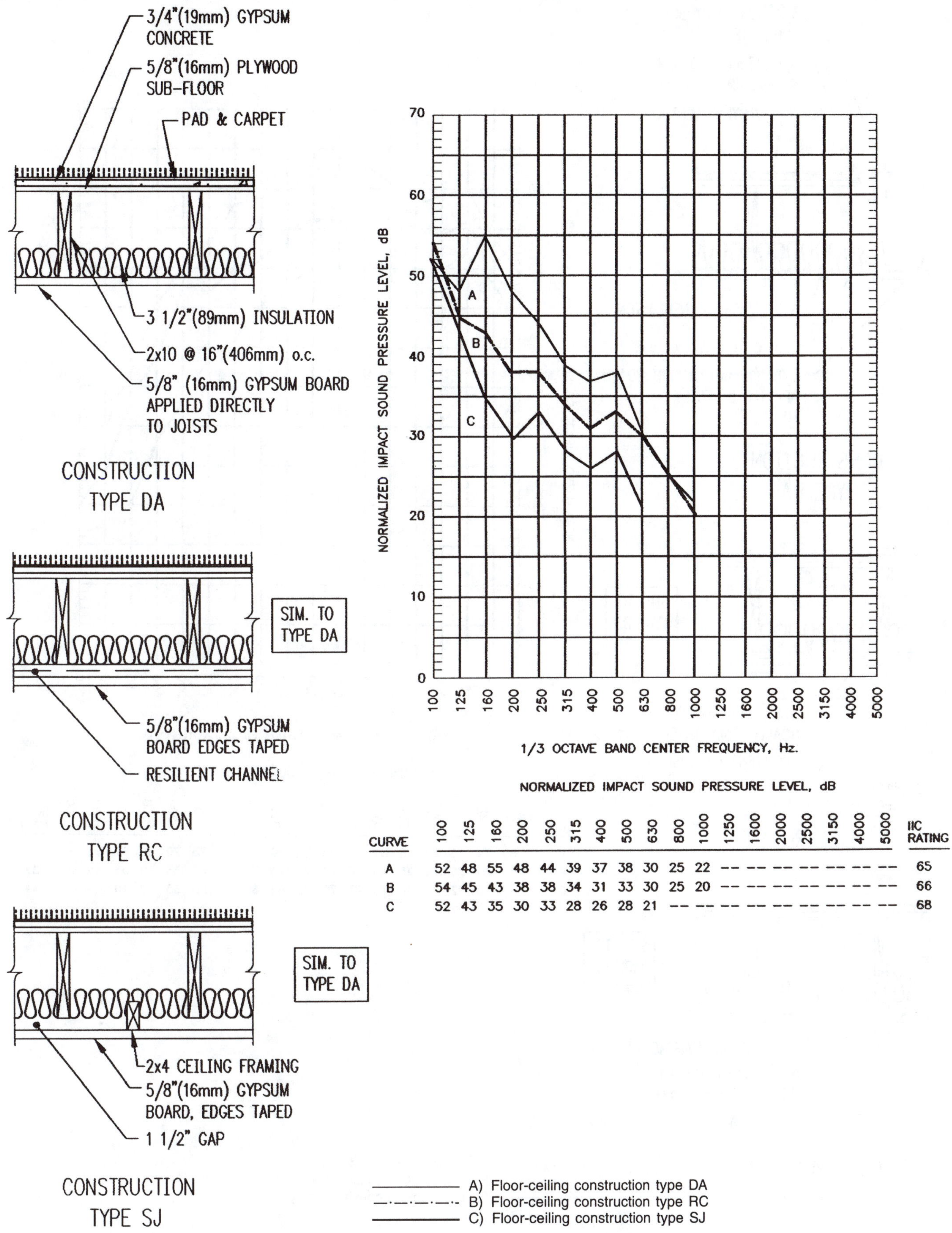


NORMALIZED IMPACT SOUND PRESSURE LEVEL, dB

| CURVE | 100 | 125 | 160 | 200 | 250 | 315 | 400 | 500 | 630 | 800 | 1000 | 1250 | 1600 | 2000 | 2500 | 3150 | 4000 | 5000 | IIC RATING |
|---|---|---|---|---|---|---|---|---|---|---|---|---|---|---|---|---|---|---|---|
| A | 52 | 48 | 55 | 48 | 44 | 39 | 37 | 38 | 30 | 25 | 22 | -- | -- | -- | -- | -- | -- | -- | 65 |
| B | 54 | 45 | 43 | 38 | 38 | 34 | 31 | 33 | 30 | 25 | 20 | -- | -- | -- | -- | -- | -- | -- | 66 |
| C | 52 | 43 | 35 | 30 | 33 | 28 | 26 | 28 | 21 | -- | -- | -- | -- | -- | -- | -- | -- | -- | 68 |

**Figure 3-29.** Impact sound insulation of wood frame floors, carpet surface.

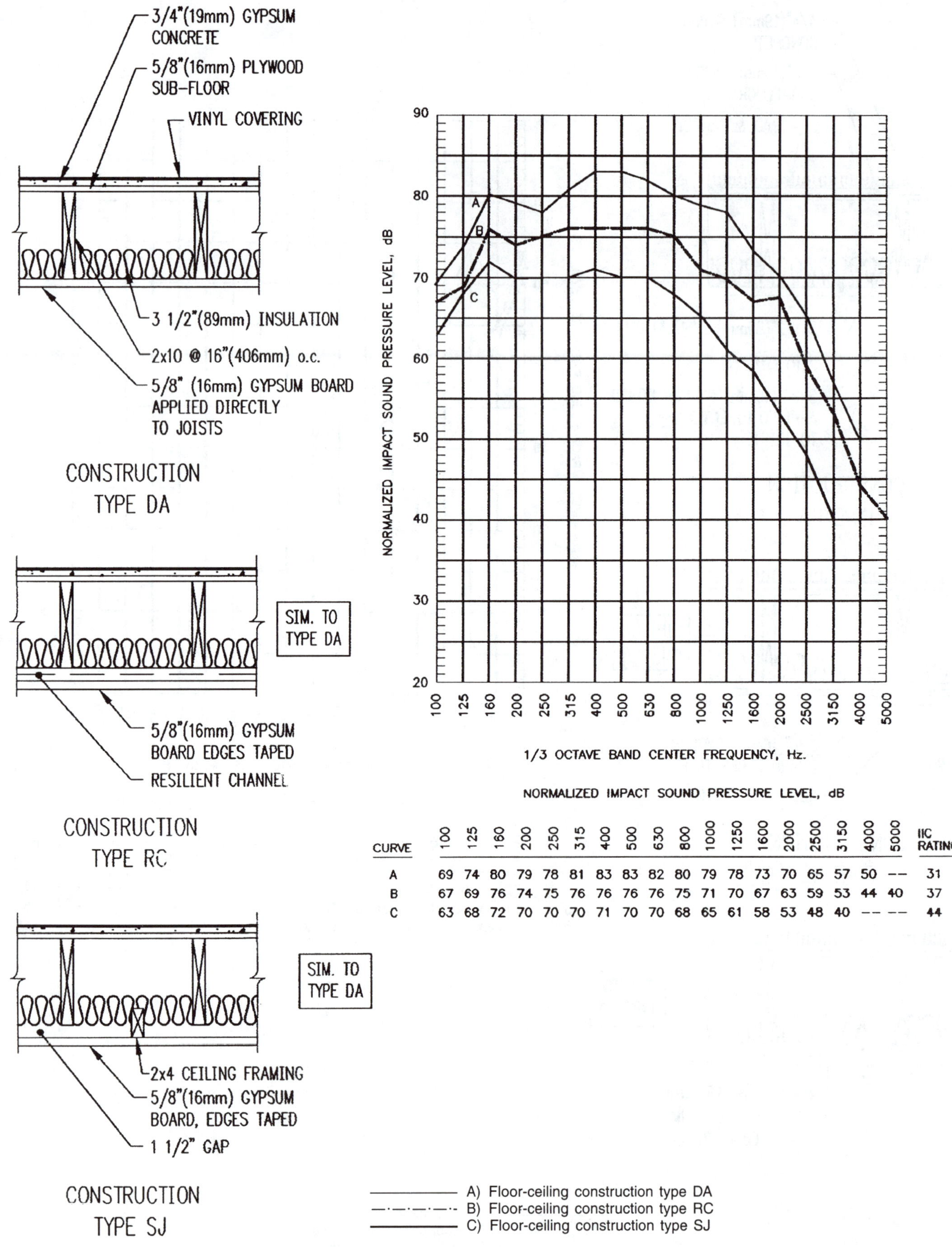


NORMALIZED IMPACT SOUND PRESSURE LEVEL, dB

| CURVE | 100 | 125 | 160 | 200 | 250 | 315 | 400 | 500 | 630 | 800 | 1000 | 1250 | 1600 | 2000 | 2500 | 3150 | 4000 | 5000 | IIC RATING |
|---|---|---|---|---|---|---|---|---|---|---|---|---|---|---|---|---|---|---|---|
| A | 69 | 74 | 80 | 79 | 78 | 81 | 83 | 83 | 82 | 80 | 79 | 78 | 73 | 70 | 65 | 57 | 50 | -- | 31 |
| B | 67 | 69 | 76 | 74 | 75 | 76 | 76 | 76 | 76 | 75 | 71 | 70 | 67 | 63 | 59 | 53 | 44 | 40 | 37 |
| C | 63 | 68 | 72 | 70 | 70 | 70 | 71 | 70 | 70 | 68 | 65 | 61 | 58 | 53 | 48 | 40 | -- | -- | 44 |



**Figure 3-30.** Impact sound insulation of wood frame floors, vinyl surface.

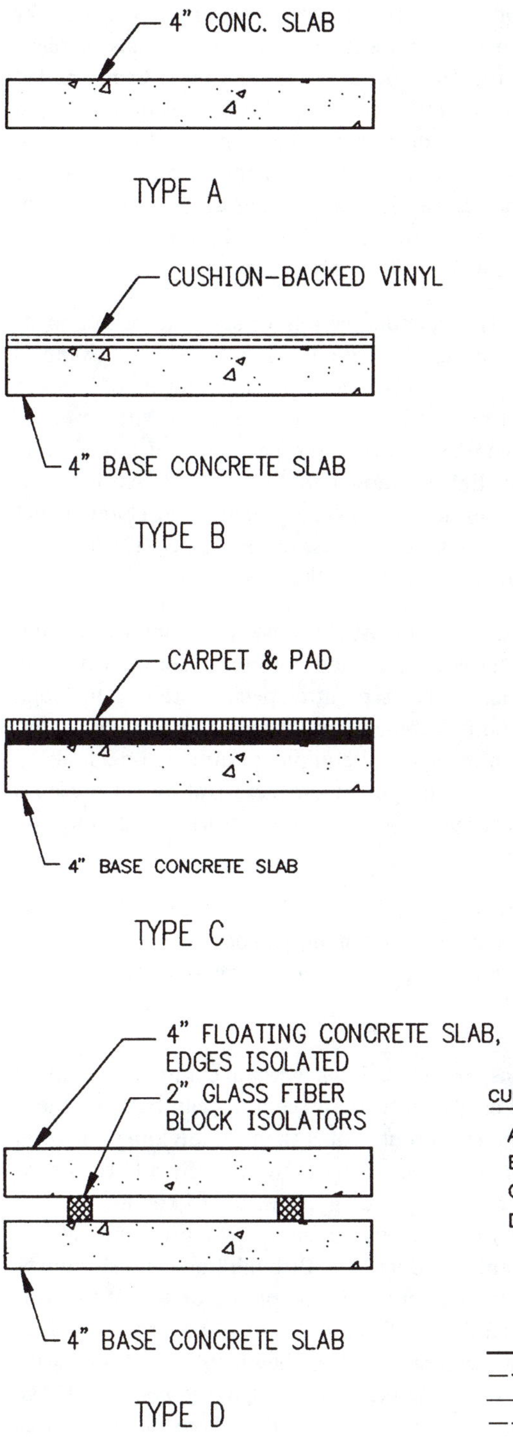

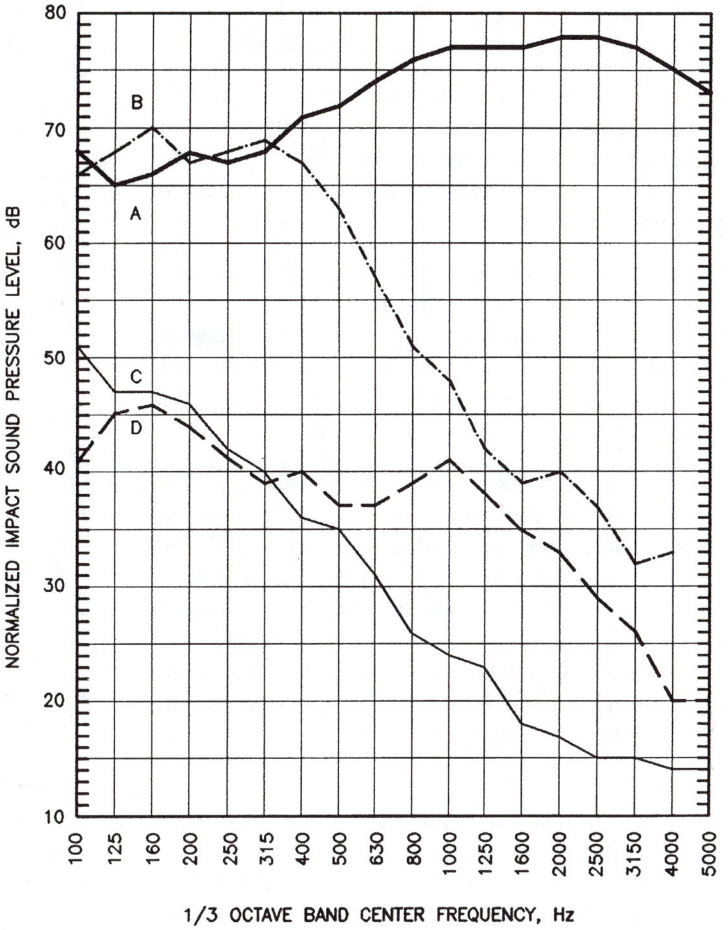

4" CONC. SLAB

TYPE A

CUSHION-BACKED VINYL

4" BASE CONCRETE SLAB

TYPE B

CARPET & PAD

4" BASE CONCRETE SLAB

TYPE C

NORMALIZED IMPACT SOUND PRESSURE LEVEL, dB

1/3 OCTAVE BAND CENTER FREQUENCY, Hz

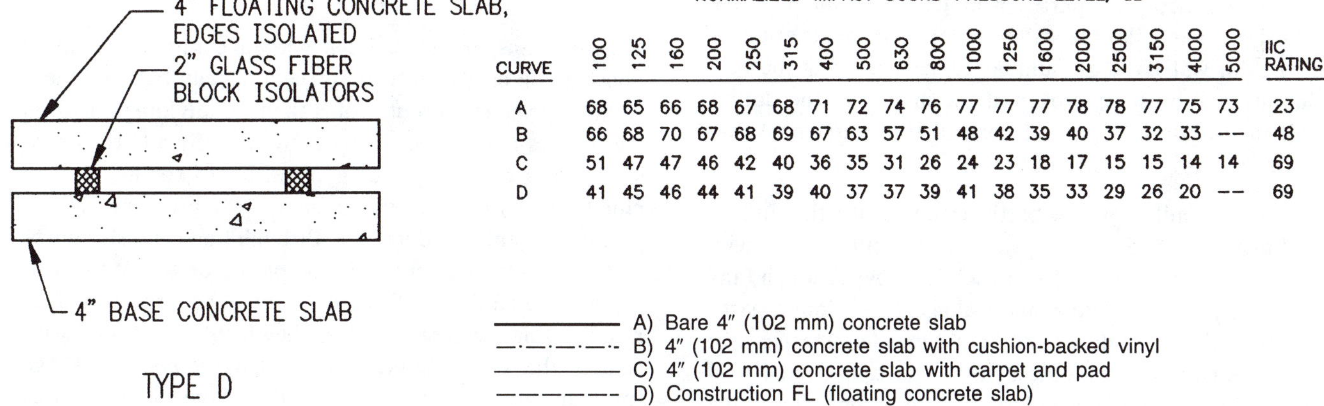


NORMALIZED IMPACT SOUND PRESSURE LEVEL, dB

| CURVE | 100 | 125 | 160 | 200 | 250 | 315 | 400 | 500 | 630 | 800 | 1000 | 1250 | 1600 | 2000 | 2500 | 3150 | 4000 | 5000 | IIC RATING |
|---|---|---|---|---|---|---|---|---|---|---|---|---|---|---|---|---|---|---|---|
| A | 68 | 65 | 66 | 68 | 67 | 68 | 71 | 72 | 74 | 76 | 77 | 77 | 77 | 78 | 78 | 77 | 75 | 73 | 23 |
| B | 66 | 68 | 70 | 67 | 68 | 69 | 67 | 63 | 57 | 51 | 48 | 42 | 39 | 40 | 37 | 32 | 33 | -- | 48 |
| C | 51 | 47 | 47 | 46 | 42 | 40 | 36 | 35 | 31 | 26 | 24 | 23 | 18 | 17 | 15 | 15 | 14 | 14 | 69 |
| D | 41 | 45 | 46 | 44 | 41 | 39 | 40 | 37 | 37 | 39 | 41 | 38 | 35 | 33 | 29 | 26 | 20 | -- | 69 |

A) Bare 4″ (102 mm) concrete slab
B) 4″ (102 mm) concrete slab with cushion-backed vinyl
C) 4″ (102 mm) concrete slab with carpet and pad
D) Construction FL (floating concrete slab)

**Figure 3-31.** Impact sound insulation of concrete floors.

will result in realizing the full acoustical potential of the construction.

*Partitions.* Wood framing should be dimensionally stable and not subject to shrinkage or warping. Gypsum board should always be type X to insure assumed density. Where two or more layers of gypsum board are applied to one side of a wall, always stagger board joints. Unless otherwise noted, any sound-rated partition must extend to structure and be sealed to it. Watch stud spacing for the STC rated partitions—usually 16″ (406 mm) for wood framing, 24″ (610 mm) for metal framing. Also be aware of metal stud gauge for single stud partitions; heavy gauge studs will not perform as well as light gauge.

Be sure to follow manufacturer's recommendations for installation of resilient channels. Always apply the channels directly to the framing member. Never apply the channels between layers of gypsum board. It is redundant to apply channels to double stud partitions.

Insulation installed inside a partition should be identified as sound batts. Density should not be less than 2.5 lb/cu ft (40.0 kg/cu m). Install so as to prevent sagging. Loose or blown-in insulation is not recommended as it settles. Never use foam-in-place insulation unless acceptable sound transmission loss data are furnished.

To prevent sound leakage, caulk continuously all around with a nonhardening caulking compound. Acceptable compounds are sometimes identified as "acoustical sealants." Where a partition meets an irregular surface, such as metal deck, the flutes of the deck must be filled in. All penetrations, such as conduit, piping, ductwork, trusses, etc., must be sealed. Electrical, telephone or TV outlets should not be back-to-back; separate by at least one stud space. Caulk all around outlet boxes. Remember: Any air leak is a sound leak.

Pipe, conduit or duct runs within, or penetration through a double partition must not rigidly tie the two sides together. Any interconnecting elements must be vibration isolated, either by means of a flexible connection or by packing around the penetration on one side with sponge neoprene, so that no direct contact is made with that half of the partition.

Where a demising partition abuts a continuous partition (i.e., a party double wall between apartments abutting a corridor wall), the gypsum board surface along the continuous wall must be broken at the middle of the double wall intersection to prevent flexural transmission. A saw cut is generally a sufficient break.

In office buildings, the partition joint with the outside wall may include a closure panel, often joining with a window mullion. Be sure the panel is heavy enough and sealed in place to avoid sound leakage. Never join a partition directly to window glass. Sometimes heating pipes or ducts are run between offices in the space under the windows and the partition is left open. It is important that such penetrations be properly sealed with acoustical sealant and insulation as needed.

CMU partitions should be constructed using blocks with dense aggregate, and core spaces can be grouted solid for added weight. Seal all exposed surfaces with two generous coats of paint. For double wythe walls, inner surfaces should not be sealed. Increasing wall separation will improve low frequency performance. Install sound batts in the airspace.

*Floors.* Stiffness of wood frame floors should exceed minimum building construction standards to avoid low frequency noise from footfalls. Use deeper joists or place them closer together. Be sure the subfloor is firmly nailed to the joists to avoid creaking. Walk the subfloor, and wherever creaking occurs, add more nails. Use coated or serrated nails. A 1″ to 2″ (25 mm to 51 mm) topping of gypsum concrete or lightweight aggregate concrete provides added weight. In areas with hard floor surfaces, a resilient underlayment is required.

A resiliently suspended ceiling or separate ceiling joists will add substantially to the TL. Apply resilient channels directly to the floor joists, never between layers of gypsum board. Select the brand of resilient channels with care; use only tested channel. "Float" the perimeter of a resiliently suspended ceiling as shown in Figure 3-32. Avoid screw penetrations through the gypsum board and channel that contacts the joists. Always use minimum 3″ thick sound control blanket between the floor joists.

Concrete floors will usually have sufficient weight and stiffness to provide adequate airborne sound isolation. Floating concrete floors require special detailing for edge conditions and plumbing penetrations, such as floor drains. Do not mount mechanical equipment directly on a floating floor. Mount the equipment on piers resting on the concrete base slab, using appropriate vibration isolation and seismic restraints.

Avoid hard surfaces for any type of floor without some type of resilient underlayment. Carpet is by far the best treatment for control of mid and high frequency impact noise.

*Ceilings.* Suspended ceilings with a common air space above, often used as return air plenums, can transmit more sound between adjacent rooms than the dividing partition which does not extend to structure. Usually a barrier consisting of a single layer of gypsum board (or material of equal weight) is sufficient to correct the problem. Return air must be entirely ducted or through lined transfer ducts. In some cases, blanket insulation placed on top of the ceiling will be sufficient. Blankets cannot be used over fire-rated ceiling assemblies unless they have been tested with exactly the same blanket. Watch out for sound leakage through light fixtures and air grilles. Some light fixtures are rated for sound transmission.

Fluorescent lights and some other special lights have ballasts which can radiate objectionable noise. Ballasts for fluorescent lights are rated A, B, C, etc., with "A" being the least noisy. However, over a period of time, quiet ballasts may become noisy. For this reason, incandescent lights should be used in recording studios and other situations where this noise would be unacceptable. Electronic ballasts are quiet but they are more expensive. The lighting

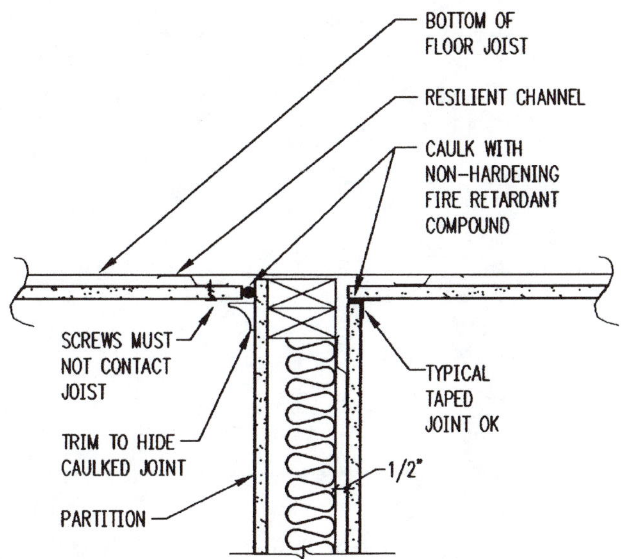


The intent is to maintain the floating suspended gypsum board around the entire perimeter of a room. Some revision may be required to meet fire codes.

**Figure 3-32.** Floating edge detail for resiliently suspended ceilings.

engineer should carefully evaluate the requirements of any noise-sensitive project.

*Doors, Operable Partitions.* Sound leakage is usually the cause of doors or operable partitions not performing as rated. Seals must be adjusted to just make continuous contact with the mating surface. If they exert too much force, it will be difficult to close the door all the way, and the seals will be ineffective. Magnetic seals for metal doors require minimal maintenance over time. Threshold closures do no provide a good seal on carpet.

Close the space above an operable partition if it is open, with a gypsum board barrier or other material of comparable weight. Seal around penetrations through the barrier. To insure optimum performance of a critical door or operable partition, require a field noise reduction test of the completed installation. The field NIC requirement should not be more than 8 points less than the laboratory STC rating.

*Mechanical noise.* Avoid mechanical rooms along side or above critical spaces. Even with adequate wall separation, duct penetrations may compromise sound isolation. Acoustical absorption on room surfaces will reduce reverberant noise build-up in mechanical rooms.

Provide proper vibration isolation for mechanical equipment, with seismic restraints or "snubbers" if appropriate. Do not support equipment on isolators if it has adequate internal isolation. Duct, plumbing and electrical connections to mechanical equipment should incorporate vibration isolation devices.

Design for laminar flow in air distribution systems. Avoid high air velocities in critical applications, as such systems are invariably noisy.

*Plumbing Noise.* All plumbing, both supply and waste, running next to or through noise sensitive spaces, must be isolated from the building structure to the maximum extent possible. Use resilient materials such as sponge neoprene around pipe penetrations through building materials, or between pipes and any structural element. Refer to Chapter 7, Figure 7-11. Plastic waste pipe is not recommended; cast iron is much quieter.

Avoid toilet rooms, bathrooms, kitchens or other fixtures requiring plumbing attachments over noise sensitive areas such as classrooms, living areas and, in particular, sleeping rooms. If unavoidable, plumbing penetrations must be totally enclosed in the space below the floor penetration.

Select quiet fixtures and valves. Flush valve toilets are much noisier than tank-type. Avoid excessively high water pressures. Provide antiknock devices in supply water lines.

# Chapter 4

# Sound in Enclosures

The behavior of sound in rooms must be understood to satisfactorily design spaces for their intended use. When sound energy is generated within a room, it travels unimpeded until it encounters a room surface or object within the room. At such an encounter the sound may be reflected in a new direction, absorbed, or diffused, usually a combination of all three. The sound may also be transmitted into an adjacent space. The previous chapters have described the acoustical characteristics of building materials, and with proper application, the designer can control how sound behaves in a room—or travels between rooms—and not be left to chance.

The subjects of sound reflection and sound absorption have been treated previously in Chapter 1, Section 1.4.2, and Chapter 2, Section 2.1, respectively. Some of that material is reviewed in the following sections of this Chapter, with emphasis on the special characteristics of sound behavior in enclosures. In particular, the concept of reverberation is discussed in some detail, with respect to its beneficial as well as harmful effects on speech and music.

## 4.1 Early and Late Reflections; Echoes

Reflected sound in a room can be either beneficial or disruptive, depending on the arrival time at the listener compared to the arrival time of the direct sound. If this time difference is less than about 30 milliseconds, recalling that sound at normal room temperature travels slightly more than 1 ft (1/3 m) in one second, the reflection will be a positive reinforcement of the direct sound. As the delay increases beyond 30 milliseconds, the combined sound starts to lose quality, and beyond about 50–60 milliseconds the reflection will destructively interfere with the direct sound.

Figure 4-1 shows an idealized impulse response in a room, where multiple reflections arrive at a listener position following arrival of the direct sound. The early useful reflections arrive within the first 30 milliseconds, followed by more closely spaced reflections gradually decreasing in level. If a particular reflection is significantly higher in level compared to adjacent reflections, and the time delay is great enough, the reflection will be perceived as an echo.

A second consideration of the destructive effect of reflected sound, in addition to time difference, is the level of the reflected sound compared to the direct. Delays longer than 50 milliseconds can be tolerated if the level of the reflection is well below that of the direct sound. The relationship between time delay and relative levels has been well documented, and is termed the "Haas effect." Figure 4-2 shows the percent of listeners disturbed as a function of both the difference in level and time delay between the direct and reflected sound (based on speech delivered at a normal rate of speaking). If the level of the reflected sound is 10 or more dB below the direct sound, the amount of time delay is not significant in terms of listener disturbance.

It is important in rooms requiring sound clarity that the direct and reflected sound paths be carefully studied to avoid destructive reflections. Particularly annoying are delayed reflections that are loud enough to be heard as a distinct sound, or echo. Echoes are frequently encountered in large arenas or sports stadiums, where delays of several hundred milliseconds are not uncommon. (See also Chapter 5, Section 5.5.1.)

If an impulsive sound is produced in a room between surfaces which are opposite, parallel and hard, a series of closely spaced (in time) reflections may occur, called a "flutter echo." Such a condition can be very disconcerting to a speaker or musician, and can be avoided by designing nonparallel surfaces, or by some treatment to absorb or diffuse the sound (see the following 4 Sections, particularly Room Modes).

Another important consideration for reflected sound is the direction from which the sound arrives at the listener. Directional realism (i.e., the sound must come from the visual source) is very important in many spaces, particularly theaters for the performing arts. Closely related to directional realism is the importance of lateral reflections in music halls, creating a sense of spaciousness or envelop-

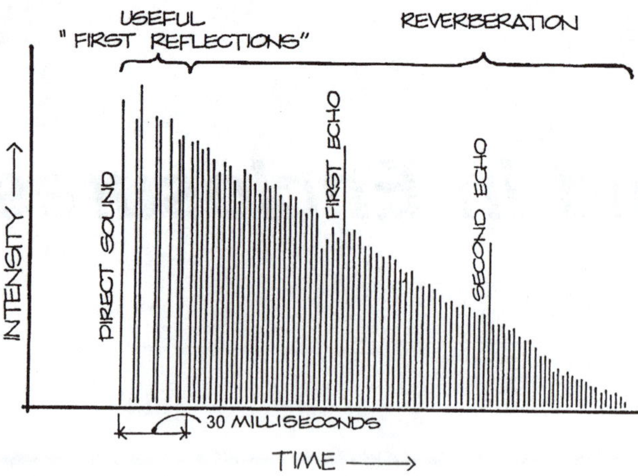


Continuity of reflections is important, otherwise lack of clarity, or even echos may result.

**Figure 4-1.** Idealized room impulse response.

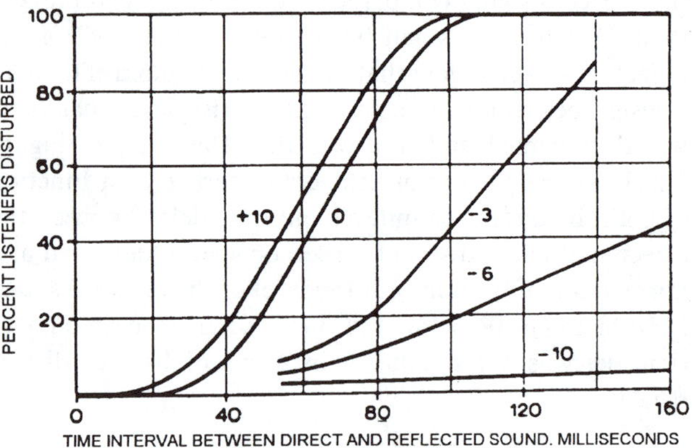


The numbers printed next to the curves are the level differences between direct and reflected sound. As the sound level difference increases, listener disturbance decreases.

**Figure 4-2.** Disturbance of delayed reflected sound.

Focusing causes uneven distribution of sound and can be very annoying.

**Figure 4-3.** Focusing caused by concave surfaces.

ment. These topics are discussed in more detail in Chapter 5.

## 4.2 Focusing

While convex curved surfaces used as sound reflectors can be useful for dispersing sound to obtain uniform sound distribution, concave surfaces can have just the opposite effect. Depending on the location of source and receiver, and the amount of curvature, the sound can focus causing very uneven distribution. For this reason, concave surfaces should be avoided in spaces intended for hearing. Figure 4-3 shows the effect of a domed ceiling and Figure 4-4 shows the problems associated with having a concave rear wall in an auditorium.

If concave surfaces cannot be avoided, or if encountered in an existing building and causing problems, it is usually possible to treat the surface with convex "bumps" to improve diffusion (recalling that the bumps must be large enough to reflect the lowest frequency of interest; see Chapter 1, Section 1.4.2), or with an absorptive material to reduce the amplitude of the reflections. The use of absorptive materials will also reduce reverberation, which may be undesirable in halls used for musical performances. In such cases, a combination of diffusing and absorbing surfaces may be the best solution.

## 4.3 Diffusion

Sound diffusion in a room is the condition of having a uniform sound level distribution throughout the room. This condition is desirable in all rooms used for speech or music, but is particularly important in music performance and rehearsal rooms, and any space where sound recordings are made. If a space lacks diffusion, the sound will have an "uneven" quality, making it difficult for musicians or students to clearly hear one another, or for locating a microphone to obtain a satisfactory recording.

Sound diffusion can be achieved by the use of irregular surfaces, splayed panels at different angles, convex surfaces or specially designed QRD diffusors (see Chapter 2, Section 2.5.2). Some diffusion can also be achieved by alternating between absorptive and reflective surfaces. Applications of such treatments are discussed in Chapter 5, Sections 5.1 and 5.2.

## 4.4 Diffraction

The ability of objects or surfaces in an enclosure to reflect sound depends on their size compared to the wavelength of the impinging sound. If the surface is large compared to the wavelength, it will reflect most of the sound striking it, in a direction controlled by the angle (or angles) of the surface. Objects or surfaces much smaller than the wave-

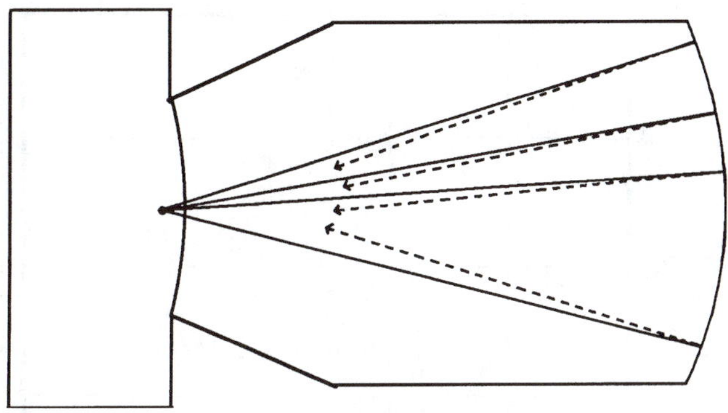

Focused reflections off the rear wall cause a concentration of sound at the front with unnacceptable time delay.

**Figure 4-4.** Auditorium with a concave, reflective rear wall.

length will have little effect on the sound as it passes by (see Chapter 1, Figure 1-10).

Diffraction must be considered when designing surfaces for reflecting sound, whether they be panels for sound reinforcement, used extensively in auditorium design, or for sound diffusion, used in music rehearsal rooms and stage enclosures. For example, if a panel is to reflect sound at a frequency of 250 Hz and above, the panel dimensions must not be smaller than about one wavelength, or 4.5 ft (1.4 m).

## 4.5 Room Modes

As sound travels around a room and reflects off room surfaces, the incident and reflected waves combine with each other, causing either reinforcement or cancellation. This process creates what are called "standing waves," which occur at a set of frequencies unique for each room, depending on the room shape and dimensions, and each frequency within this set is called a room mode.

Standing waves are stationary in space, and are characterized by large variations in sound level as a listener moves through the room. These variations are most noticeable when the sound is a musical tone or single frequency, and must be avoided in rooms used for music performances or rehearsals, recording studios, and other similar critical uses.

There are three types of room modes: axial, tangential and oblique, involving two, four and six surfaces, respectively, in rectangular rooms. The frequency distribution for each type of mode consists of a fundamental, and a series of integral multiples of the fundamental called harmonics. The frequencies at which room modes occur are determined by

$$f = \frac{c}{2}\left[\left(\frac{n}{L1}\right)^2 + \left(\frac{n}{Lw}\right)^2 + \left(\frac{n}{Lh}\right)^2\right]^{1/2}, \text{ Hz}$$

where f = mode frequency, Hz
c = speed of sound, ft/sec (m/s)
L = room dimensions (*L*ength, *W*idth and *H*eight), ft (m)
n = integral number of mode

A particular mode is specified by listing the mode numbers n for each of the three principal room axes. For example, the mode designation (1,1,0) would indicate the fundamental frequency of a tangential mode determined by the length and width dimensions of a room.

To facilitate visualizing the distribution of modal frequencies in a room, it is convenient to determine only the axial modes, designated by (n,0,0), (0,n,0) and (0,0,n). If two of the terms within the square root expression are zero (indicating an axial mode), the preceding equation simplifies to

$$f = \frac{cn}{2L}, \text{ Hz}$$

The frequencies for axial modes in two rectangular rooms are shown in Figure 4-5. The modes are plotted on a horizontal frequency scale. The left scale is for a room 40 × 20 × 10 ft (12.2 × 6.1 × 3 m). Notice the coincidence of room modes at 28 Hz (2 modes, 2,0,0 and 0,1,0), 56 Hz (3 modes, 4,0,0, 0,2,0 and 0,0,1), 85 Hz (2 modes, 6,0,0 and 0,3,0) and 113 Hz (3 modes, 8,0,0, 0,4,0 and 0,0,2). The room would tend to be more responsive at these frequencies, an undesirable condition for any type of musical performance or rehearsal space.

The right scale in Figure 4-5 is for a room 42 × 17 × 9 ft (12.8 × 5.2 × 2.7 m). For this room the modal distribution is spaced along the frequency scale in a reasonably uniform manner, and there are no coincident modes. This room would be preferred to the room represented by the left scale.

While standing waves cannot be completely eliminated, their effects can be minimized in rectangular rooms by choosing room dimensions which cause the modes to be uniformly spaced along the frequency scale. A good ratio of room dimensions (height, width and length) is $1:2^{1/2}:3^{1/2}$, or 1:1.4:1.7. The shaded area within Figure 4-6 defines preferred ratios of room dimensions to minimize the detrimental effects of room modes. These preferred ratios apply primarily to rectangular rooms such as music practice and rehearsal rooms and recording studios; they are not as important in larger auditoriums where extensive shaping in both plan and section minimizes problems with room modes.

Rooms with coincident dimensions, or with one dimension exactly twice another, such as 1:1:2, should be avoided, because the modes are more pronounced and are not uniformly spaced. A cubical room is a worst case example, since the modes in all three dimensions would coincide in

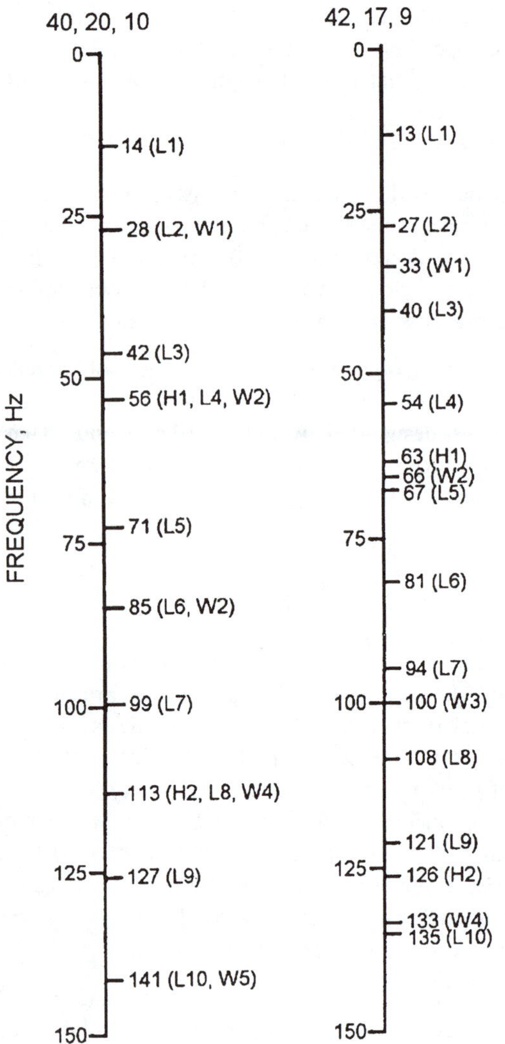


Notice the coincidence of axial modes for the room whose dimensions are integral multiples of 2. The modes are not coincident and more evenly spaced for the second room.

**Figure 4-5.** Axial modes with room dimensions (L, W, H, ft).

frequency. Such modes are sometimes called degenerate modes.

Standing waves can be minimized by designing nonrectangular rooms, thus avoiding opposite, parallel, reflecting surfaces. (Such surfaces may also produce flutter echoes; see Section 4.1.) Alternatively, such surfaces may be treated with diffusing elements or acoustical absorbents. The subject of treatments to control all types of undesirable reflections for specific building types is presented in Chapter 5.

## 4.6 Reverberation

When a steady sound source is activated within a room, the sound quickly builds up to a constant level where the rate at which acoustic energy is introduced into the room

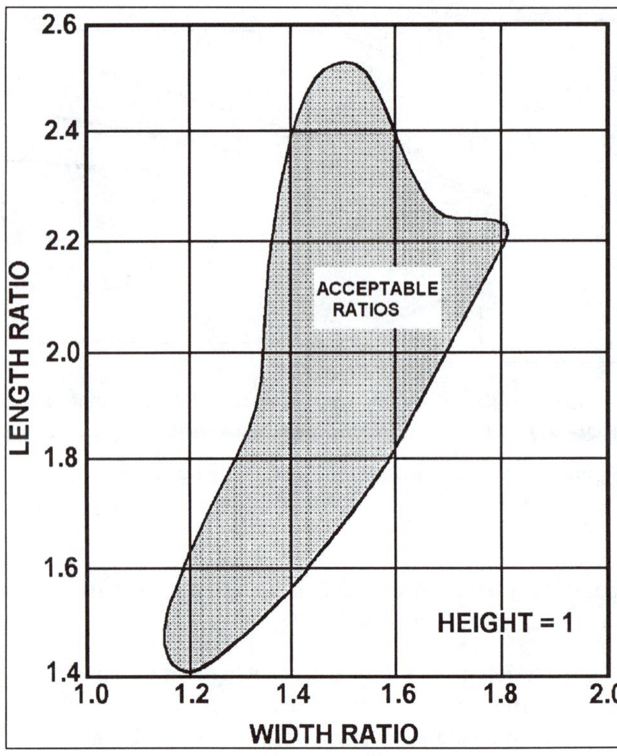


Preferred ratios of dimensions for essentially rectangular rooms which provide reasonably uniform distribution of room modes.

**Figure 4-6.** Optimum room ratios.

is equal to the rate at which it is absorbed. If the sound is turned off, the rate of absorption remains the same, so the sound level gradually diminishes. If most of the room surfaces are reflective, it will take a relatively long time for the sound to disappear, or "decay," using the terminology of the profession. Introducing more absorption in the room will increase the rate of decay, and the sound disappears more rapidly.

The persistence of sound within a room after the source has been turned off is called *reverberation*. The time it takes for the sound level to decrease 60 decibels is the *reverberation time*, symbolized $T_{60}$. The reverberation time is a useful measure of the suitability of a space for particular functions. In the following sections, reverberation time criteria are developed for specific uses, followed by room design features that control reverberation, and how reverberation times are calculated.

### 4.6.1 Reverberation Criteria for Speech and Music

As a general rule, programs that consist primarily of speech favor short reverberation times. (A discussion of speech intelligibility is contained in Chapter 1, Sections 1.8.1 and 1.8.2.) Excessive reverberation interferes with the progression of speech syllables, and intelligibility suffers. On the other hand, music is enhanced by some reverberation, depending on the type of music. In Figure 4-7, mid-frequency reverberation times are recommended for a variety of speech and music programs. The recommended range of times shown for a particular use are related to the cubic volume of the room. Shorter reverberation times are appropriate for smaller volumes, and vice versa. In multipurpose auditoriums, where both speech and music must be accommodated, a compromise reverberation time may be appropriate, or some means of achieving adjustable reverberation. Churches featuring choir and organ music are best served by long reverberation times, and this condition can be accommodated if a properly designed sound reinforcement system is used for speech (see Chapter 6).

Reverberation is controlled primarily by the volume of a room and the amount of acoustical absorption contained within it (including air absorption at the higher frequencies; see Chapter 2, Section 2.5.4). To a lesser extent, reverberation is also controlled by the shape of the room and the placement of the absorption. All of these factors are frequency-dependent, so reverberation times vary with frequency.

Because the ear is less sensitive to the lower frequencies compared to the middle and higher frequencies, the reverberation time at 63–125 Hz should generally be about 1.5 times longer than the reverberation at 500–1000 Hz, as shown in Figure 4-8. The times at 2000–4000 Hz should be the same as 500–1000 Hz, but this is difficult to attain because the absorption of air increases with frequency, and some common materials such as carpet have their highest absorption at the higher frequencies.

It is important that the reverberation time be related to the size of the space. The visual impression should be consistent with the acoustical conditions. A small space such as a classroom used for speech should have a short reverberation time, while music halls and cathedrals should have much longer times.

Reverberation at low frequencies gives music a feeling of warmth, and is particularly important in spaces for musical performance. Wood finishes in concert halls are thought to add warmth to the music, but thin wood paneling can have just the opposite effect because of low frequency flexural absorption (see Chapter 2, Section 2.5.2). This absorption can be avoided by using paneling at least 1 1/2" (38 mm) thick, or by applying wood veneer to heavy plaster or concrete walls.

### 4.6.2 Near/Direct/Reverberant Fields

When sound is generated in a room, the initial sound paths from source to room surfaces are considered the near and direct fields. The near field is closest to the source, and its extent depends on the size and radiation characteristics of the source. Within distances comparable to the dimensions of the source, the sound level may decrease, be constant

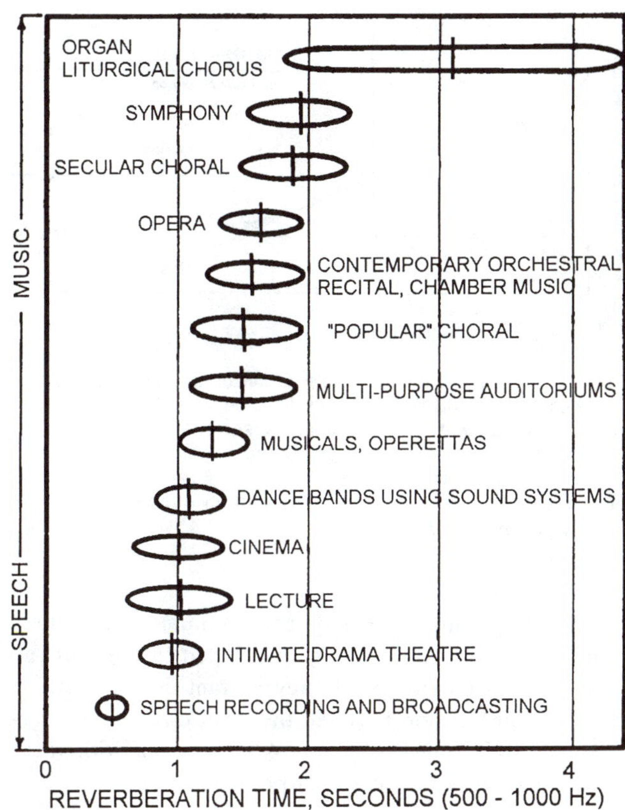


The range indicated for each use is a function of volume, the larger spaces permitting longer times.

**Figure 4-7.** Optimum mid-frequency reverberation times for various uses.

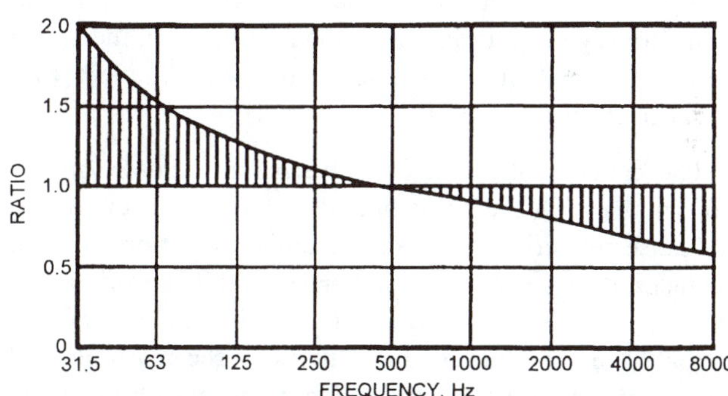


For music performances, reverberation times should be longer at the low frequencies. Air absorption usually limits high frequency reverberation.

**Figure 4-8.** Ratio of reverberation times at various frequencies.

or even increase with distance. Prediction of sound levels within the near field requires detailed information on source radiation characteristics.

The direct field is between the limits of the near field and room surfaces. Since no reflection has taken place as yet, the sound behaves as if in open space, decreasing at a rate of 6 dB per doubling distance (See Figure 1-8 in Chapter 1).

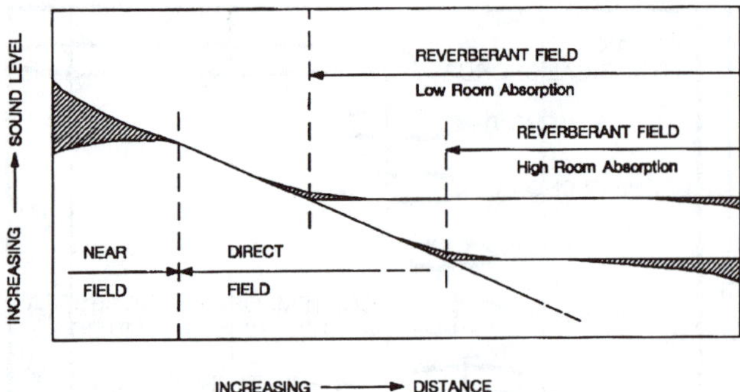


Close to the source, the near field is difficult to predict. The direct field decreases 6 dB per doubling distance. The reverberant field is steady throughout a room with low absorption, but will decrease with distance in a room with high absorption.

**Figure 4-9.** Variation in sound level in an enclosure with distance from the source.

Once the sound has undergone a number of reflections, it becomes more evenly distributed throughout the room. This is considered the reverberant field, or diffuse field. The interaction between the near, direct and reverberant fields is shown in Figure 4-9, for rooms with varying amounts of acoustical absorption.

The near and direct fields are unaffected by room absorption, but the level of the reverberant field is inversely proportional to the absorption. In a highly reverberant room, therefore (characterized by low absorption), the space where the direct field is predominant is small compared to a room with high absorption and less reverberation. In very large rooms with high absorption, the reverberant field will not be uniform, but will diminish with distance from the source.

The concept of direct and reverberant fields is important in assessing the value of adding absorption to a room for noise reduction purposes. Adding absorption to room surfaces affects *only* the reverberant field. If a worker in a shop is stationed close to a noisy machine, the worker is probably in the near or direct fields of the machine, and adding absorption to room surfaces will have little or no affect on the worker's noise exposure. Other workers in the same shop further away from the machine may be in the reverberant field, and would benefit by wall or ceiling treatments.

The amount of noise reduction (NR) achieved in the reverberant field by adding sound absorption is given by

$$\text{NR} = 10 \log \frac{\text{Existing plus added absorption}}{\text{Existing absorption}}, \text{dB}$$

where the absorption values are expressed in sabins (U.S. or metric)

For example, if the total absorption is doubled, the NR is 10 log 2, or 3 dB. A practical limit of achieving reverberant field noise reduction by adding absorption is 6 to 8 dB. See Chapter 9 for additional methods for achieving noise reduction in industrial applications.

### 4.6.3 Reverberation Time Calculation

Reverberation times may be calculated by using the Sabine equation, as follows:

$$T_{60} = \frac{0.049\text{V}}{\text{A}} \text{ (U.S.), or } \frac{0.161}{\text{A}} \text{ (metric)}$$

where $T_{60}$ = Reverberation time for a 60 dB decay, seconds
V = Volume of the room in cu ft (cu m)
A = Total absorption in the room, U.S. sabins (metric sabins)

The total absorption A is the sum of all the acoustical absorption in the room, and is obtained by summing the area times the absorption coefficient for each material in the room, plus the absorption of objects within the room, and air absorption (refer to Sections 2.3.1 and 2.5.4). "A" may be expressed as $S\bar{\alpha}$, where S is the total area of all room surfaces, and $\bar{\alpha}$ is the average sound absorption coefficient of these surfaces. Because the absorption varies with frequency, the calculation is done for each of the 6 standard test frequencies, and may be extended to lower frequencies for music halls if sound absorption data are available.

For example, a room 30 ft (9.14 m) × 75 ft (22.9 m) with a 10 ft (3.05 m) ceiling height will have a volume of 22,500 cu ft (637 cu m) and a surface area of 6600 sq ft (613 sq m). If the *average* sound absorption coefficient $\bar{\alpha}$ for all room surfaces is 0.05 (at a specified frequency), and the room has no other absorption present, the reverberation time is given by

$$T_{60} = \frac{0.049 \times 22500}{6600 \times 0.05} \text{ or } \frac{0.161 \times 637}{613 \times 0.05} = 3.34 \text{ seconds}$$

If A is increased to 0.20, the reverberation time then becomes (U.S. units only)

$$T_{60} - \frac{0.049 \times 22500}{6600 \times 0.2} = 0.84 \text{ second}$$

The Sabine equation for calculating reverberation times assumes that a reasonably diffuse sound field exists throughout the decay. In highly absorbent rooms, particularly if the absorption is not uniformly distributed over all room surfaces (such as carpeted rooms with acoustical ceilings but mostly hard walls), this condition may not be fulfilled. In such cases, another procedure developed by Norris-Eyring is available which seems to provide results that agree better with measurements.

**TABLE 4-1.** Values of $-2.3 \log(1 - \bar{\alpha})$ as a function of $\bar{\alpha}$

| $\bar{\alpha}$ | $-2.3 \log(1 - \bar{\alpha})$ | $\bar{\alpha}$ | $-2.3 \log(1 - \bar{\alpha})$ | $\bar{\alpha}$ | $-2.3 \log(1 - \bar{\alpha})$ |
|---|---|---|---|---|---|
| .01 | .010 | .15 | .162 | .29 | .342 |
| .02 | .020 | .16 | .174 | .30 | .356 |
| .03 | .030 | .17 | .186 | .31 | .371 |
| .04 | .041 | .18 | .198 | .32 | .385 |
| .05 | .051 | .19 | .210 | .33 | .400 |
| .06 | .062 | .20 | .223 | .34 | .415 |
| .07 | .072 | .21 | .235 | .35 | .430 |
| .08 | .083 | .22 | .248 | .36 | .446 |
| .09 | .094 | .23 | .261 | .37 | .462 |
| .10 | .105 | .24 | .274 | .38 | .477 |
| .11 | .116 | .25 | .287 | .39 | .494 |
| .12 | .128 | .26 | .300 | .40 | .510 |
| .13 | .139 | .27 | .314 | .41 | .527 |
| .14 | .151 | .28 | .328 | .42 | .544 |

Methods for performing reverberation time measurements are presented in Appendix A.10.

The Norris-Eyring equation considers not the sound absorption but the sound reflected back from the room boundaries $(1 - \bar{\alpha})$, and $\bar{\alpha}$ utilizes the concept of the "mean free path" that sound travels in a room between reflections. This results in a different expression for the denominator of the Sabine equation given above, by substituting the term $S[-2.3 \log(1 - \bar{\alpha})]$ for A (or $S\bar{\alpha}$). Reverberation times may be calculated using the Norris-Eyring method by substituting the value of $-2.3 \log(1 - \bar{\alpha})$ for $\bar{\alpha}$ in the Sabine equation. Table 4-1 shows the value of $-2.3 \log(1 - \bar{\alpha})$ as a function of $\bar{\alpha}$. Notice at small values of $\bar{\alpha}$ the Norris-Eyring and Sabine equations yield virtually identical results.

# Chapter 5

# Designing Acoustical Spaces

The purpose of this Chapter is to offer suggestions on methods for incorporating good acoustical design for the types of uses most often encountered by the building design and construction professions, particularly those uses most subject to development of acoustical problems. Not surprisingly, this category covers most spaces occupied by people.

It may be useful at this point to review the acoustical design attributes that have been presented in other Chapters (particularly Chapters 3 and 4), that have to be considered to provide satisfactory acoustical conditions for virtually all types of spaces. Whether designing a large auditorium or classroom, a sports stadium or swimming pool, an office building or condominium, achievement of good acoustics demands that these attributes be considered.

1. **Reverberation.** The reverberant condition must fit both the size of a space and its use. For performing arts facilities, evaluating the program emphasis on speech and/or music is particularly important. This Chapter includes detailed reverberation time calculation procedures for a multiuse auditorium. A similar method is used for all other types of spaces. See Section 5.9.
2. **Reflections.** Reflected sound should provide useful reinforcement and uniform coverage for all listening areas. Early frontal reflections will help reinforcement as well as directional realism. Care must be taken that there are no excessively delayed reflections, echoes or focusing. See Chapter 4, Section 4.1.
3. **Diffusion.** Diffuse reflections improves the sound quality in spaces for music rehearsal and performance, and is critical in spaces where recordings are to be made. This attribute is less important for spaces designed only for speech. See Chapter 4, Section 4.3.
4. **Room Dimensions.** Particular proportions of length, width and height can be detrimental if the room modes are coincident and not well distributed. This is an important consideration for studios and music rehearsal rooms. See Chapter 4, Section 4.5.
5. **Sound Reinforcement.** Large volume spaces—particularly with a low ceiling—or spaces with high background noise levels, may require a sound reinforcement system to ensure good articulation. See Chapter 6.
6. **Sound Isolation.** Transmission of sound between spaces must not interfere with critical activities. Typical noise sources are other activities, mechanical equipment within a building or sources outside the building. Transmission paths can occur in the architectural, structural or mechanical systems. Background noise levels must be compatible with the use of the space. See Chapters 3 and 7.

## 5.1 Performing Auditoriums

Performing auditoriums, regardless of size, probably rank as the most critical of acoustical spaces. As the very name states, the space is for hearing, and if it fails in that requirement, it has failed its fundamental purpose. It is certainly prudent for anyone designing (or renovating) an auditorium for the performing arts to engage the services of an acoustical consultant. All too often, however, for whatever reason, a consultant is not involved, but the designer should at least have an understanding of the basic acoustical design parameters. The purpose of this Section is to provide those parameters.

### 5.1.1 Concert Halls and Opera Houses

It is beyond the scope of this handbook to present in great detail the design methods of major projects in this category. The requirements of reverberation control, reflection patterns and delays, and noise control are so critical that a qualified acoustical consultant must be involved from the start. Acoustical design methods will probably involve studies of physical or computer models, and the acoustical requirements are likely to determine the size, shape and interior surfaces of the facility. Even so, the architect fortunate enough to have a commission for the design of a major performing arts facility will benefit from the material presented in this Section.

### 5.1.2 Recital Halls and Drama Theaters

Recital halls and drama theaters are usually at the smaller end of the size range of performing arts facilities, with seating capacities in the 200 to 500 range. The small size allows for an intimate sound quality, a very desirable feature for such uses.

Acoustical intimacy is achieved with a strong direct sound signal, reinforced by reflections arriving within about 30 milliseconds of the direct sound (see Chapter 4, Section 4.1). These reflections should be in the vertical plane between actor and audience, thus the reflector surfaces are located in the ceiling. Because the reflections are in the vertical plane, the source directionality is also reinforced at the listener.

In drama theaters, it is difficult to locate ceiling reflectors at effective positions because of requirements for catwalks and numerous lighting positions. Usually reflectors must be located just below catwalks to be out of the way of lighting. Intimacy is more often achieved in drama theaters by plans which shorten the actor-to-audience distance, such as wide fans (with due consideration for sight lines), thrust stage, arenas (surround seating) and balconies. It is usually necessary to add acoustical absorption to minimize reverberation. The theater rear wall is a good place to locate absorption, where it also prevents undesirable reflections returning to the front of the theater.

The attribute of intimacy for recital halls—or perhaps clarity is a better term in this case—is to facilitate critical listening to student or faculty performances. (Most recital halls are associated with academic institutions.) Because lighting requirements are much less imposing than for drama theaters, overhead reflectors are easily incorporated. Also, some reverberation is desirable, and provisions should be made for lateral reflections. (The importance of lateral reflections is discussed in the next Section.)

### 5.1.3 Multipurpose Auditoriums

By far the most common type of auditorium is that which must accommodate a variety of program types. Only the largest cities can support single-use performing arts facilities. Smaller cities and most academic institutions must accommodate their varied programs in a single, multiuse facility.

The multiuse feature inherently must cope with the conflicting acoustical requirements of speech versus music. For maximum clarity, speech should be supported with a strong direct signal and early overhead reflections (see the discussion of drama theaters in the preceding Section) with minimal reverberation. Music, on the other hand, is enhanced by some reverberation—the amount depending on the type of music—and by lateral as well as overhead reflections. The requirement for variable reverberation can be partially resolved by some type of adjustable absorption, or settle for a compromise, nonadjustable design.

This Section outlines the major acoustical design considerations for a multipurpose auditorium. Section 5.9 at the end of this Chapter provides a detailed, step-by-step design procedure, including drawings demonstrating many of the design features discussed here.

*Program.* This important first step must define the categories of users of the facility. It is particularly important to determine if there is an emphasis on speech or music, though many owners are unable to provide specific information. It is advisable to engage the services of a program consultant, who will conduct a thorough survey of potential users. From such data, the designer is better equipped to properly accommodate the requirements of the principal users.

*Shape and Volume.* Shaping in plan will be influenced by an emphasis on either speech or music programs. Wide fans are preferred for speech, since side wall reflections are less important than for music, and minimizing the distance from speaker to audience is desirable. On the other hand, an emphasis on music would dictate a longer and narrower "shoe box" plan, which is better able to provide lateral reflections (see the following section, Reflectors). Auditorium widths are generally in the range of 80 ft (24.4 m) to 100 ft (30.5 m).

In a multipurpose facility where both speech and music programs have equal emphasis, a modified fan shape in plan with both ceiling and sidewall reflectors represents an acceptable compromise. Excessive length of a hall leads to delayed coordination between visual movement and sound arrival. The suggested maximum length of a hall is about 120 ft (36.6 m). If the required seating capacity results in a distance between stage and rear row exceeding this value, a balcony should be considered. The acoustical design of halls with seating capacities greater than about 2000 presents many difficult problems, particularly for achieving clarity and adequate loudness for unreinforced speech.

For a given plan, the cubic volume is determined by the height of the ceiling surface. Reverberation times are affected by volume, as explained in a following Section on Reverberation Control, and demonstrated in Section 5.9 (see also Chapter 4, Section 4.6). For a typical auditorium section (no balcony), a minimum height for the ceiling might be determined in the following manner:

| Location | Elevation, ft | (m) |
|---|---|---|
| Floor at front of seating | 0 | (0) |
| Stage floor | +3 | (+1) |
| Top of proscenium | +28 | (+8.5) |
| Catwalks | +33 | (+10) |
| Bottom of roof structure | +41 | (+12.5) |

If the roof structure is open to the auditorium (a common design feature), this space could add 6 ft (1.8 m) or more to the overall height. It is not satisfactory to rely on the roof deck for sound reinforcement if the structure is exposed, as the beams, trusses and catwalks will interfere with the reflected sound, the deck surface will be too high for useful early reflections, and may distort the sound because of the flutes. Suspended cloud reflectors (such as under the catwalks) or a full reflective suspended ceiling is preferable. The presence of a balcony also will increase proscenium and catwalk heights, increasing the cubic volume.

Balconies have the advantage of adding seating without increasing sightline distances. Deep balcony overhangs must be avoided, however. The depth of a balcony should not be much greater than the opening under the balcony, and definitely less than twice the opening height; see Figure 5-1. Someone seated in the rear row under the balcony should be able to see all of the first ceiling reflector, usually located just outside and at the top of the proscenium opening (the sound system loudspeakers also should be located here). Also, balcony seats are costlier than main floor seats. It has been estimated that the cost per balcony seat is about 1.5 times the cost per main floor seat, due mainly to increased structural costs and added requirements for an elevator, stairs and circulation space.

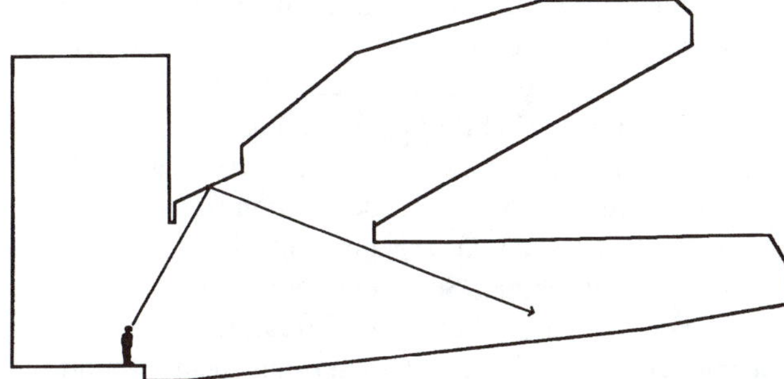

People seated under the balcony toward the rear receive no reflections from the ceiling and are deprived of many early reflections and a strong reverberant field. A distributed sound system with delayed signal can help (see Chapter 6, Figure 6-3).

**Figure 5-1.** Auditorium with a deep balcony overhang.

*Reflectors.* Passive reinforcement of the direct sound from source to listener is an important aspect of auditorium design. Ceiling, sidewall and rear wall reflecting surfaces are all possible, but each location has special reflection attributes.

Ceiling reflectors are important for both speech and music programs. Their location and orientation must be carefully integrated with the requirements for catwalks, associated lighting instruments, and HVAC diffusors. A location directly beneath the catwalks usually works well with lighting requirements, as shown in Figure 5-2.

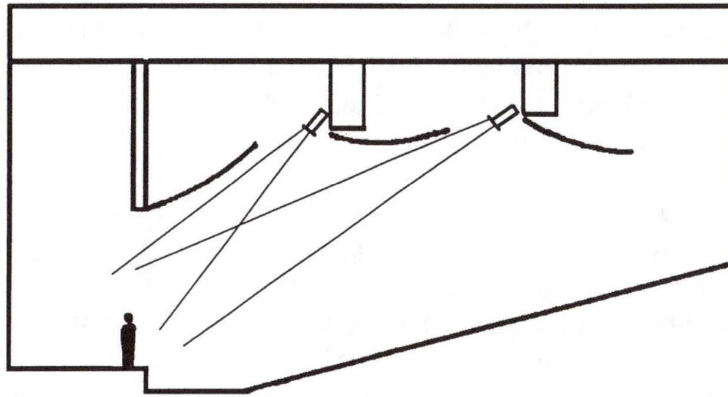

The reflectors are bowed slightly to increase the angle of coverage. Catwalks may be located above the reflectors, but sufficient space must be left between reflectors to avoid obstructing light from instruments clamped to the catwalk structure.

**Figure 5-2.** Auditorium ceiling reflector design.

Horizontal ceiling reflectors at the proper angle will redirect sound in a vertical plane containing both source and listener. In this case, the arrival time of the sound at both ears of the listener is the same, or, more significantly, is exactly in phase at each ear. This condition provides a strong sense of source directionality. See Chapter 1, Section 1.5.2. It is important that the arrival time of the reflected sound be within about 30 milliseconds of the direct sound, but in no case more than about 50 milliseconds. This may be a problem in large auditoriums, particularly with a balcony, which requires that ceiling reflectors be at a higher elevation above the main floor audience.

The upper rear wall may be tipped out with the top towards the proscenium, to provide reflections for the rear few rows of seats. This is not always possible the full width of the auditorium because the control room is usually at this location in the middle. Make sure that the reflected sound does not return too far forward, because the reflected path compared to the direct path may exceed the time delay limitation stated above.

Sidewall reflectors designed to produce lateral reflections play a different role in auditorium acoustics than do ceiling reflectors. If sidewall reflectors are designed to produce strong lateral reflections, the arrival time and phase of the sound at the two ears are not the same, producing a sense of envelopment or being surrounded by the sound. This is a very desirable attribute for music programs, and does not adversely affect speech.

If there are side aisles, sidewall reflectors can be placed starting at about 8 ft (2.4 m) above the aisles, and should be large enough to reflect low frequency sound. Height and width dimensions should not be less than about 6 ft (1.83 m). It is desirable to have at least four reflectors on each sidewall. The reflector tops should not extend all the way to the ceiling reflectors, and should be tipped in about 10° towards the middle of the auditorium to return the sound back down to the middle seating areas. Also, the reflectors should be slightly bowed (radius of curvature about 25 ft (8 m) to increase the angle of coverage, and prevent any possibility of source localization from a side wall. Figure 5-3 shows an arrangement of sidewall reflectors having these attributes. Some degree of lateral reflections can be realized with other sidewall treatments (see the following paragraph), but the designer must bear in mind the limitations imposed by reflector dimensions compared to sound wavelength, in achieving low frequency reflections. See Section 1.4.2.

Some acoustical designers prefer sidewall diffusing surfaces rather than discrete reflectors as described above. This approach may be satisfactory for shoebox plans, but should not be used for a fan-shaped plan. The lateral component will not be strong enough to fully develop the sensation of envelopment. Furthermore, many diffusing surfaces exhibit some degree of sound absorption, which may be undesirable in attaining the desired reverberation.

*Seating.* In the past, styles of seating layout were classified as standard and continental. Standard seating required aisles so that there was never more than a certain number of seats between a seat and an aisle. For continental seating, with greater row-to-row spacing, aisles were required only at row ends, and a row could consist of any number of seats. Current codes merge the two methods, so that the row-to-row spacing increases as the row length becomes longer, and aisles are required only at the row ends. From an acoustical standpoint, there is only a small difference in sound absorption within the range of typical row-to-row spacings. Manufacturers can provide seats of various widths, usually ranging from 19″ to 22″ (483 mm to 559 mm). A row may have seats of several widths to conform to aisle width code requirements, stagger seats from row to row for better sight lines, and keep the row ends reasonably well lined up.

The total sound absorption of the seating area may be calculated by either multiplying the seating coefficients by the seating area (see Section 5.9, and Chapter 2, Section 2.5.4), or by multiplying the sabins/seat by the total number of seats. Different absorption values apply to occupied and unoccupied seats. The difference can be minimized by using cushioning that will be approximately equal to the absorption of the occupant, who will cover most of the seat absorption. The absorption of seats will vary with the type of padding and covering. *Caution*: When hard seats are occupied, there will be a change in both reverberation times and sound levels. Rehearsals and sound system settings in an empty hall with hard seats may not be valid for the occupied hall. This problem is avoided by using upholstered seats.

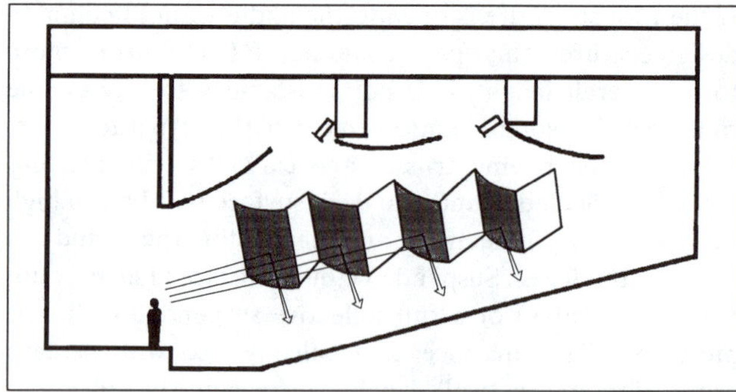

The reflectors are bowed and tipped in at the top so that the reflected sound projects down slightly and arrives at the audience from a lateral direction. Such reflections are particularly desirable for music.

**Figure 5-3.** Auditorium sidewall reflector design.

The rake of the seating area must provide satisfactory line of sight viewing of the stage, but must also follow the requirements of applicable codes. Seating rake also improves sound propagation by raising the sound path above the people seated in front of the listener. Staggering seats from one row to the next is another way of improving sight lines, though there is no significant acoustical effect. There are strict code requirements governing the slope of aisles without steps, as well as the design of steps if they are used.

*Stage House and Orchestra Pit.* The stage house may be the same height as the auditorium with minimal storage capacity for sets and drapes, or it may have a full fly loft with a gridiron over twice the height of the proscenium opening. About the only acoustical requirement of the stage house is to ensure that it does not have a reverberation time longer than the auditorium. Usually there is enough drapery, sets and other equipment stored on stage or in the fly loft to provide adequate acoustical absorption, so that excessive reverberation is not a problem. The owner should understand, however, that if this absorption is not present, the stage house may have unacceptable reverberation.

An orchestra pit is a virtual necessity for musical comedy and operatic productions. Without a pit, members of the orchestra at the usual location in front of the stage will interfere with sightlines, because the musicians will project above stage floor height.

The floor of the pit should be large enough to accommodate 20–25 musicians, including their instruments. Most pits project under the leading edge of the stage floor. The depth of the recess should not be more than 6 ft (1.8 m), otherwise some musicians are "buried" at the back of the pit and will have difficulty being heard by other musicians, let alone by the audience or performers on stage.

Blending of sound within the pit can be improved by installing diffusing panels on some of the pit wall surfaces. QRD-type panels are useful for this purpose (see Chapter 2, Section 2.5.2). Also, it is desirable to incorporate some adjustable acoustical absorption on other pit wall surfaces. Sectional drapery, such as heavy velour, is a good solution. The adjustable feature allows the absorption to be positioned close to the louder instruments, such as the brass and percussion.

The front of the pit should not have any openings that would cause the front rows to hear the orchestra at excessive sound levels. Sufficient *quiet* ventilation should be provided. Access doors should be big enough to accommodate the largest instruments. A lift may be required for wheelchair access. Lighting should not be seen by the audience. A cover for the pit, consisting of removable panels, will serve as a thrust stage when use of the pit is not required.

*Stage Acoustical Shells.* For any type of musical performance, a stage acoustical shell is very important. A shell has three functions. It must conserve sound to keep it from being "lost" in the fly loft, it must project sound into the auditorium, and it must provide feedback so that performers may hear each other.

Acoustical shells usually consist of reflective panels attached to towers on casters that may be moved on and off stage, and to suspended overhead frames that can be raised in the fly loft for storage in a vertical position. Some shells form a tight enclosure with very little sound leakage into the stage house, while others (usually in smaller auditoriums) are more open. The tighter shells must incorporate lighting instruments in the ceiling reflectors, while the others rely on separate lighting units attached to pipes located in openings between ceiling units. A typical shell configuration is shown in Figure 5-4.

There are several types of materials used for panel construction, such as plywood, hardboard, and plastic laminate sheets either single ply or as a facing with a foam core. While fairly heavy panels are desirable for low frequency reflectance, excessive weight is a safety consideration, as the shell is often times handled by students or other nonprofessionals. Panel material weighing about 1.5 lb/sq ft (7.32 kg/sq m) is acceptable. The panels are preferably bowed to a radius of curvature of 3 ft to 6 ft (0.9 m to 1.8 m) to achieve a more uniform distribution of sound within the shell. Also, bowing will increase the stiffness of the panels, thus providing better low frequency reflectance without excessive weight. Reminder: If the shell is to be stored in a side stage area, be sure there is adequate height to accommodate the towers, and make certain there are no light fixtures or air ducts below the top of the towers. The tallest towers are typically proscenium height. Manufacturers of acoustical shells are listed in Appendix A.12.

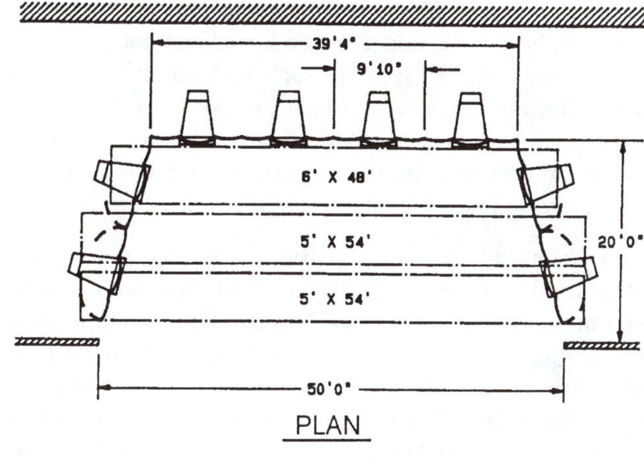


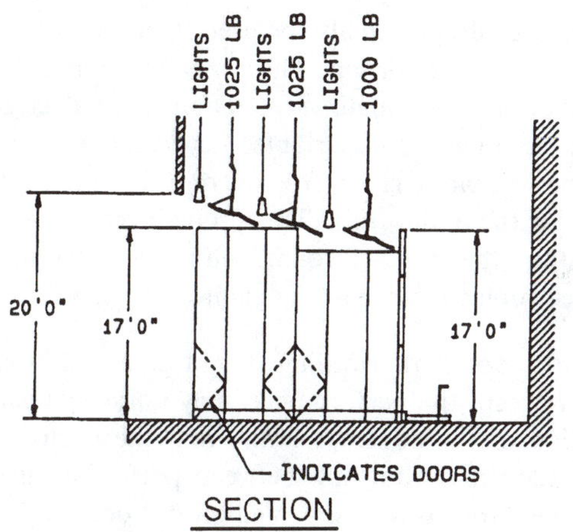


Panels should be bowed for better sound diffusion, and may be constructed of plywood or a composite of fiberglass and foam or honeycomb core. (Courtesy of Wenger Corporation.)

**Figure 5-4.** Typical acoustical stage shell.

*Reverberation Control.* Recalling the reverberation time calculation equation given in Chapter 4, Section 4.6.3, the two main factors affecting reverberation are cubic volume and acoustical absorption. Reverberation times increase with volume because the average path length between successive reflections becomes longer. (Most acoustical absorption takes place at the room boundaries.) Increasing the acoustical absorption will lower reverberation times; less sound energy is returned to the room following reflection.

The audience is the main source of acoustical absorption in most medium-sized multipurpose auditoriums, and will limit mid-frequency reverberation times from 1.25 to 1.5 seconds. Times in this range represent a reasonable compromise for accommodating both speech and music. If more reverberation is desired, the volume will have to be increased, being careful not to add any more absorption than is necessary. If less reverberation is desired, absorption will have to be added to room surfaces not providing useful reflections, such as the rear wall and ceiling above any reflectors. Rear walls which are not sloped must be treated with a broad-band absorbent and/or diffusing panels to prevent undesirable reflections, or focusing if the wall is curved (see Chapter 4, Figure 4-4).

As mentioned earlier, provisions can be made for adjustable absorption, to better accommodate both speech and music. Heavy drapery such as a velour functions well for this application. It can be located on the rear wall, or on the side walls above the area providing useful reflections. There should be sufficient drapery area to affect the mid-frequency reverberation times by at least 0.2 second, preferably 0.4 to 0.5 second.

A large drape can also be used to close off the rear portion of an auditorium. The drape will visually create a smaller, more intimate theater, and will reduce reverberation, both desirable attributes for drama. However, such a drape will block vision from a control room located at the rear of the auditorium. This problem has been solved by locating the control room behind a sidewall reflector, which projects out into the auditorium above the seating.

*Sound Isolation.* Nothing should be heard in an auditorium except the performance. Any noise that calls the attention of the audience away from the performance has degraded that vital link between performer and listener. The key word is *intrusion*, and once it occurs, the communication link is never fully restored.

Listed below are methods by which the most common noise sources in auditoriums can be controlled.

Lobby noise. Provide vestibules between the lobby and auditorium. This will not only control noise, but light intrusion as well. Absorptive treatment for both walls and ceiling helps to reduce noise transmission, and gives a subtle message to the entrant that "Now I am entering the auditorium and I must be quiet." The doors on the auditorium side of the vestibule should not have latching hardware, as the hardware noise of the door closing will be distracting. Latching hardware should be on the lobby side of the vestibule.

Scene shop noise. Locate a buffer zone between scene shop and stage, such as a corridor or storage room. Stage doors that open directly into the shop (or into any other noisy space, inside or out) must have a high acoustical rating for sound isolation, STC 50 or more. These are very expensive doors, in that they must be large enough to pass scene sets, grand pianos or other large items.

HVAC, plumbing and electrical equipment noise. Avoid locating mechanical rooms adjacent to the auditorium or stage house, and never on top of them. Even with double wall construction, structural connections can become vibration transmission paths. Furthermore, air duct penetrations directly from mechanical room to auditorium will almost certainly be a noise transmission path.

HVAC systems serving auditoriums usually require quiet and low-vibration fan units, acoustically lined plenums or banks of sound traps, lined ductwork, low air velocities and low noise air diffusers. Return air systems must receive the same careful attention as the supply systems. See Chapter 7.

Like mechanical rooms, toilet rooms should not be adjacent to the auditorium. Flush valve toilets are very noisy and create high levels of water turbulence. If water pipes supplying toilets are in contact with auditorium wall or floor surfaces, plumbing noise will very likely be audible. A common location for toilet rooms is below the seating riser structure at the rear of the auditorium. This location is acceptable *only* if the toilet room construction consists of a heavy, solid ceiling, and does not contact the riser structure at any point—particularly the plumbing. Be particularly careful if the space below the seating risers is used for return air (which means openings through the riser structure).

Pipes from roof drains should never be exposed in the auditorium. Even if enclosed in chases, the pipes should not contact the chase walls. If the pipe is slightly off vertical, water will run down the sides of the pipe and dripping will be avoided.

Noise from electrical equipment could include transformers and dimmer boards. While the noise from this type of equipment is not particularly loud, the character of the noise is highly annoying. If such equipment is located close to the auditorium or stage, the equipment must be installed using appropriate vibration isolation devices, and the room construction must provide adequate sound isolation. Conduit and air duct penetrations must be carefully sealed to avoid sound leakage. Never have a dimmer room door opening directly into the auditorium or stage.

Exterior noise. Avoid auditorium or stage doors opening directly to the outside, particularly if the facility is located in a noisy area. If such doors are unavoidable (code-required exits, for example), sound-rated doors equipped with acoustic seals should be used.

The roof is vulnerable to noise penetration from aircraft and other sources above roof height. Adequate sound isolation may require a heavy roof and/or gypsum board applied to the bottom of the roof structure. Smoke hatches will probably be required for the stage house roof. Sound-rated hatches are available from manufacturers listed in Appendix A.12.

*Sound Reinforcement and Effects.* All multipurpose auditoriums will require an electroacoustic system for reinforcement of live sound and reproduction of prerecorded material, and for special effects. Sound reinforcement of live speech is usually required to provide uniform sound coverage and sufficient loudness throughout the audience area. Reinforcement of live music is usually not required unless the system provides for special signal processing. This subject and a more detailed description of electroacoustic systems are presented in Chapter 6.

Microphone inputs should be provided at various locations around the auditorium, including several on stage, the orchestra pit, catwalks and a few in the house. The control room will contain an input mixer console, tape and CD players, equalizers, amplifiers and possibly special processing equipment. It is desirable to provide a small portable mixing console that may be plugged into a location in the auditorium seating area, so that the operator can make system adjustments while experiencing the same sound as the audience.

The system loudspeaker should be located just above the middle of the proscenium opening (see Chapter 6). Usually an array of several loudspeakers, covering the full frequency range necessary for speech and music, is required to provide uniform coverage for all seating areas. The speakers may be inside a tubular frame covered with an open weave fabric, but should not be covered with perforated metal, wood slats or similar material which may cause selective filtering of the sound. The loudspeakers should be far enough forward from the location of stage microphones to avoid feedback.

Notice that a monophonic system has been described. Stereophonic systems in properly designed auditoriums are not particularly effective, because the ceiling and wall reflectors are designed to provide uniform sound coverage, and the stereo separation is defeated by this provision. Feeding a monophonic signal into a stereo pair of loudspeakers causes disturbing interference patterns in the audience area. Furthermore, a stereo system will almost double the cost of a monophonic system, since twice the number of all components are required. A high quality monophonic system will best serve the multipurpose auditorium.

An auditorium may have seating areas that cannot be adequately covered by the proscenium loudspeaker cluster, such as under a deep balcony overhang, or in convertible classroom areas at the rear of the auditorium. In this case, ceiling loudspeakers located above these areas with time delayed input are required. The delay must be set so the sound from these loudspeakers arrives at the listener just a few milliseconds after arrival of the sound from the proscenium speakers, or live sound from the stage. Because the ear locates the source by the first arriving sound—even if the sound from above is higher in level—the correct directional information is preserved. See Figure 6-3, Chapter 6.

The basic auditorium sound system as described can be enhanced with special processing equipment. Such equipment can add delay and/or reverberation to the signal, both of which are adjustable over a wide range of settings. The enhanced signal is fed to "surround" loudspeakers located unobtrusively in the walls and ceiling, giving the listener the impression of being in a much larger auditorium or concert hall. Indeed, such a system can be a powerful teaching tool, and can provide the sound operator with the same kind of creative capability enjoyed by the lighting designer.

*Ancillary Spaces.* Most multipurpose auditoriums are surrounded by supporting spaces such as a lobby, dressing rooms, toilet rooms, a green room, scene and costume shops, and possibly a rehearsal room or a smaller arena-type "black box" theater. The major acoustical concern is to prevent noise generated by these activities from entering the auditorium. Location of these spaces with buffer zones between them and the auditorium can reduce the need for costly double-wall constructions.

Black box theaters are usually fairly small, typically about 50 ft (15.2 m) square, and are equipped with little more than a lighting grid, modest control room and raised platform. Seating is often portable. The theater is used mostly for rehearsal and small, intimate performances. Acoustically, the ceiling and about 50% of the wall surfaces should be treated with highly absorbent panels. The result is very little reverberation and virtually no reinforcement by reflected sound, but the actor-to-audience distances are short enough to make reinforcement unnecessary.

## 5.2 Instructional Spaces

Acoustics plays a vital role in the functioning of rooms used for instruction. To the extent that speech is difficult to understand, that music is weak or distorted, or that intruding noise causes a distraction, the educational process is hampered. The purpose of this section is to guide the designer through the methods by which acoustical problems in schools can be avoided.

Music rehearsal rooms, lecture rooms and auditoriums usually receive the most attention for acoustical design, but classrooms, gymnasiums, cafeteria/kitchens, laboratories, shops, commons and administrative offices also have acoustical requirements. Some of these uses are covered in later sections in the Chapter.

### 5.2.1 Classrooms

The design of typical lecture classrooms, 1000 to 2000 sq ft (92.9 to 186 sq m). has historically received little attention with regard to acoustics. A suspended acoustical ceiling and perhaps a carpeted floor is considered adequate by most school designers. Yet the spoken word is at least as important a part of the educational process as the visual portion. Guides and standards for lighting levels and air supply quantities are well defined, but the acoustical environment is seldom referred to at all, or if it is, with such indefinite statements as "avoid noise intrusion" or "provide good acoustics."

This condition of neglect with respect to classroom acoustical design may finally be entering into a phase of correction. In a recently issued paper by the American Speech-Language-Hearing Association (ASHA) entitled "Position Statement and Guidelines for Acoustics in Educational Settings" (November 1994), the failure of classroom design in this country to consider acoustics is well documented. The paper makes the salient point that achieving appropriate classroom acoustics is particularly important considering the emphasis of the Americans With Disabilities Act of 1990 (Public Law 101-336) on removing barriers and improving accessibility of educational facilities. Poor acoustics is a barrier to accessibility, just as curbs or lack of ramps to upper floors.

The ASHA paper offers several guidelines to improve the acoustical design of classrooms, as follows (see Figure 5-5):

1. The background noise level must not be above RC-30. This will be difficult to achieve in classrooms with unit ventilators.
2. The mid-frequency reverberation time should not exceed 0.5 second. Achieving this will require acoustical wall panels (see item 4).
3. Provide reflecting surfaces to reinforce the instructor's speech. This can be accomplished with a sloped ceiling at the front and rear of the room, and angled side walls at the front. See Figure 5-5. (The sloped front ceiling is particularly useful when the instructor is facing the black board.)
4. Apply tackable acoustical wall panels to the rear wall and the side wall opposite the windows (although much of the acoustical absorption will be lost for panels covered with heavy paper or posters).
5. A portion of the ceiling extending from the front wall to the middle of the room should be left hard (reflective). The remainder of the flat, horizontal ceiling should consist of absorptive treatment. Additional absorption may be required on the upper side walls.

These guidelines will provide a significant improvement in the ability of students to hear and understand the instructor's speech, particularly at the rear of the room. A number of other design considerations are presented below.

A plan that is slightly off square, such as 30 ft by 35 ft (9.1 m by 10.7 m), has some acoustical advantage by avoiding degenerate modes (see Chapter 4, Section 4.5). At

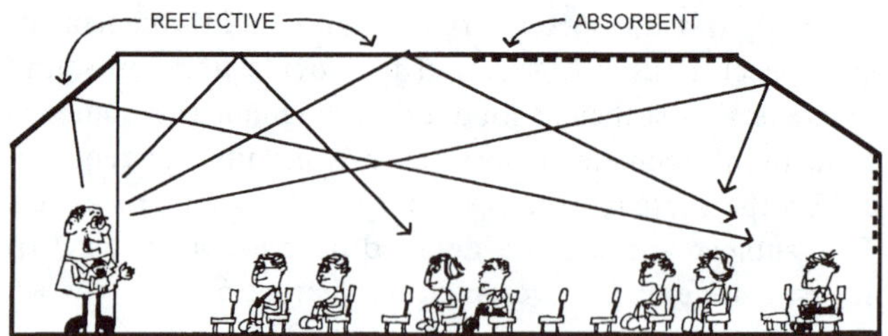


Some acoustical wall panels are also required (not shown), best placed on the wall opposite the windows.

**Figure 5-5.** Classroom design incorporating shaped, reflective surfaces for speech reinforcement.

least a partial acoustical ceiling is required, and a carpeted floor helps reduce noise from scraping chairs or desks.

Acoustical wall panels are particularly important for classrooms that will be used for music instruction, audio recording, or live audio (and video) pick-up for transmission to other locations. Classrooms of this latter type are becoming popular on branch campuses at the university level for live transmission and/or reception with another campus. These classrooms require more acoustical absorption than normal rooms, with the absorption distributed on all room surfaces. Also, it is particularly important to avoid excessive background noise from the air supply system, which should not exceed an RC-30 curve as previously recommended (see Chapter 1, Section 1.7, and Appendix A.4).

Noise control requirements between classrooms are dependent on the teaching philosophy of the school district. Some districts prefer a fairly open plan to permit maximum flexibility in the arrangement of teaching stations, quite often having a cluster of teaching spaces opening into a central team project area. Some degree of acoustical privacy can be achieved with barriers (fixed or movable) between teaching stations, but open to the project space. A highly absorbent acoustical ceiling and judiciously placed acoustical wall panels are also helpful. If the cluster is opened up for large group instruction, a sound reinforcement system will be required to achieve adequate loudness throughout the seating area. Such a system can be zoned to cover an individual classroom or a group of rooms sharing a common program, and can receive inputs from audiovisual program material.

A word of caution concerning open plan schools should be made. Many school districts have experimented with open plans, only to retreat to enclosed classrooms after a less than satisfactory experience. In most cases the problems stem from a poor open plan design, or failure of the administration and teaching staff to coordinate teaching activities in a manner that minimizes interference. Open plan schools have been remarkably successful where these problems have been resolved, but open plan should not be attempted by districts unwilling to undertake the necessary planning.

Other school districts prefer a high degree of acoustical privacy between classrooms, usually requiring fixed partitions extending from floor to structure, rated at about STC 50. Alternatively, if the space above a suspended ceiling is used as a return air plenum and the ceiling consists of lay-in mineral fiber panels, the sound transmission between rooms will not exceed an STC of 35 to 40 maximum. This condition represents a substantial compromise when used in conjunction with a dividing partition rated at STC 50. If the plenum space is shallow, sound transmission between rooms can be reduced by 3 to 5 dB by placing a 4 foot wide band of acoustical blanket along either side of the partition on both ceilings. This method will be ineffective for ceilings rated below about STC 35 unless the entire ceiling over each room is covered.

Another method for achieving both flexibility and acoustical privacy is to use operable partitions for separating classrooms. Common complaints regarding operable partitions are that they are difficult to operate and that serious sound leakage occurs because of panel misalignment or the acoustical seals have become damaged. While operable partitions have laboratory STC ratings in the low 50's, performance in the field can be 10 or more points below this value (see Chapter 3, Section 3.3.5).

### 5.2.2 Lecture Halls

Lecture rooms can be considered small auditoriums and the same acoustical principles can be applied. The room usually has a steep rake to provide excellent sightlines to a demonstration desk at the front of the room. This design feature has acoustical merit for the same reason: an uninterrupted sound path from lecturer to student.

If properly designed, most lecture rooms can perform acoustically without the need for sound reinforcement of the lecturer, though a sound system will be required for audiovisual material. Several room surfaces lend themselves to providing useful reflections if properly shaped, as shown in Figure 5-6. Sound reflections to the rear seats are particularly important when the instructor is talking at a blackboard, facing away from the students. The ceiling above and just in front of the lecturer's position should be sloped to redirect the sound to the rear portion of the room. The same can be done on the upper rear ceiling, being careful to limit reflections from this surface to the rear rows of seats. The side walls at the front can be angled to reflect sound towards the middle of the seating; this feature also avoids opposite, parallel reflecting surfaces which could produce undesirable flutter echoes for the lecturer.

Although it is important to keep reverberation to a minimum with ample absorptive treatment, the reflecting surfaces shown in Figure 5-6 should remain hard. Surfaces not providing useful reflections can be treated with absorption, particularly the vertical portion of the rear wall, which could produce undesirable reflections. Appropriate reverberation times depend on the size of the hall and can be selected from the recommended reverberation time graph (see Figure 4-7 in Chapter 4).

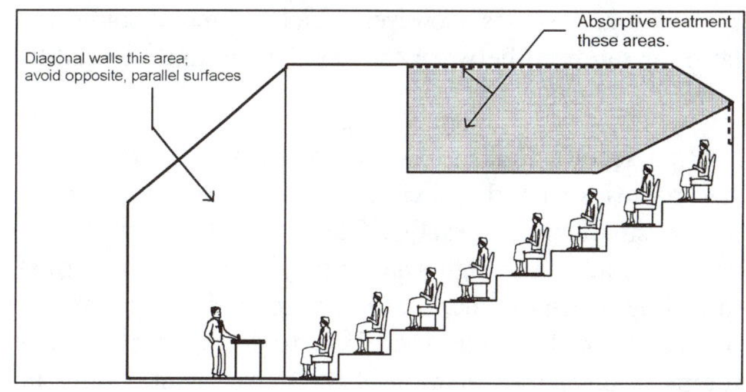


Lecture rooms typically have steep-raked seating, helpful for both sight lines and acoustics. Shape surfaces for natural reinforcement. Apply asbsorptive materials on surfaces which do not provide useful reflections.

**Figure 5-6.** Lecture room acoustical design.

Lecture halls should be free of intruding noise, whether from adjacent activities or the air distribution system. Partitions surrounding the hall should be full height, and entrances preferably should be through vestibules. The air supply and return systems must be laid out to eliminate communication between spaces through the ducts. System noise should not exceed an RC-30 curve (see Chapter 1, Section 1.7).

A sound reproduction system will be required for audiovisual programs. If projection equipment is to be located at the rear of the seating, it should be at least partially enclosed to keep equipment noise from students seated nearby. If the sound system is monophonic, the loudspeaker should be located directly above the lecturer, being careful not to interfere with projected images to a screen. A stereophonic system will require two loudspeakers separated by about one-half the length of the hall, remembering that each channel must cover the entire seating area. Like the single loudspeaker, the pair must be mounted at ceiling height to provide even sound distribution throughout the seating area.

### 5.2.3 The Music Department

Because of the high sound levels that are generated in music departments, they are often located in well separated areas of the school complex, and sometimes in separate buildings. If the school includes an auditorium, it may want to be located close to the music department, as the rehearsal rooms are often used as a staging area, as well as for rehearsals. Proximity also reduces the distance for transporting large musical instruments for performances, such as grand pianos. However, a high degree of sound isolation is required between the department and the auditorium.

High schools and universities will generally have separate rooms for band and choral rehearsal, and for individual or small group practice. The practice rooms in universities and many high schools may be used at all hours, including at times when both band and choral rooms are in use. It is desirable, therefore, to locate the practice rooms in a separate section of the department. Under no circumstances should practice rooms open directly off of band or choral rehearsal rooms.

Both sound isolation and background noise are important acoustical considerations in the music department. Requirements for sound isolation are discussed for room types in the following sections. Background noise from the air distribution systems serving rehearsal rooms should not exceed an RC-30 rating (see Chapter 1, Section 1.7).

### 5.2.4 Band Rehearsal

A common problem in the design of band rehearsal rooms is not providing sufficient cubic volume. High sound levels are generated and if the volume is too small, the level of sound will be uncomfortably high. Indeed, if an instructor teaches several classes a day, there is a possibility of permanent hearing damage over a period of time if the sound levels are excessive.

A good rule of thumb for determining an optimum volume for a band room is to allow about 500 cu ft (14.2 cu m) per student. For example, if a room is being designed for 75 students, a volume of about 37,500 cu ft (1061 cu m) is desirable. Allowing about 25 sq ft (2.3 sq m) per student with instrument for the floor area, the result is 1875 sq ft (174 sq m). Being careful to avoid room dimensions that are equal or integral multiples (see Chapter 4, Section 4.5), plan dimensions of 39 ft by 48 ft (11.9 m by 14.6 m) and a height of 21 ft (6.4 m) would be a good choice. The resulting cubic volume would be about 39,312 cu ft (1112 cu m), thus meeting the 500 cu ft/student requirement. This ceiling height may seem excessive, but low ceilings not only limit volume but make it difficult to achieve uniform distribution of sound throughout the room.

In addition to having adequate volume, band rehearsal rooms require a considerable amount of broad-band acoustical absorption, and will benefit from incorporating some diffusing elements. Both of these objectives can be achieved by alternating between absorptive and reflective surfaces. For example, a suspended, T-grid ceiling consisting of about 2/3 glass fiber lay-in panels and 1/3 gypsum board cut to fit in the grid, laid out in a checkerboard pattern, is quite effective. Even more diffusion can be obtained by replacing the gypsum board with inverted pyramidal lay-in panels, available from several manufacturers (see Chapter 2, Section 2.5.1, category "Solid Plastic Panels"). Lay-in light fixtures should be counted in the reflective 1/3 of the ceiling area. The airspace behind the suspended ceiling is important for low frequency absorption. See Figure 5-7.

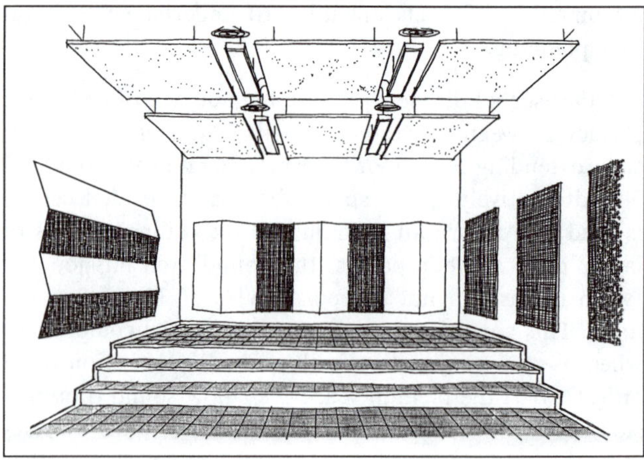

The ceiling is treated with suspended absorptive "clouds," which sound strikes from both bottom and top. Walls have splayed (plain) and absorptive (hatched) panels for both absorption and diffusion. Notice the wall treatment is at the same elevation as the students, and must be impact-resistant.

**Figure 5-7.** Music rehearsal room acoustical design.

Acoustically absorptive wall panels are also required for band rooms. The quantity will depend on the room volume, occupancy and desired reverberation times. Optimum reverberation times are in the range 0.6 to 0.8 second for all frequencies, with minimum occupancy (a jazz trio, for example). The room must not be excessively reverberant; bands perform in spaces where there are few or no reflections, such as out-of-doors, and if the students rehearse in a room which provides too much reverberant assistance, they will experience difficulties in other, less reverberant environments.

Absorption should be placed on all four wall surfaces, and could be in the form of individual acoustical wall panels, at least 2″ (51 mm) thick. Again, alternating between absorptive and hard surfaces helps provide sound diffusion. Even more diffusion and added low frequency absorption can be achieved if the hard surfaces consist of splayed or bowed plywood panels. If there are instrument storage cabinets across the rear wall, sometimes the panel doors can made with a perforated material backed with glass fiber insulation board to provide some additional absorption. Figure 5-7 shows a typical music rehearsal room with acosutical treatments of the type discussed.

Carpet on the floors is a matter of owner preference. It adds considerable sound absorption but mostly at the high frequencies, tending to throw the reverberation time curve somewhat out of balance. Carpet also causes a sanitary problem when students clear their instruments of saliva.

On the other hand, it is much quieter when chairs and music stands are moved around. Another consideration is the thought that exposed wood risers assist in the production of tones for some instruments in direct contact with the floor, such as violas and cellos.

Because of the high sound levels produced in band rooms, sound isolation is an important design consideration. Within the music department, it is good design to provide buffer zones between the band room and other rehearsal or practice spaces, such as corridors, offices or instrument storage rooms. Even so, doors to the band room should be equipped with acoustical seals. The band room should never have doors opening directly to another rehearsal room. Without intervening buffers, double wall construction is required, consisting of masonry, heavy frame or a combination of the two. Whatever the wall construction, it must be completely sealed all around, particularly at the top. If the wall abuts or supports a corrugated metal deck with the flutes running perpendicular to the wall, the flutes must be carefully packed—both below and above—with dense insulation or special neoprene inserts to avoid sound leakage. Undamped metal deck running continuously from one room to another also may be a structural transmission path.

Multichannel sound recording and playback systems are usually required for the band room. Because the room is not very reverberant, stereo separation is fairly well maintained throughout the room.

Interconnecting ductwork can be a flanking sound transmission path. If the music department is served by a single air distribution system, the supply and return ducts for each room must have long runs of lined ducts with several turns before joining with ducts from other rooms. If there is insufficient space to permit long runs, in-duct sound traps may be required. The noise from the lighting fixture ballasts can be a serious problem and remote class A minimum noise or solid state ballasts should always be specified.

### 5.2.5 Choral Rehearsal

Many of the acoustical requirements for band rehearsal rooms apply to choral rooms as well. Principal differences are that choral rooms do not require as much cubic volume per student, and may have somewhat more reverberation.

The rule of thumb for choral room volume is 350 cu ft (9.9 cu m) per student. Optimum room dimensions are determined in the same manner as for band rooms. Mid-frequency reverberation times should be in the range of 0.8 to 1.2 seconds. Somewhat longer times are permitted at the low frequencies, since the singing voice generates very little acoustic energy at these frequencies.

Choral instructors often prefer some adjustability of reverberation. This can be achieved by covering one wall with a heavy drape (such as 25 oz velour), which can vary the reverberation time by several tenths of a second depending on the area exposed. Otherwise, absorptive and diffusing treatments for the room wall and ceiling surfaces are much the same as for band rooms, as are the requirements for sound isolation.

### 5.2.6 Practice—Ensemble

The suite of individual practice rooms may consist of only a few rooms to as many as two dozen or more. It has been common practice to lay out the rooms so that two opposite walls are out of parallel to avoid standing waves (see Chapter 4, Section 4.5), and while this plan has some merit, it is not really necessary if the room is adequately treated with absorptive/diffusing wall materials. An upright piano is often placed along one sidewall. If the other sidewall and the adjacent wall opposite the door have had acoustical treatment, such as a combination of absorptive and diffusing panels, and the ceiling consists of suspended acoustical panels, the room will perform adequately.

Double walls with an STC rating of at least 55 are required between practice rooms for adequate sound isolation. *Caution*: Some single stud/gypsum board walls have STC ratings this high, but they will not provide sufficient isolation at the low frequencies. The walls may be either masonry or frame, but care must be taken not to tie the two wall components together. The room-to-corridor wall may be single frame or masonry with an STC rating of about 45. If a practice room wall is common with a band or choral room wall, an STC rating of at least 60 is required, insuring that the wall has adequate TL at the low frequencies.

It is desirable to have all partitions extend from floor to structure, and all joints well sealed, but this is not always practical for practice rooms if in the same building with larger rehearsal rooms with higher ceilings. In such cases, the ceilings must provide good sound isolation to prevent transmission between rooms in the common space above the ceiling. Typically, one or two layers of 5/8″ (15.9 mm) gypsum board with an acoustical blanket above will be sufficient. Acoustical treatment is then glued or otherwise attached to the underside of the gypsum board. If there is sufficient height, a suspended acoustical ceiling can be used.

Ceilings must never be continuous between adjacent rooms. If the room walls support the ceiling, do not bridge from room to room. If the ceiling is suspended from above, extend the walls at least 6″ (152 mm) above the plane of the ceiling.

Practice room doors opening into a corridor should have an STC rating of about 45, comparable to the rating of the wall. As mentioned earlier, under no circumstances should a practice room door open directly into another rehearsal space; there should always be an intervening buffer zone.

Because practice room doors may be used frequently, the seals should be of a type that do not require adjustment after installation. Pressure-type seals require the locking hardware to be heavy-duty, and automatic door closures may not operate to ensure a good seal. For these reasons, a magnetic-type seal is preferable. Acceptable types of sound seals and installation procedures for sound-rated doors are further discussed in Chapter 3, Section 3.3.5.

Relights in either the door or corridor wall are usually required for surveillance. If the doors have windows in them, the STC rating must be for a test door that includes a relight of comparable size and construction. Wall-installed relights should have STC ratings comparable to the wall.

Cross-talk between practice rooms through the air supply and return duct systems can easily be a problem because the rooms are so small. The problem can usually be avoided by incorporating several lined elbows between outlets for adjacent rooms. If the return air is not ducted but uses the space above the ceiling as a plenum, the air may be vented from the room through lined elbows with maximum separation from one room to the next.

Ensemble rooms may be designed for as many as 6 students, and require a higher degree of sound diffusion than do individual practice rooms, to insure that each student can hear the others. To achieve this diffusion, bowed or splayed diffusing reflectors should be spaced on the walls between the absorptive panels. Other acoustical design considerations are the same as for the individual practice rooms.

The instructor's office may also double as a studio/practice room and should be designed similar to an ensemble room, though wall space may be limited because of a desk, book shelves and other fixtures. The door and relight design should be the same as for the other rooms.

Practice and ensemble rooms which are prefabricated then assembled on site from modular panels are also available. The modular panels allow for rooms of different sizes, are both absorptive and reflective, and special panels incorporate the door, relights, ventilation equipment, lights and electrical service. Such rooms perform very well for interior acoustics and room-to-room sound isolation, and have the advantage of being disassembled and relocated to another area at a later date if desired.

A more recent development in modular practice rooms incorporates a microphone to pick up the sound produced by a student, which is then processed and fed back into the room through hidden loudspeakers. The signal processing adds simulated early reflections and different amounts of reverberation to create a variety of acoustical environments. A multiposition switch in the room allows the user to select a particular environment, which may range from a classroom to a large concert hall. These rooms serve to provide students with an opportunity to practice in a variety of performing environments, and to provide instructors with a unique teaching tool.

Manufacturers of prefabricated rooms are listed in Appendix A.12, numbers 1, 33 and 81.

### 5.2.7 Multipurpose Rooms

In addition to serving as a performance auditorium, multipurpose rooms often must also function as a cafeteria ("cafetorium") or gymnasium. The latter two uses require flat floors, so the auditorium function is already hampered by poor sight (and sound) lines. A slight improvement can be accomplished in a cafetorium by stepping up the floor in increments with increasing distance from the stage.

Sound absorbent ceilings are usually required in multipurpose rooms for noise and reverberation control, while a shaped, reflective ceiling would best serve the auditorium function for the propagation of sound from the stage to the rear seating. In a cafetorium, where the ceiling may be quite low compared to the room length, a distributed sound reinforcement system is a virtual necessity (refer to Chapter 6, Section 6.1). If the room doubles as a gymnasium, the ceiling will be higher, but impact resistance becomes an important consideration. With a high ceiling, a high level, single cluster sound reinforcement system may be satisfactory.

Because of the requirement for acoustical materials to control noise, the reverberation times are likely to be quite short. The room, therefore, is better suited to speech programs rather than music. It is desirable to select one end of the room as the performing area if there is no dedicated stage. If possible, slope the ceiling over the performing area to reflect the sound to the rear of the room. Sloping the upper rear wall to provide additional reflections to the rear seating is also desirable.

### 5.2.8 Gymnasiums

Gymnasiums are basically intended to serve athletic instruction and sports events, but many are also used for school assemblies and community functions. Some even have an elevated stage opening at one end for school music or drama performances, and can be classed as a multipurpose room as discussed earlier. Control of noise and reverberation is, therefore, very important, and most functions will require a sound reinforcement system.

Two frequently encountered problems are excessive reverberation, caused by lack of sufficient acoustical absorption, and high noise levels caused by industrial-type air handling units hung from the ceiling or otherwise directly exposed to the space.

Reverberation control should be taken care of by both acoustical ceilings and absorbent wall panels. Desirable mid-frequency reverberation times depend on the size of the room, but typically should be about 1.0 second for small gyms to not over 1.5 seconds for larger gyms.

There are a number of suitable materials available for acoustical treatment of the ceiling, such as lay-in acoustical panels, spray-on insulation and perforated metal deck. Whatever is selected, impact resistance from thrown balls is an important consideration. Probably the perforated metal deck is best in this regard, but most decks are deficient in high frequency absorption. Select with care.

Acoustical wall panels also must have impact resistance. In large gyms, it is desirable to locate the panels low on the walls if possible, as this location will help prevent disturbing echoes as well as reduce reverberation. If there is a stage at one end of the gymnasium, the opposite wall should be treated with acoustical absorption to prevent echoes reflecting back to the stage.

Care must be taken if an acoustical material is applied to an exterior wall. The thermal insulation provided by the material may cause moisture condensation on the wall surface or within the material in cold weather (see Chapter 2, Section 2.4). The material may become moist which can cause both fungus and odors. For this reason, all acoustical materials applied to exterior surfaces must be furred out from the surface to provide ventilation behind the material. The furring on the walls should be vertical, as horizontal members can interfere with the vertical air circulation. The heating and ventilating engineer can help resolve this problem by controlling the relative humidity and by providing the required amount of thermal insulation in the walls and on the roof deck.

Many space heating units cannot be enclosed or connected to silencers or ducts because they will not tolerate even small increases in static pressure losses. The only satisfactory solution to this problem is to provide separate mechanical rooms or enclosures and use fan systems which can tolerate noise control treatments.

For many uses in a gym a sound reinforcement system will be required. A centrally located loudspeaker cluster is best for announcements during sports events, or for someone speaking from the floor. If speakers or performances are at one end of the gym, a second loudspeaker cluster at this location allows the sound to be coming from the direction of the source. In-wall cabinets for the electronic equipment allows an operator to effectively select microphones or make gain adjustments, since the operator is actually in the gym and experiences the sound everyone is hearing. Care should be taken to acoustically treat wall surfaces that can cause echoes from loudspeaker reflected sound.

### 5.2.9 Shops

Wood, metal and automotive shops require highly absorbent acoustical ceilings for noise and reverberation control. The shops are usually located in a separate area of the school complex to avoid noise transmission to other school functions. Shops should never be located over classroom areas, as impact sound transmission will be very difficult to control.

A noise problem commonly encountered in wood shops is caused by the blower in the dust removal system. The blower should not be located in the shop itself, but rather outside on a side of the building where it will not cause problems for adjacent buildings. A partial enclosure may be required for noise control.

Air compressors in shops are a source of noise that requires attention. The air intake should have a noise suppressor, or the air should be taken in from outside. If the compressors are in a separate room or provided with enclosures, proper ventilation must be supplied to control the temperature.

Automotive shops have the potential for creating community noise problems. During nice weather the shop doors are often left open, and engine testing and body repair can create objectionable noise if there are residences nearby. The architect should be aware of such problems if the school is located in a residential neighborhood.

Drafting classrooms are often located within the shop area. Special care should be taken to isolate the classroom from shop noise. Recent advances in teaching industrial arts utilize laboratory-like facilities where students work in teams, with extensive use of computers to explore design and construction techniques. Like many other laboratories, such areas usually do not require acoustical treatments other than an absorbent ceiling.

## *5.3 Office Buildings*

In recent years many building projects—office buildings in particular—have combined the design and construction functions into a what is called a "design-build" procedure, where the architect and contractor collaborate from the start of the project. The first phase of such a project usually consists of constructing the structural shell, exterior enclosure, toilet rooms, elevator core and building-standard mechanical and electrical distribution systems. Interior finishes may be limited to a suspended ceiling, basic lighting and a finished floor. The second phase, called "tenant improvements," consists of a partition layout for a particular tenant's needs.

Acoustical problems sometimes arise when the building-standard systems cannot readily or easily accommodate special needs such as confidential privacy for offices, conference rooms, board rooms, teleconferencing rooms, etc. Some examples: If the space above the suspended ceiling is used as a return air plenum, it may not be possible to easily achieve confidential acoustical privacy by extending partitions through the previously installed suspended ceiling, to avoid a flanking sound transmission path in the open plenum. Perimeter heating systems that extend continuously around the building may present difficulties in obtaining an airtight seal where the system penetrates sound-rated partitions. The floor-to-floor spacing may be inadequate to contain air ducts above the suspended ceiling large enough to keep air velocities sufficiently low to avoid generation of excessive noise. Terminal HVAC boxes above the suspended ceiling may be too noisy if the tenant wishes to place a large conference room directly below. Penthouse mechanical rooms may not have sufficient noise and vibration isolation to permit noise-sensitive executive office suites directly below. It may not be possible to install a satisfactory sound masking system for open plan office spaces if no provisions were made for such a system in the initial design.

Problems such as listed above can be avoided if the basic design can easily be upgraded to accommodate the special conditions that some tenants require. It is particularly important for design-build projects that these potential problems be considered before construction has progressed to the point where it is very costly or impossible to achieve the desired goals.

### 5.3.1 Private Offices

The expectation in private offices is that some degree of acoustical privacy can be realized. Many offices do not require complete privacy, while others, such as counseling and interview rooms for doctors or attorneys, will usually require a condition of confidential privacy. Patients or clients will be reluctant to discuss personal problems if they believe they can be heard in adjacent spaces. Their clue that this may be happening is when they hear intruding speech. The following discussion applies particularly to offices requiring confidential acoustical privacy.

Acoustical privacy is essentially a problem of achieving the proper signal-to-noise ratio at the listener. The intent is to have the intruding signal "buried" in the background noise, to the extent that the speech is unintelligible. At the same time, the background noise must not be so loud that it is disturbing. Problems occur when the intruding signal is too high and/or the background noise—or masking noise—is too low.

The level of speech at the listener is affected by the voice level of the speaker, and losses in the transmission path(s) between speaker and listener. For normal voice levels and background noise from the HVAC system in the RC-35 to 40 range, confidential speech privacy can usually be achieved with a metal stud (light gauge) partition consisting of an extra layer of gypsum board on at least one side, and insulation in the stud space. The partition should extend from floor to floor and not terminate at the suspended ceiling. This partition will have an STC rating of about 50, while a typical mineral fiber suspended ceiling will have a room-to-room ceiling attenuation class (CAC—see Section 2.2) rating of about 40. Over the ceiling sound transmission, therefore, is the controlling transmission path. The common practice of placing batt insulation on top of the ceiling may raise the CAC rating to 45, an improvement but still less than the partition STC rating of 50.

If the space above the suspended ceiling is used as a return air plenum (not recommended for high privacy offices), there should be no openings in the partition extension above the ceiling. The air should be returned to the space above the corridor ceiling with a lined transfer boot through the corridor partition (see Chapter 7, Section 7.4). If both the supply and return air is ducted, care must be taken to avoid crosstalk between rooms. Fan-powered VAV boxes located above the ceiling may produce excessive noise, and should be located in adjacent noncritical spaces if possible.

Another common sound leakage path between offices on exterior walls occurs at the joint between the partition and the outside wall. Many times this joint occurs at a window mullion utilizing a thin closure panel between the partition and mullion. Because of its low mass it can transmit considerable sound, and the panel may not be well sealed in place, causing sound leakage. Under no circumstances should a partition join the outside wall at mid-window. In the same area, low wall heating coils may extend from room to room leaving a large opening, constituting a significant leakage path.

### 5.3.2 Open Plan Office Areas

Open plan office design can be an effective way to combine reasonable acoustical privacy between work stations with a highly flexible space layout. Even though the lack of full height partitions and doors results in a significant loss of sound privacy, freedom from distraction and annoyance in open plan offices can be achieved. It is essential, however, that certain basic factors be effectively incorporated into the design, or the system will not perform satisfactorily.

As discussed for private offices, acoustical privacy in the open plan environment is essentially a signal-to-noise ratio problem. Unlike private offices, the sound reduction between speaker and listener—termed the "interzone attenuation"—is limited because of partial high barriers instead of full height partitions. Controlling the background sound level, therefore, is extremely important if acoustical privacy is to be achieved.

The factors that must be considered in open plan office design are (refer to Figure 5-8) as follows:

**Sound source characteristics.** In most cases the source is speech for someone assumed to be in a seated position, but also could include ringing telephones and office equipment.

**Barrier or screen construction.** The screen should not be less than 5 ft (1.52 m) in height and extend to the floor. Both screen surfaces should have a highly absorptive treatment to reduce reflections. There must be a solid noise barrier set in the panels which will provide a sound reduction of at least 20 decibels above 250 Hz.

**Separation between work stations.** The interzone attenuation increases with increasing distance between work stations. Open plan areas requiring a higher degree of acoustical privacy should have more separation between work stations.

**Reflective surfaces.** The area layout must avoid hard wall or window surfaces that can reflect sound to adjacent work stations around the end of the screen. Problem areas are where separating screens are perpendicular to a hard surface but do not butt up against it. Outside walls that are predominantly windows are particularly troublesome. The floor must be carpeted.

**Ceiling treatment.** Use only glass fiber ceiling panels that are specially manufactured for open plan areas, having interzone attenuation ratings of NIC'20 or more. Most mineral fiber panel ceilings will not perform satisfactorily. Hard ceilings such as gypsum board are unacceptable for open plan areas. Light fixtures with flat lenses will cause troublesome reflections. Parabolic type fixtures are an improvement, but task lighting at individual work stations is best.

**Background noise.** Even if all the architectural surfaces in an open office area are installed as discussed above and screen heights are adequate, satisfactory acoustical privacy will not be achieved if the background noise is too low. The characteristics of an ideal background noise are quite restrictive. The noise must be broad-band such as random noise (see Section 1.1.2) but with reduced energy at the high frequencies to avoid sounding "hissy." The level of the noise must be within the fairly narrow range of 46 to 48 dBA. If it is not loud enough it does not mask out intruding sound, and if it is too loud it becomes annoying.

It is important that the noise level be uniform throughout the area. A person walking through the area will become aware of the noise if there are noticeable changes in level, and this is not a desirable condition. Uniformity is achieved by locating the loudspeakers above the suspended ceiling, properly spaced and pointing up so that the sound must undergo at least one reflection before reaching the top of the ceiling. Open slots or grilles for return air can be leakage paths causing noise "hot spots" directly beneath.

Another problem with noise distribution occurs at the boundaries of the open plan area. A very gradual transition is required to avoid detection by people entering or leaving the area.

The major sound transmission paths between work stations are diffraction over the top of the barrier and reflection off a ceiling with incorrect material. There also may be reflections around the end of a barrier, particularly at outside walls with windows. The loudspeakers above the suspended ceiling provide essential background masking noise.

**Figure 5-8.** Open plan acoustical considerations.

Private offices opening directly into an open plan area should have similar background noise but about 5 dBA lower. This masking will also serve to improve confidential acoustical privacy in the office.

Noise from the air distribution system is not satisfactory for open plan masking. It is too difficult to design and control system noise to meet the strict requirements for level, uniformity of distribution and spectrum. Also, background music is not suitable for masking because the level is not steady with time, and it has information content which can be distracting. Every effort must be made to render the masking noise as unobtrusive as possible.

The degree of acoustical privacy between work stations can be evaluated by determining the sum of the interzone attenuation and the background sound level, both measured in A-weighted decibels (dBA). Figure 5-9 shows this sum for a variety of acoustical treatments and the degree of satisfaction (or dissatisfaction) associated with each treatment configuration. It is interesting to note that only one condition has a satisfactory rating, the condition that achieves *all* of the design parameters described in the foregoing text.

The ASTM test method designations for determining the interzone attenuation and masking sound characteristics for open plan are as follows:

E 1111: Interzone Attenuation of Ceiling Systems, Measuring.

E 1573: Evaluating Masking Sound in Open Offices Using A-Weighted and One-Third Octave Band Sound Pressure Levels.

E 1375: Interzone Attenuation of Furniture Panels Used as Acoustical Barriers, Measuring.

E 1376: Interzone Attenuation of Sound Reflected by Wall Finishes and Furniture Panels, Measuring.

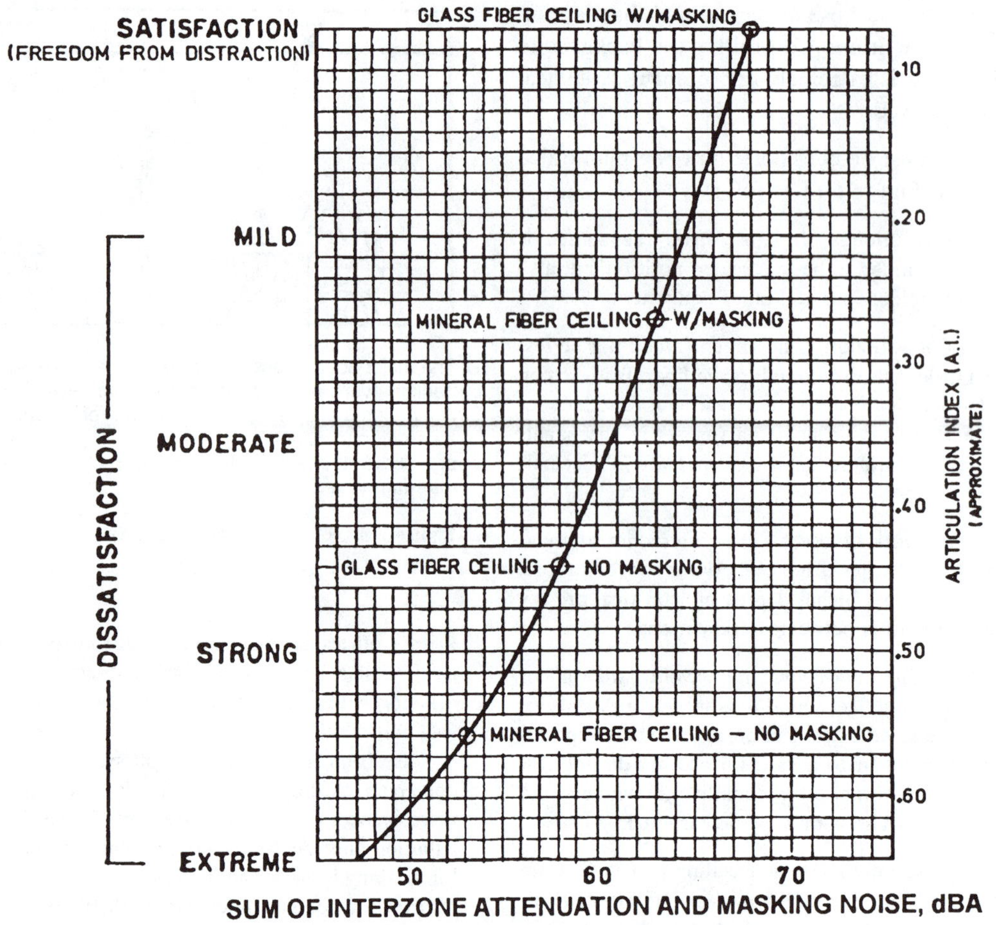


Assumes 60 inch (1.5 m) absorbent screens, no reflections.

**Figure 5-9.** Open plan satisfaction ratings for various treatments.

### 5.3.3 Conference Rooms

Large conference and board meeting rooms have all the acoustical requirements of private offices, with additional considerations for speaking over distances that may exceed 50 feet (15.2 m), audiovisual presentations, recording capabilities and teleconferencing. The room may be so large that speech reinforcement is required. Such reinforcement systems have very demanding requirements to preserve sound quality and avoid feedback. See Chapter 6, Section 6.1.

Because of the extraordinary speech communication and recording requirements, conference rooms can benefit from several specialized acoustical treatments. Leaving the center of the ceiling reflective (such as gypsum board, plaster or wood) will assist in voice transmission from one end to the other. Absorptive treatment should then be located in a peripheral band. Acoustical wall panels are required to increase the total absorption to a value to avoid excessive reverberation or flutter echo, both of which will seriously degrade the quality of recordings. The absorptive wall panels should be located on at least two adjacent wall surfaces, preferably at an elevation between table top and door height.

Another requirement for obtaining good listening conditions and quality recordings is low background noise. There should be no intruding noise from adjacent activities, and the HVAC noise level should not exceed an RC-25 to a maximum of RC-30.

## *5.4 Residential*

Most people want to live in a place that is private and reasonably free of intruding noise. One of the conditions that should be considered when selecting a location in which to live is the exterior noise environment. Obviously locations near highways or airports will be noisy at all times, but many commercial or industrial noise sources, or even school activities, may not be present on weekends when house or apartment hunting is commonly done. It is wise to be aware of zoning or planned projects in the immediate area, as to what potentially noisy uses are permitted or under consideration. Such uses could include car washes, 24-hour minimarts, fast food outlets (often a gathering place for youth), etc.

The construction of the house or building under consideration is important in excluding exterior noise, and in controlling noise transmission within.

### 5.4.1 Private Residences

Usually very little attention is paid to acoustical privacy in the design of single-family residences. The bedrooms are often in a separate section away from the living spaces, though the openess of some comtemporary designs does not provide much acoustical privacy.

To help reduce exterior noise from entering the residence, consider the following:

1. **Orientation** on the site can be of significant benefit. The side of a building facing away from a highway or aircraft flight path will receive up to 10 dBA of shielding from the building itself. Try to have patios and building penetrations, such as doors and windows, on the quiet side. See also control of highway noise in Chapter 8, Section 8.3.2.
2. **Windows** are usually the primary noise transmission path from out to in. Insulated glass which is heavier and thicker than what is customarily used will improve sound isolation. Sound-rated windows for very noisy environments are available (see Chapter 3, Section 3.3.4). Operable units must have air tight seals. Storm windows with maximum obtainable separation from the primary window can also help.
3. **Doors** should weigh at least 5 lbs/sq ft (24.4 kg/sq m) and be equipped with air tight seals. Enclosed entrance vestibules can virtually eliminate noise transmission through doors. For sliding glass doors, airtight seals are a requirement to prevent sound leakage. Secondary storm doors are also a benefit.
4. **Roofs** can be a significant path for transmitting noise if there is no attic or similar buffer space between the outside and the room. In such cases consider increasing the mass of the roof to reduce noise transmission.
5. **Walls** may require an extra layer of gypsum board or paneling inside to reduce noise transmission. In extreme cases, such as within 100 ft (30.5 m) of a heavily traveled highway, a veneer layer of brick or CMU will greatly decrease noise transmission.
6. **Heat pumps** are noisy and if not properly located can be a problem for both owner and neighbors. Most competent installers will provide suggestions on where to locate the unit, obviously not directly below a window or opposite a window in the adjacent house. Partial enclosures can act as a noise barrier, but proper allowance must be made for air circulation around and through the unit.

There are a number of methods available for the control of noise generation and transmission within the house. The selection of quiet appliances is a start. Such information is available from some manufacturers, and from publications such as *Consumer Reports*. While rating methods vary, the most commonly used descriptor is A-weighted decibels (dBA). To help evaluate the rating, refer to Chapter 1, Figure 1-5, listing the A-weighted level for a number of common noise sources. The use of isolation pads or other devices under clothes or dish washers to reduce vibration transmitted to the house structure should be done with caution, as such action may void the manufacturer's warranty.

If greater than normal acoustical privacy is desired for certain rooms, consider the use of staggered stud partitions, solid core doors (acoustical seals will be of considerable help), and for floors, suspended gypsum board ceilings using resilient channels. Sound isolation performance data on these and other constructions are presented in Chapter 3. The addition of acoustical absorption to rooms is usually not necessary if the rooms have carpet and upholstered furniture.

### 5.4.2 Apartments and Condominiums

Intrusion of exterior noise for apartments and condominiums is controlled in much the same manner as that presented in the preceding section for private residences. For buildings to be located in a noisy environment, many building codes and lending agencies will require that if certain noise levels are exceeded, the building design must incorporate appropriate noise reduction features. The Federal Department of Housing and Urban Development (HUD) has defined exterior noise environments that require the building exterior construction meet higher noise reduction values than provided by "normal" construction (see Chapter 8, Section 8.1.1).

It is the interior acoustics of multifamily buildings that takes on more importance than for single family residences. Walls and floor/ceiling systems that separate living units must provide an acceptable level of acoustical privacy for both airborne and impact noise.

Appendix Chapter 35 of the Uniform Building Code specifies minimum performance standards for airborne and impact sound isolation of walls and floor-ceiling assemblies between dwelling units of multifamily housing, or guest rooms of hotels or motels. The airborne sound transmission loss is set at STC 50 for laboratory tests, and FSTC 45 for field tests. The same numerical values are specified for impact sound isolation, namely IIC 50 and FIIC 45 for laboratory and field tests, respectively. Experience has shown that performance just meeting these requirements is still subject to numerous complaints, and that minimum performance values be 5 points higher (STC/IIC 55, FSTC/FIIC 50). For "luxury" units, performance 5 points higher yet are not unreasonable.

A comprehensive discussion of airborne and impact sound isolation can be found in another HUD publication, "A Guide to Airborne, Impact, and Structure Borne Noise Control in Multifamily Dwellings," prepared for the Federal Housing Administration (FHA) and published in September 1967. The acoustical performance recommendations specified in these guidelines are probably more realistic compared to the UBC standards in terms of occupant satisfaction. The following Table 5-1 summarizes these recommendations with some modifications; the reader is referred to the original document for a complete listing.

**TABLE 5-1.** Guidelines for Airborne and Impact Sound Isolation Between Dwelling Units

| | STC/IIC | |
|---|---|---|
| Location of Partition or Floor-ceiling* | Quiet** | Noisy** |
| Bedroom—Bedroom | 58 | 54 |
| Bedroom—Living Room | 58 | 54 |
| Bedroom—Bathroom or Kitchen | 60 | 56 |
| Bedroom—Corridor (partition only)*** | 55 | 52 |
| Living Room—Living Room | 55 | 53 |
| Living Room—Bathroom or Kitchen | 57 | 55 |
| Living Room—Corridor (partition only)*** | 53 | 50 |
| Bathroom or Kitchen—Bathroom or Kitchen | 52 | 50 |

*For vertical separation, the lower room is listed first.

**Exterior noise environment. Quiet means surburban not close to a highway or airport. Noisy means urban close to a highway or airport.

***Corridors are not recommended over bedrooms or living rooms.

Special situations to be aware of are particularly noise-sensitive rooms adjacent to noisy areas in adjoining units, such as bedrooms adjacent to bathrooms, kitchens or laundry rooms. Bathrooms above bedrooms or living rooms are another situation requiring special attention. The plumbing penetrations through the party floor must be boxed in below, and pipe runs down through walls must be carefully isolated from the structure (see Chapter 3, Section 3.8). Use cast iron drain pipe instead of plastic. Cast iron does not radiate as much noise. Where bathrooms are back to back, be sure to continue walls behind bathtubs all the way to the floor using a double thicknesses of water-resistant gypsum board. Kitchen cabinets should not be attached directly to party walls or adjacent to a bedroom. Impact noise from closing doors and drawers is readily transmitted through the walls.

Wood frame floors are subject to low frequency "thumping" noise and creaking from footfalls. The thumping is aggravated by a floor structure that lacks stiffness. It is advisable to use heavier or more closely spaced framing than the minimum requirements of building codes. For example, use 2″ × 12″ wood joists (51 mm × 305 mm) instead of 2″ × 10″ (51 mm × 254 mm) joists (nominal dimensions), or space the joists 12″ o.c. (305 mm) instead of 16″ o.c. (406 mm). Floor creaking can also be minimized by using heavier or more framing, and by insuring that the subfloor is in intimate contact with the joists. After the subfloor has been applied to the joists, walk the floor and wherever creaking occurs, apply additional coated or serrated nails. Refer to Chapter 3, Section 3.8 for precautions to be taken for noise control in general.

### 5.4.3 Hotels—Motels

Hotels and motels are often located in noisy areas such as near airports or highways, consequently preventing noise intrusion is a major acoustical concern. Preventing intrusion of noise should follow the same procedures as recommended for apartments and condominiums located in similar environments. Locations near railroads may have additional concerns of excessive low frequency noise and vibration, and train whistles if there is a crossing nearby.

The mass of heavy masonry exterior walls will provide more low frequency attenuation than will frame construction. Room doors opening directly to the outside can be avoided with interior corridors. Windows may require special glazing such as heavier and\or laminated glass. Avoid operable windows if possible. Air conditioning units installed in outside walls constitute noise leakage paths. Baffling with absorptive treatment between the unit and the room will reduce both intruding and unit-generated noise, although care must be exercised so as not to restrict air flow.

Interior noise problems also have similar solutions to those for apartments and condominiums. Acoustical privacy between guest rooms is a major concern, and in most situations will require double wall construction with an STC rating of at least 50, and preferably 55 for luxury facilities or those in particularly quiet locations. Rooms with interconnecting doors should have a door in each side of the double wall construction with maximum obtainable separation between the two doors. The doors must be solid core with acoustical seals all around. Additional sound reduction can be achieved if a side of one door (facing into the enclosed space) is treated with a highly absorbent acoustical panel.

Room-to-corridor partitions can usually be rated about 5 STC points less than the partition between rooms. The room entrance door should be solid core with acoustical seals.

Plumbing noise is a common source of complaints in hotels and motels. Both supply and waste water piping should be isolated from the building to the maximum extent possible. Methods for accomplishing this are presented in Chapter 7, Section 7.5. See also the discussion of plumbing noise in the preceeding Section 5.4.2.

Hotels and motels specializing in hosting conventions have large rooms for meetings and banquets which can usually be subdivided with operable partitions. Because so many seals are required between individual panels and between the partitons and adjacent surfaces, the sound reduction of these partition systems is very dependent on the quality of installation. Always insist that the product being bid is the same as that for which performance testing is submitted for approval. Field testing of the completed installation is the only sure way of achieving acceptable performance. Installation and testing procedures are discussed in Chapter 3, Section 3.3.5.

There are two additional precautions for the installation of operable partition systems which are occasionally overlooked. First, some systems are suspended from the overhead roof structure. Heavy partitions may cause deflection of the structure, in turn causing the partitions to have excessive drag on the floor. This may be a particular problem where partitions are mounted under roof systems where there are heavy snow loads in the winter, causing added deflection. The support of the partitions should be independent of the roof structure in such cases. Second, The space above the partition (and above the suspended ceiling) may be open, creating a serious sound leakage path over the partition. A barrier of at least one layer of gypsum board should extend from the top of the partition track to the roof structure, and seal to it. Be wary of duct penetrations.

Convention facilities will in most cases require a sound reinforcement system. The system will very likely be a distributed one (see Chapter 6, Section 6.1), and will have to be zoned to serve separate rooms individually or one large room when the operable partitions are opened. It is important that the loudspeaker layout and radiation patterns be considered to achieve uniformity of coverage.

## 5.5 *Sports—Entertainment*

A fundamental acoustical requirement for most sports facilities is good speech communication, whether for instruction, announcements or crowd control. This requirement in turn means reasonably quiet background noise and, for almost all the facilities, a functional sound reinforcement system.

For enclosed facilities, reverberation and echo control must be attained to preserve speech intelligibility. For outdoor activities, high background noise levels may be encountered from aircraft, surface traffic or spectator noise, all of which can cause problems hearing or understanding the sound system.

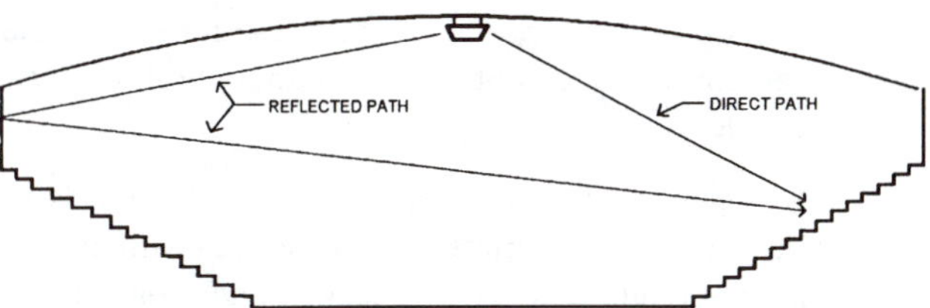


Surfaces that can produce long delayed reflections—or echos —must be treated with an absorbent. If the path length difference between direct and reflected sound exceeds about 60-70 ft, the reflected sound must be minimized.

**Figure 5-10.** Delayed reflection in a large arena.

### 5.5.1 Arenas—Stadiums—Field Houses

Acoustical problems encountered in large sports facilities typically are the result of too much reverberation, long-reflected sound paths that cause echoes, concave shaping (such as a dome or barrel ceiling) that may cause focusing, poor sound system design, and excessive mechanical noise from air handling units installed within or immediately adjacent to the main space. Obviously the larger the space the more critical the acoustical design. The potential problems for spaces seating more than a few thousand are so demanding that an acoustical consultant must be employed.

The cure for excessive reverberation is adequate ceiling and wall sound absorption. Except for the largest stadiums, mid-frequency reverberation times should be kept to 2-3 seconds, and not more than about 3-5 seconds at the low frequencies. Use caution if the main acoustical absorption is to be provided by a spray-on insulation, as such materials in thicknesses less than 2 inches have little low frequency absorption. Suspended baffles or clouds can have excellent broad-band absorptive characteristics.

It is important to identify surfaces from which reflected sound can cause echoes. It is usually sound from the sound system loudspeakers that causes problems, but the combination of a rock band on the floor near the center of the building and a focusing ceiling can produce surprising results in the form of low frequency flutter echoes. Offending surfaces must be treated with an acoustical material having the proper absorption versus frequency characteristics. A situation where a reflection would be trouble in a large arena is shown in Figure 5-10.

There are frequently problems of reflections from unoccupied, upturned seats with the sound from the loudspeaker system reflecting to the opposite side of the building, or down onto the floor where the sound is heard as a strong echo. This problem occurs only when the seats are unoccupied. A possible method of minimizing the problem is to apply absorption to the bottom of the seats, such as a heavy, fire-rated carpet or other abuse-resistant material. A perforated seat bottom is another method if the seat is padded.

Good design of the sound reinforcement system is essential in stadiums, as unreinforced speech will not be nearly loud enough. In smaller spaces a single central cluster of loudspeakers will be adequate, but in larger houses a circumferential arrangement will be required to keep the loudspeaker-to-spectator distance within acceptable limits. If this distance is too great, the reverberant sound field masks the direct sound and intelligibility suffers. To avoid this problem the loudspeaker(s) must either be moved closer to the spectators, or the level of the reverberant field must be lowered by adding absorption. Simply increasing the volume of the sound system will not help; the reverberant field is energized by the sound system, and will increase in level by the same amount the sound system volume is increased. The *ratio* of direct to reverberant sound remains unchanged.

It is important that the sound system be designed so the sound is directed just to the seating areas. If possible, the system should be zoned so that only those seating areas that are occupied receive sound. See Chapter 6 for a more complete discussion of sound reinforcement systems.

### 5.5.2 Movie Houses

Movie houses are frequently constructed with two or more theaters in a cluster, side by side or back to back, so that sound transmission from one to another must be controlled. A well-designed sound system is capable of producing very high levels, particularly at the low frequencies. The most critical situation occurs when a battle scene is taking place in one theater while a tender love scene is playing in the adjacent theater. Dividing walls should be rated at STC 65 minimum, but one-third octave band TL data should be consulted to insure adequate low frequency attenuation.

To achieve an STC 65, the dividing wall will have to be of double construction. In theaters with surround sound, loudspeakers will be mounted on the side walls. In this case, it is essential that double wall construction be used, to avoid sound transmission directly through the wall to the adjacent theater.

Modern movie houses are typically designed with extensive use of sound absorbing materials to minimize reverberation and room effects, so that all sound effects are provided through the theater's sound system. It is important that the acoustical materials used have adequate low frequency absorption, because of the low frequencies produced by the sound system.

It is not uncommon to install the theater's air handling units directly on the roof over the middle of the theater, with a hole punched in the ceiling for return air. In this situation fan noise is excessive; the system should be designed not to exceed NC-30. Furthermore, a mid-span location is the worst place in terms of vibration, particularly if the mechanical unit contains a compressor. A better location is over the projection booth, or in a separate mechanical room. Each theater should have its own mechanical system, to avoid the possibility of crosstalk through the air ducts.

Many movie houses now, such as IMAX and Cinerama, are designed to meet or exceed THX performance criteria, which establish strict requirements for sound isolation, reverberation, background noise and sound system performance. Design of theaters to meet these strict requirements should not be attempted without the help of an acoustical consultant.

### 5.5.3 Outdoor Amphitheaters

Attending a music or drama performance out of doors can be a very enjoyable experience, but the acoustical aspects of an outdoor amphitheater require careful study. For music, lack of an enclosing structure deprives the audience of early reflections and reverberation so necessary for enhancement of the program, as well as not providing a barrier for intruding noise. There are, however, some methods available for minimizing these problems.

Obviously an amphitheater should not be located close to an airport, freeway or other noisy facility. Even so, street traffic of some type is likely to be close by. Earth berms between the traffic or parking areas can provide some acoustic shielding. See Chapter 8, Section 8.3.2. There is no way of providing shielding from aircraft; the only solution is to locate away from airport flight paths. Also, be aware of buildings or other structures close enough that could cause annoying delayed reflections from the sound system or performing group.

The performing area requires, at a minimum, some type of shaped enclosure to provide feedback for the musicians, and project as much sound as possible toward the audience. The enclosure must also provide sufficient blending of the sound so that microphones required for the sound system are receiving a well-balanced signal.

A sound reinforcement system is required to generate levels loud enough to be clearly heard by the audience, particularly in the presence of background noise levels, likely to be considerably higher than would be experienced in a concert hall or theater. Most modern systems will incorporate additional side loudspeakers and signal processing for time delays to simulate sidewall reflections, and for added reverberation. The quality available from state-of-the-art systems can create surprisingly realistic concert hall sound effects.

Spillover from the sound system into adjacent residential areas can be a problem. Even with the use of highly directional loudspeaker arrays, some sound will propagate beyond the limits of the amphitheater. The sound system designer must calculate the probable sound levels at adjacent residences to ensure that the levels are not excessive.

### 5.5.4 Swimming Pools

Swimming pools are used for both recreation and instruction. Without acoustical treatment the reverberation times may be so long that it is hard to understand the instructor at any distance, particularly if there is loud conversation or considerable activity in the water. The problem is caused by the fact that water is almost 100% reflective of sound and just putting acoustical material on the ceiling may not be adequate.

Acoustical wall panels can provide the required additional absorption. It is desirable to locate the panels as low on the wall surface as possible, allowing for the fact that walkways are hosed down for cleaning. *Caution*: Acoustical panels against outside walls in cold climates can produce condensation between the panel and the wall surface. See Chapter 2, Section 2.4. A similar problem may occur on the ceiling if there is no plenum space above, or if there is, not being properly ventilated.

A sound reinforcement system will be required if the pool is used for competitive events, and also can be used for water exercise classes. An underwater sound system is required for underwater ballet, but is of little benefit for surface swimmers.

## 5.6 Medical Facilities

Medical facilities command special acoustical attention because the people utilizing such facilities are to some degree in distress, and should not have to endure a noisy environment in addition to their other problems. Furthermore, the confidentiality of doctor-patient conversations must be ensured, and the medical staff should be able to function without excessive noise. Vibration-sensitive imaging equipment will not yield satisfactory results if the motion of the surfaces or structures to which they are attached exceeds acceptable vibration levels. In short, noise should not add to the unavoidable stress that the people associated with these facilities must endure.

### 5.6.1 Clinics

Clinics are much like private offices in their acoustical requirements (see Section 5.3.1). Acoustical privacy is very important for counseling and examination rooms; patients will be inhibited from open and frank discussion if they believe they can be heard in adjacent spaces. The same applies to persons waiting for an appointment if they can overhear a doctor-patient conversation coming from an adjacent room. Acoustical seals on doors opening into waiting areas are required. Low level background masking noise also may be used in waiting areas, similar to the masking noise used for open plan offices (see Section 5.3.2). Supply and return air must be done in a way to avoid crosstalk problems.

Rooms used for diagnosing speech disorders or screen testing of hearing acuity should have lower HVAC background noise (RC-25) and acoustical wall panels as well as an acoustical ceiling. The acoustical requirements for hearing threshold testing are even more restrictive, but such testing is usually done in specially designed audiometric booths. Manufacturers of audiometric booths are listed in Appendix A.12, numbers 1 and 33.

### 5.6.2 Hospitals

Private offices and exam rooms in hospitals have the same acoustical requirements as similar rooms in clinics (preceeding section). However, hospitals have additional functions that are both noise and vibration sensitive, and will have patients admitted for at least one night or longer.

Acoustical privacy between patient rooms should be sufficient to prevent disturbing a patient trying to rest or sleep in one room from a patient in the next room who may be in pain, talking with a doctor or nurse, listening to TV or talking with visitors. This condition will require a partition with an STC rating of 45 to 50. Sound leakage can be a problem because the partitions often contain equipment for testing, oxygen supply piping, and recessed casework for cabinets, etc.

Acoustical seals on doors are seldom used. Indeed, patient room doors are often left open for surveillance by the staff, and are closed more often for visual privacy than for acoustical privacy. Patient rooms near nurses stations and elevators are more vulnerable to noise intrusion than are rooms located further away.

Cleanliness is a major concern in hospitals, and since most acoustical materials consist of fibers, they are vulnerable to infestation by bacteria and vermine. If the material is protected by some type of washable film, there can be an unacceptable loss of sound absorption, particularly at the high frequencies. Operating rooms are of particular concern, and are often quite reverberant because acoustical materials cannot be used. Fiberous air duct lining insulation to control transmission of noise also may be prohibited from use. There are specially treated sound traps that may be acceptable for such use.

Special instruments, such as electron microscopes and magnetic resonance imaging equipment, have special requirements for vibration isolation. Vibration of the supporting structure, which can result from mechanical equipment or footfalls, can cause serious problems with such instruments. The manufacturer of the instruments should provide maximum permissible vibration levels for their instruments, and the acoustical consultant working in concert with the structural engineer can then design the required supporting structure and vibration isolation to meet the criteria.

Noise and vibration from emergency helicopters serving hospitals can be a problem, and their landing pads should not be located near or above noise-sensitive areas.

## 5.7 Miscellaneous Spaces

Virtually all spaces occupied by man have acoustical requirements. In the foregoing sections of this Chapter, specific categories of occupied spaces have been discussed. In this Section are presented acoustical design considerations for those additional uses where acoustics is (or should be) a major design concern. Indeed, the acoustical requirements for some of the uses discussed in this Section are so critical that design should not be attempted without the services of an acoustical consultant. The information given here outlines these requirements, but should not be considered sufficient for a complete design.

### 5.7.1 Libraries

Libraries are traditionally quiet, and the sound of conversation or the shuffling of books and papers can be disturbing to people reading or studying. The problem is aggravated when there are reflective, untreated areas such as a hard ceiling, or the background noise is either too low or too high. Acoustical ceilings are required.

The presence of low level masking noise in libraries can be beneficial in achieving some degree of acoustical privacy. Background noise may be provided by the air handling system, but this source is unreliable even in carefully designed systems. If the system is too quiet, no masking is provided, and if it is too noisy, that just compounds the problem. A good design objective would be an RC-35 curve.

A more reliable source of masking noise can be achieved with an electronic system consisting of a noise generator, equalizer, and loudspeakers above the suspended ceiling. See Chapter 6, Section 6.3, Sound Masking Systems.

Work areas may contain noise sources that have to be acoustically isolated from study and reading areas. Also, many libraries now have media rooms where groups of students may be viewing (and listening) to recorded program material, that must similarly be isolated. Where computers are used in the general library reading and study areas, the printers should be laser-type. Impact printers should be located in a separate room.

### 5.7.2 Churches

The acoustical requirements of churches are as varied as they are in size, which range from small, intimate chapels to auditoriums seating 5000 or more. To some degree all denominations involve both music and the spoken word, with the attendant conflict in desirable reverberation characteristics. Indeed, there are few other building categories with such widely varying acoustical requirements for the same space. For speech the reverberation time is ideally about 1 second, while for choral and particularly organ music, times in excess of 3 seconds are often preferred. Such a range is well beyond what could reasonably be expected from a variable absorption system.

Fortunately, recent advances in the design of sound reinforcement systems and available hardware permit good speech intelligibility in churches which are quite reverberant. Chapter 6 provides basic information on state-of-the-art sound systems and how they may be successfully integrated into the architectural design of churches. What this means is that the passive or "natural" acoustics of the space can permit the reverberant enhancement of choral and organ music, while relying on a properly designed sound system to achieve acceptable speech intelligibility.

A visual problem that sometimes is difficult to resolve is the location of the loudspeaker cluster serving the congregation area. The ideal location is usually high above the altar, a point that is also a visual focal point for the entire church. Architects and consultants alike must be prepared to work out a design that can successfully meet both requirements.

The acoustical problems are somewhat different in the larger churches (seating 2000 or more) that have become popular in recent years. Many rely on sound reinforcement for both speech and music, and the sound systems can provide signal processing for added reverberation. In such cases it may be advisable to add sufficient acoustical absorption to favor speech intelligibility (since the cubic volume is probably quite large), and have the sound system provide added reverberation for music.

Typical sound isolation problems in churches are cry rooms, adjacent meeting or sunday school rooms (which can be below the main church) and mechanical rooms. The methods for controlling such noise are discussed in other sections of this handbook, particularly Chapters 3 and 7.

### 5.7.3 Courtrooms

Excellent speech intelligibility is a prime acoustical requirement of courtrooms. Complicating this requirement are witnesses that because of intimidation or emotional stress do not speak loudly or clearly, and a growing tendency to rely on the use of electronic sound systems for audio reinforcement, recording or transmission to remote locations. If an audio recording is to become the official court record, the quality of the recording cannot be compromised by poor room acoustics or excessive background noise.

Room acoustics are not a major influence on speech recordings if the speaker-to-microphone distance is within a foot or two (about 0.5 m). At greater separations, reflections from room surfaces or reverberation (multiple reflections) in a room with inadequate acoustical treatment can seriously degrade the quality of the recording. It is difficult to maintain small speaker-to-microphone distances when attorneys move around the courtroom, or when a witness may leave the witness box to approach a large exhibit. Lavaliere microphones are not considered practical by most courts. However, recording quality can be achieved through careful placement of acoustical absorbents on both the walls and ceiling to minimize room effects such as flutter echos and excessive reverberation.

Another important acoustical factor in courtrooms where audio recordings are made is background noise, whether from the HVAC system or adjacent activities. HVAC system noise should not exceecd an RC-25. Because public lobbies can be noisy, courtroom entrances should be through vestibules. Entrances for the judge and supporting staff are usually from secure corridors and not a problem. Holding cells for prisoners waiting to enter the courtroom, however, should be well isolated acoustically, particularly if a single holding space serves two courtrooms. Acoustical treatments in holding areas must be rugged and abuse-resistant. Perforated metal or hard plastic over insulation is a commonly used treatment for such spaces.

Confidential acoustical privacy is a requirement for the judge's chambers, law clerks, attorney conference rooms and offices for supporting services such as pretrial and probate. Occasionally a room will be provided for remote witness testimony, for use by children who would be intimidated by the courtroom environment. Such a room must accommodate both send and receive live audio (and video), with the same acoustical requirements as the courtroom, only on a smaller scale.

### 5.7.4 Radio and TV Studios

The control of sound transmission, reverberation control within the room and low background noise are major design concerns for radio and TV studios. The demands are complicated enough that an acoustical consultant should be employed.

Studio entrances should be through vestibules or at least open into quiet, controlled access corridors. Sound isolation between critical spaces will require multiple wall construction and a high degree of structural isolation. Relights must be double glazed with heavy, laminated glass and a wide separation.

Mechanical equipment should not be located close to studios. The air distribution system must attenuate equipment noise before entering the studio. Air velocities through the ducts and at the diffusors must be low to avoid noise regeneration. Air ducts and signal cables between studio and control must be done without creating sound leakage paths. Chapter 7 addresses noise and vibration control in mechanical systems.

The studio interior will most likely be highly sound absorbent with both ceiling and wall treatments to minimize reverberation and undesirable reflections which can interfere with "clean" sound recording. Indeed, room effects in recorded sound should be minimized, as the final program audio will most likely be the product of postrecording signal processing and mixing. Even the traditional cyclorama or background set is becoming a thing of the past as virtual backgrounds are created electronically.

One of the more important rooms acoustically is where postrecording audio mixing is done, sometimes called "sound sweetening." Many times this is done in the control room. The person doing the mixing must hear the recorded signals without the presence of pronounced room interference patterns caused by the combination of direct and reflected sound. At the same time, the room cannot be anechoic (removal of all reflected sound) as this results in a very unnatural listening environment. Reflected sound, therefore, must be highly diffused. Specially designed quadratic diffuser reflectors are available for applications of this type. Refer to Chapter 2, Section 2.5.2.

Quite often a "voice-over" recording booth will be associated with either live audio pickup or postrecording operations. The booth must be acoustically isolated from the adjacent studio or control room, but at the same time have a visual connection requiring specially designed, double-glazed relights. Such booths can also be obtained from the same manufacturers that provide audiometric testing booths and small music practice rooms. (See Appendix A.12, manufacturers 1 and 33.)

### 5.7.5 Recording Studios

Many different methods are employed for making audio recordings in studios. It is common practice for small bands to record each performer on a separate channel with mixing and signal enhancement done later in post-production sessions. A studio functioning in this manner should have very little reverberation, and may even have partial barriers between peformers.

If the natural reverberation of the studio is desired, the studio must have sufficient cubic volume to create the required reverberation, and methods of diffusion for sound uniformity. Recordings of large symphony orchestras are usually made in concert halls where, if the hall is properly designed, all desirable sound attributes can be recorded with judicious placement of microphones and postrecording mixing.

Modern digital recording equipment is capable of achieving signal-to-noise ratios approaching 100 dB. Such a capability requires that the background noise from any source be extremely low (RC-15 for example; see Chapter 1, Section 1.7). It is evident that the acoustical requirements of recording studios are very demanding, and the design should not be attempted without the services of an acoustical consultant.

### 5.7.6 Laboratories, Testing Facilities and Clean Rooms

High resolution imaging devices, computer chip manufacturing facilities and other highly vibration-sensitive uses must be designed with great care. It is usually a requirement that this type of equipment be mounted on ground floors with concrete slab isolation from other parts of the building. If such equipment must be located on upper floors, massive concrete structure is preferable to frame construction. Special attention is required in the design of air distribution systems for such areas to insure against low frequency noise, as this can induce unacceptable vibration in building components (see Chapter 3, Section 3.7). Vibration from footfalls also must be considered in the structural design of such facilities.

Air bags can be used when maximum vibration control is required at the very low frequencies. These bags also accurately control the height of test benches, etc. In earthquake zone areas, seismic movement restraints must be provided by suitable snubbers. Design for these situations must not be attempted without the assistance of engineers competent in this field.

Many laboratories have fume hoods which can be very noisy when operating at maximun exhaust CFM. Multi-speed fans permit operation at lower CFMs resulting in less noise. The use of sound traps or duct lining in the exhaust ductwork may not be permissible because of corrosive fumes.

There are frequently other requirements such as shielding from electromagnetic radiation which must be considered along with noise and vibration control.

## *5.8 Rehabilitation of Acoustical Spaces*

Correcting acoustical problems in existing buildings is a basic service provided by acoustical consultants. Many older buildings were designed without the knowledge of architectural acoustics principles. That cannot be an acceptable excuse for poor acoustical design in new buildings, but it happens with disturbing frequency. Another major reason for having to make acoustical modifications is when the use of a building changes. Sometimes proposed new uses are incompatible with remaining existing uses, such as attempting to locate a disco or live music in a hotel with guest rooms directly above, or placing a gymnasium over classrooms in a school. It is surprising how often acoustical consultants are presented with problems of this type.

One advantage in correcting acoustical problems in an existing space is that many times acoustical measurements can be made to better define the nature of the problems. Measurements can determine reverberation times, the presence of undesirable reflections or echos, background noise, and sound isolation between spaces, including identification of various sound and vibration transmission paths. Methods for performing various types of measurements are presented in Appendix A.10

A number of precautions need to be observed in acoustical rehabilitation. If the problem is sound isolation, it cannot be solved simply by adding acoustical material to the receiving space. It is necessary to determine the sound transmission path(s) and treat each in the order of its contribution to the transmitted sound. Sound leakage through holes and cracks is usually obvious and can be corrected by sealing with caulking or some other suitable treatment. Following this, solving the problem may require heavier or double construction for partitions, and modification of floor/ceiling systems, doors or windows/relights. Be aware of flanking transmission through adjoining structure, interconnecting air ducts, over suspended ceilings common to adjoining rooms, etc. As mentioned above, measurements are very useful for identification and rank-ordering of sound transmission paths.

Many times the source of noise is plumbing. Supply water piping contains a lot of turbulence and resultant vibration which, if transmitted to other building components, can result in noise. Selection of quiet valves can help, but the most effective treatment is to isolate the pipes from the building to the maximum extent possible (see Chapter 7, Section 7.5). This may be very difficult in existing buildings if the piping is not accessible, but if the problem is serious enough the only solution may be opening up a wall or floor to reach the plumbing. Obviously, solution of such problems is best achieved during construction.

Similar problems can be caused by waste lines, particularly if plastic pipe was used. Avoiding contact with the surrounding building construction again is the best solution. If the pipe is accessible, wrapping it with glass fiber blanket then covering it with barium-loaded vinyl will reduce radiation from the pipe. There are also damping compounds that can be applied directly to the pipe, but such treatments are usually less effective than wrapping.

Change in use of a room often requires acoustical correction. For example, if a room is to be converted to a musical rehearsal room, there is likely to be a need for both improved control of sound transmission to adjacent spaces, and room interior treatment. Added acoustical absorption will probably be required, and care must be taken to insure that the added absorption functions in the proper frequency ranges. Many acoustical materials do not have significant absorption at the high and, in particular, the low frequencies. See Chapter 2 on absorption characteristis of acoustical materials.

If an auditorium with a sound system has problems, a study must be made to determine if the problem is with the acoustical treatments, the sound system, or both. Common problems in auditoriums are too much (or too little) reverberation, lack of clarity, uneven sound characteristics throughout the seating area, too much background noise and a sound system that sounds unnatural.

Excessive intruding or background noise can be a problem in almost any type of building. Solving the problem will require determining both the sources of noise and primary transmission paths. A decision must be made whether it is feasible or more cost-effective to treat the source, or impose controls along the transmission path. Methods of treating various types of noise are contained in several chapters of this handbook.

A frequently encountered limitation in rehabilitating older buildings is when the buildings are placed on a historic register. When this is done the project often becomes eligible for funding, but with the caveat that the architectural character of the building cannot be significantly altered. If significant architectural modifications are not permtted even though required to correct acoustical problems, other methods will have to be considered.

## *5.9 A Practical Procedure for Designing a Multipurpose Auditorium*

The following sections outline a straightforward procedure to follow for incorporating acoustical considerations in the design of a multipurpose auditorium. The same design procedure can be applied to any large or small space for which good acoustics is desired, such as lecture rooms, recital halls, classrooms and so forth. The steps for this procedure are itemized on the room calculation sheets found at the end of this Chapter. The same forms without calculations are also included, which may be photocopied for future use. The designer may wish to use a spreadsheet (or write a computer program) to speed up the calculation procedures.

The procedure for the acoustical design of an auditorium is an exercise in coordination between drawings and calculations. A computer or drafting table will be necessary to produce the drawing. As the design progresses, the drawing will begin to modify the calculations and vice versa. Starting points for both the drawings and the calculations are only the beginning; they will undergo modification until a satisfactory design is achieved.

The design begins with the directives contained in the user program relating to acoustics. As the physical characteristics take form, they must accommodate the program acoustical requirements in both the space and time domains. For example, reflections must be directed in three-dimensional space, and they must reach the listener at an appropriate time interval with respect to the direct sound and other reflections.

Having developed an initial shape in both plan and section, the designer then performs preliminary reverberation time (RT) calculations. The two steps of calculating RTs and shaping the room are continually refined until satisfactory results are achieved. These concepts are covered in detail in Section 5.1.3, "Multipurpose Auditoriums," and the reader is encouraged to refer to this Section as the design progresses. The following exercise involves shaping for reflections and reverberation time calculations only. Additional design considerations include sound isolation, noise from mechanical systems and design of sound reinforcement/reproduction systems.

### 5.9.1 Design Criteria

For determing the design criteria, use the ROOM CALCULATIONS STEP ONE PROGRAM SHEET.

1. Determine the uses of the auditorium. For the purposes of this exercise, assume a multipurpose auditorium that will be used primarily for speech, music and drama.
2. Determine the size of the audience. Assume the program requires an auditorium to seat 2200.
3. Select the type of seating. Standard seating will be used with an area per seat of 6.5 sq ft (0.6 sq m). The range of area per seat could be between 6 and 8 sq ft (0.56 and 0.74 sq m).
4. Calculate the floor area (excluding stage): A = 2200 × 6.5 = 14,300 sq ft (2200 × 0.6 = 1320 sq m).
5. Determine the "volume per seat" value from Table 5-2. The range for an auditorium is 200 to 250 cu ft/seat (5.7 to 7.1 cu m/seat). A middle value of 220 cu ft/seat (6.2 cu m/seat) is appropriate for a multipurpose auditorium.

**TABLE 5-2.** Recommended Cubic Volume per Person

| Use | cu ft/person | (cu m/person) |
|---|---|---|
| Movie Theater | 100 | (2.8) |
| Multipurpose Auditorium | 200–250 | (5.7–7.1) |
| Concert Hall | 200–300 | (5.7–8.5) |
| Choral Rehearsal | 350 | (9.9) |
| Band Rehearsal | 500 | (14.2) |

6. Calculate the design room volume (excluding stage house): V = 220 × 2200 = 484,000 cu ft (V = 6.2 × 2200 = 13,640 cu m).
7. Select an average width of 96 ft (29.3 m) for the hall.
8. Calculate the length of the auditorium. This step will also determine if a balcony is desirable:

L = 14,300/96 = 149 ft (1320/29.3 = 45.1 m). This length exceeds the maximum desirable length (proscenium to rear row) of about 120 ft (36.6 m). The design, therefore, will have a balcony, which will improve both the acoustics and viewing distance by limiting the distance to the furthest seat. The main floor area is determined by the permissible length times the width, 120 × 96 = 11,520 sq ft (36.6 × 29.3 = 1072 sq m).

9. Determine the design ceiling height by dividing the volume by the main floor area. This number will be the average height since the ceiling height will vary. H = 484,000/11,520 = 42 ft (13,640/1072 = 12.7 m).
10. The ratios of the dimensions should not be integer multiples of each other (see Chapter 4, Section 4.5, "Room Modes"). The dimension ratios are acceptable: 42:96:120, or 1:2.3:2.9.
11. Select the stage opening dimensions. Assume the opening to be 50 ft wide by 30 ft high (15.2 m × 9.1 m).

12 & 13. Select the design reverberation times (RTs) for 125 Hz and 500 Hz. Referring to Chapter 4, Figures 4-7 and 4-8, appropriate times would be about 1.6 seconds at 500 Hz and 2.25 seconds at 125 Hz.

14. Calculate the constant that will be used in the RT calculation:
0.049 × volume = 0.049 × 484,000 = 23,716.
(0.161 × volume = 0.161 × 13,640 = 2196).

15 & 16. Calculate the number of sabins (total sound absorption, see Chapter 2, Section 2.1) required to meet the design RTs at 500 and 125 Hz:
$0.049V/RT_{500}$ = 23,716/1.6 = 15,810 sabins.
($0.161V/RT_{500}$ = 2196/1.6 = 1464 metric sabins).
$0.049V\ RT_{125}$ = 23,716/2.25 = 10,540 sabins.
($0.161V/RT_{125}$ = 2196/2.25 = 976 metric sabins).

### 5.9.2 Drawing the Auditorium

The design dimensions of the auditorium are taken from the program sheet as a starting point. Preliminary sketches of the plan and longitudinal section are made to scale. Shape the wall and ceiling surfaces to provide reflections for uniform coverage of sound over the seating areas, with emphasis for the rear seating. Direct sound from the stage or sound system proscenium loudspeakers drops off with distance. For this reason, it is necessary to provide useful reflections from the ceilings and walls to offset this condition.

In hand drawing, a protractor is an indispensable tool to draw the angles of the reflected sound; with a CAD program, mirroring the angles of reflection off the panels is a simple operation. See Chapter 1, Figure 1-9, which illustrates that the angle of reflection equals the angle of incidence.

When plotting reflected sound, there must not be any delayed first reflections with path lengths greater than about 55 ft (16.8 m) or 50 milliseconds, and preferably less than 35 ft (10.7 m) or 30 milliseconds compared to the direct sound. If the difference is greater than 55 ft, there is a noticeable deterioration of sound quality. It is desirable to have a series of well spaced early reflections during the first 1/10th of a second, after which the reflections become very closely spaced leading to the reverberant decay. Avoid single reflections that are much higher in level compared to reflections just before or after, as they may be perceived as echos (see Chapter 4, Figure 4-1). If the surfaces producing echoes cannot be reoriented, they must be treated with absorption and/or diffusion. This also applies to opposite, parallel walls, which can cause "flutter echos" if not highly diffused or absorptive. Beware of "corner reflections," which can act as a flat surface returning the sound back towards the source.

A good procedure for shaping the walls and ceilings follows. Begin working in section.

1. Select a source point on the stage about 3 ft (1 m) upstage from the proscenium opening and about 5 ft (1.5 m) above the stage floor, as illustrated in Figure 5-11. Draw a line from the source to the front edge of the first ceiling reflector panel. Draw another line from a point about one-third of the distance from the stage to the rear row of seats, to the panel at the intersection of the first line. Set the angle of the panel by making the incident and reflected angles of the two lines the same.
2. Lengthen the reflector panel until a reflected ray also covers the rear seats. If there is a balcony, either continue extending the panel until the rear seats in the balcony are covered, or, if the panel is becoming too large, bow the reflector slightly to increase the angle of coverage. This panel will then provide a first reflection to most seats in the auditorium, and should be within the permissible time delay compared to the direct sound.
3. A second ceiling panel is then constructed in the same manner, except the leading edge reflects to about 80 ft (24.4 m) from the proscenium opening, and the far edge reflects to the last seating in the balcony. The rear seating has now been covered twice. Note: If there are catwalks above the reflector panels, the panels will have to be spaced to accommodate lighting fixtures.
4. A third reflector panel may not project to main floor seats under the balcony, so this panel can be angled to reach from the front of the balcony to the back. Additional reflections for the main floor rear seats can be obtained by angling the underside of the balcony to cover some of these rows.
5. The top portion of the balcony rear wall (and main floor rear wall above head height, if there is room) can be tipped forward to reflect sound down to the rear seats. Make certain that the reflection from the top of the angled area does not extend forward more than about

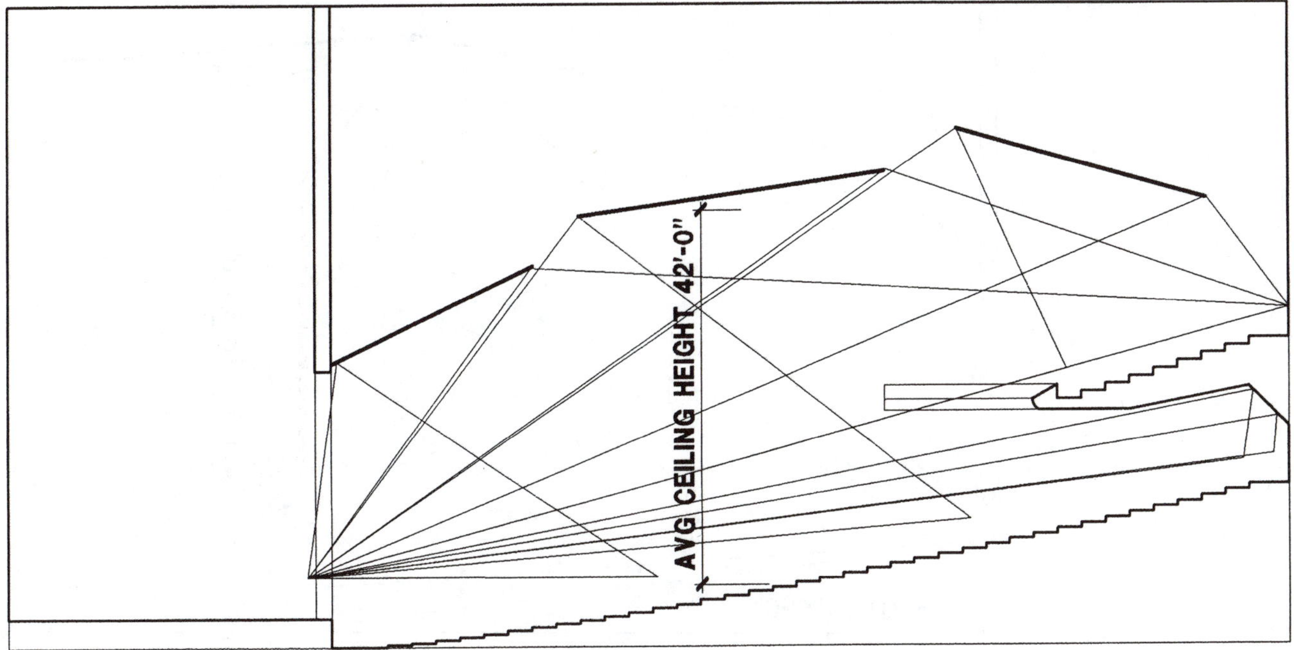


**Figure 5-11.** Auditorium exercise section.

25 ft (7.6 m) to the heads of the audience. Any space below the bottom of the angled wall down to 4 ft (1.2 m) from the floor should be treated with absorption.

6. The front of the balcony can reflect sound from the loudspeakers back to the stage if this surface is not treated acoustically or has a convex surface. An alternative can be to angle the front surface so that reflections from the source on the stage will be directed down towards the seating area, being careful that the sound path difference does not exceed permissible limits.

Continue shaping in plan. (Another approach to the design of sidewall reflectors is presented in Section 5.1.3 which is desirable for halls with an emphasis on music.) The method described here is appropriate for multipurpose halls:

1. Sidewall reflections are determined on a plan of the auditorium floor using a source point about one third the distance from one side of the stage opening, and 3 feet (1 m) behind the proscenium opening. See Figure 5-12. Reflection lines are then drawn from this point to the far wall. The first reflection from the wall next to the stage wall will come across the hall to seating in front of the source location. If the sound path difference between the direct sound and the total reflected path exceeds 55 ft (16.8 m), the angle of the sidewall will have to be adjusted until the sound path difference does not exceed the 55 ft (16.8 m) limit.
2. The wall may continue at this same angle, producing a continuous reflection across the seating area. This procedure may make the hall wider than is desired, and the angle of the wall can be decreased at any point, forming panels with less angle as desired, as long as the last panel is not parallel to the panel on the opposite side.
3. Where the sidewall intersects with the rear wall as in Figure 5-11, there can be reflections from the corner back to the stage. To avoid this condition, the corner will either have to be rounded (convex), or covered with an acoustical material. This corner effect can occur at other locations.
4. If it is desired that the rear panel be parallel to the opposite panel, then it must be covered with absorptive material or diffusing elements to eliminate a "flutter" between the parallel walls.
5. A single loudspeaker cluster should be positioned in the middle above the proscenium opening to cover the entire seating area. Distributed, delayed loudspeakers may be required under the balcony. See Chapter 6, Figure 6-3.

The drawings should also indicate the areas of reflective and absorptive surfaces. The types of materials of these areas will need to be selected for the takeoff and RT calculations. The actual values acheived in the auditorium design such as volume, area, average ceiling height and room width, should be compared with the design criteria on the programming sheet. There should be good agreement between the two if the design has been properly done.

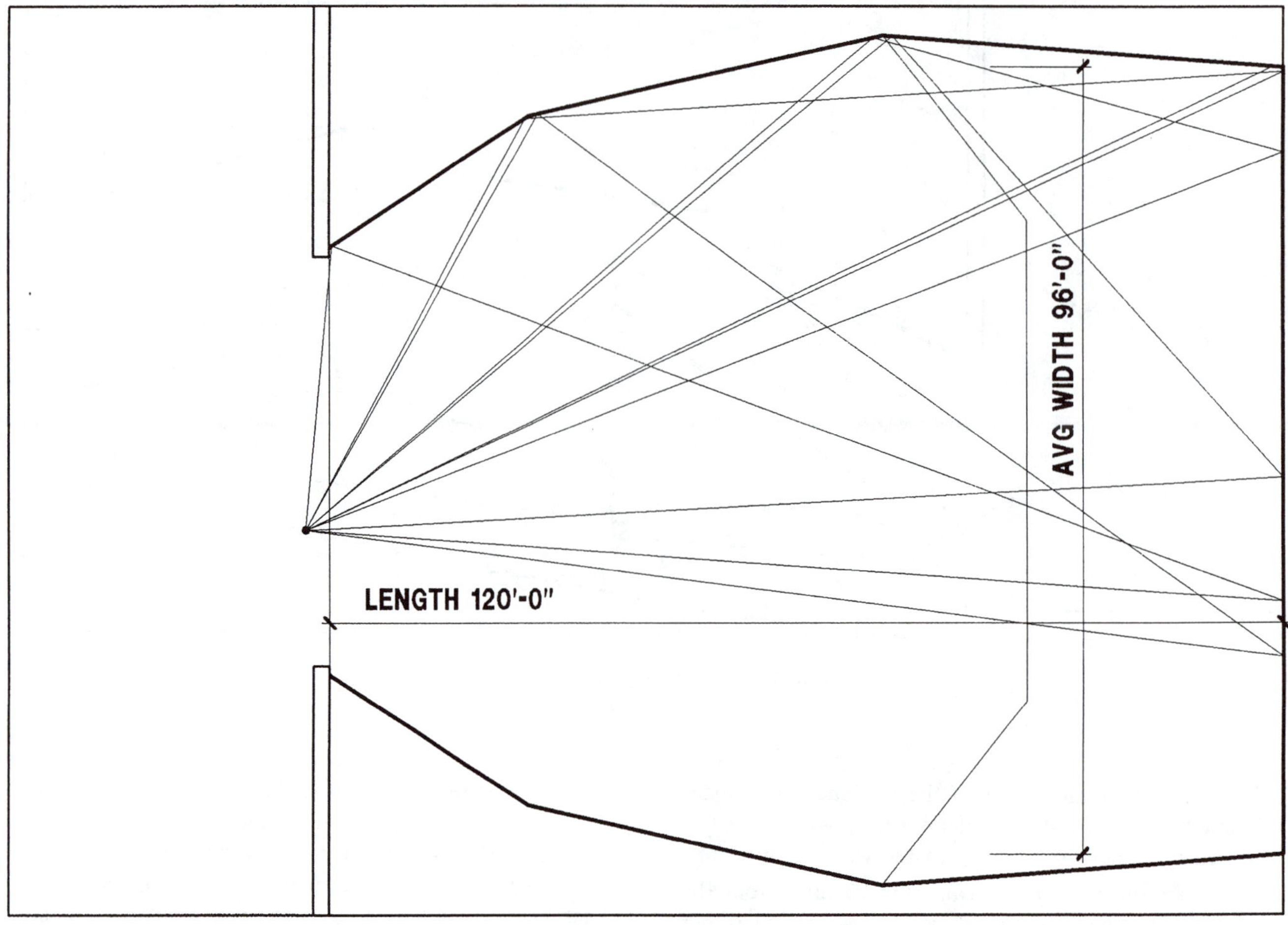


**Figure 5-12.** Auditorium exercise plan.

### 5.9.3 Take-off Information

The drawings that were created in Section 5.9.2 are used to obtain data for the RT calculations; the materials and areas shown in the drawings are "taken-off" and entered on the "Room Calculations Step Two Take-off Sheet." If an orderly procedure is used, the numbers for calculation of RT will be much simpler.

In order to compute the RT, every material and the surface area it occupies will need to be determined. If the total surface area of the auditorium is first computed, then this may be used as a cross-check to ensure that all surface areas have been accounted for. The procedure that is illustrated on sheet two and described herein is the one the authors find simplest and most accurate. The designer may prefer an alternate method.

Find the gross area for each of the following surfaces:

1. Main floor
2. Balcony floor (if applicable)
3. Main ceiling
4. Ceiling under the balcony
5. Rear wall below the balcony
6. Rear wall above the balcony
7. Front of balcony
8. Front sidewalls back to the balcony
9. Sidewalls under balcony
10. Sidewalls above balcony
11. Front wall

Total the gross areas to obtain the total surface area for the auditorium. Within each gross area there may be subareas of different materials. These subareas include materials such as wood doors, wainscotting, acoustical absorption, the proscenium opening, carpeting, drapery, etc. For example, the main floor area is concrete and has a gross area of 10,874 sq ft (1010 sq m). The carpeted aisles are a subarea of the main floor area, with an area of 2682 sq ft (249 sq m).

The next step is to subtract all subareas from the gross areas. In the example of the main floor gross area, the aisle areas are subtracted from the concrete column and added to the carpet column (see line 18, ROOM CALCULATIONS STEP TWO). This procedure is followed for each subarea. The total surface area will remain the same; this cross-check should be made to ensure accuracy.

The volume of the hall may be calculated by multiplying the floor area by the average height. If there is a balcony, use the floor area in front of the balcony multiplied by the average height for this area. Add to this the floor area under the balcony multiplied by its average height, and the floor area of the balcony times its average height. Once this is done, the actual volume should be entered on the programming sheet and compared with the volume established for the design criteria. (Note: The actual auditorium areas and volume as they appear at the bottom of the Step Two sheet were calculated using CAD program functions.)

For use in the RT calculations, multiply the actual volume by 0.049 (0.161 metric). This number is a constant that will be used for the RT calculations, and when divided by the design reverberation times at 500 Hz and 125 Hz will give the sabins of absorption required. These values are as follows:

$$\text{Sabins at 500 Hz} = \frac{0.049\ V}{1.5} = 16{,}088 \qquad \frac{(0.161\ V = 1497\ \text{metric sabins})}{(1.5)}$$

$$\text{Sabins at 125 Hz} = \frac{0.049\ V}{2.25} = 10{,}725 \qquad \frac{(0.161\ V = 998\ \text{metric sabins})}{(2.25)}$$

### 5.9.4 Reverberation Time Calculations

The air absorption is the first calculation. See Chapter 2, Table 2-1, for values of air absorption. This calculation is made on the bottom of the PROGRAM SHEET. Enter the air absorption under 1000, 2000 and 4000 Hz. The absorption of the air is significant only at the higher frequencies.

Enter the square footage for each material surface from the TAKE-OFF SHEET onto the ROOM CALCULATIONS STEP THREE REVERBERATION CALCULATIONS sheet. Select the sound absorption coefficients for each of the surfaces at each frequency of interest from the data in Chapter 2. Calculate the sabins of absorption for each surface by multiplying the area by the absorption coeficients, as shown in lines 1 through 7 in the example.

Add the absorption of the seating, occupied and unoccupied, as shown on lines 8 and 9 of the reverberation calculations form, at each of the frequencies. For design purposes, the number of occupants is usually assumed to be about two-thirds of the total number of seats. Exceptions could include auditoriums frequently used for small group lectures.

Calculate the total absorption at each frequency by summing lines 1–10, and show on line 11. The reverberation time is calculated as shown at the end of the previous Section, and entered on line 12.

### 5.9.5 Adjusting the Reverberation Times

The calculated reverberation times of the hall are adjusted to meet the design criteria established on the programming sheet. The adjustment is made on the basis of comparing the total calculated sabins with the design sabins at a frequency of 500 Hz, and is accomplished by subtracting line 11 from line 14 on the step three sheet. The difference is entered on line 15, indicating the number of sabins required to meet the design reverberation time at 500 Hz.

The next step is to select a material from those listed in Chapter 2 that will add sufficient absorption to reduce the reverberation time at 500 Hz to 1.5 seconds. Because the added material will be covering a surface that already has some absorption, the coefficient of the covered material (line 17) will have to be deducted from the coefficient of the added material (line 16). The result is the *effective* absorption coefficient of the new material being added, shown on line 18. The area of the acoustical material to be added is calculated by dividing the number of sabins needed by the effective coefficient at 500 Hz. This area is shown on line 18.

The added absorption at the other frequencies is determined by multiplying their respective effective coefficients with the area shown on line 18. The added absorption for all frequencies is then added to the sabins shown on line 11 and new reverberation times calculated. The new reverberation times should be plotted on the ROOM CALCULATIONS STEP FOUR sheet, along with the recommended reverberation times.

It is unlikely that the reverberation times will exactly agree with the recommended times except at 500 Hz. If there is a major deviation at some other frequency, a compromise may be in order. For example, if the final reverberation times at the low frequencies are too long, absorption can be enhanced by providing an airspace behind the added material (see Chapter 2, Figure 2-5). The high frequency times are likely to be lower than the criteria because of air absorption.

The process of balancing the reverberation times at the various frequencies is an exercise in judgment as to which materials to use and an appropriate area for each. This process may require the use of two or more different types of materials to secure a satisfactory balance.

### 5.9.6 Final Design Checklist

The following checklist is a guide which can be used to make sure that the design has considered all important acoustical concerns.

1. Select reverberation times appropriate for major uses.
2. Be sure volume is adequate for desired reverberation times.
3. Allow for absorption of air at selected relative humidity.
4. Provide useful reflections for reinforcement of the direct sound and uniform sound distribution.
5. Avoid opposite, parallel hard surfaces (flutter).
6. Watch out for excessively delayed reflections (echoes).
7. Avoid concave surfaces (focusing).
8. Ensure adequate space for sound and light control rooms.
9. Design a suitable sound reinforcement/reproduction system.
10. Include an orchestra pit and trap room if required.
11. Provide a stage shell for music performances.
12. Select maximum background noise levels (RC or NC values). Advise mechanical engineer.
13. Provide room absorption data for use by mechanical and sound system engineers for calculation of room noise and sound system parameters.
14. Control noise from HVAC, plumbing, elevators and dimmers.
15. Control noise from ancillary spaces (scene shop, rehearsal room, dressing rooms, lobby, green room, etc.).
16. Control noise from external sources (surface traffic, aircraft). Stage house smoke vents can be a problem.

**ROOM CALCULATIONS STEP ONE**

**PROGRAM SHEET**

**JOB NO:** ______
**SHEET NO:** *1*
**DATE:** *1997*

ROOM: *Auditorium* BUILDING *School*
PROJECT *Sample* CLIENT ______
CALC. BY *Irvine* CHECKED BY *Richards*

| | | | |
|---|---|---|---|
| 1. Uses | *Multipurpose - Speech, Music & Drama* | | |
| | | **Design Criteria** | **Final Design** |
| 2. Number of Seats | # | *2200 seats* | *2200 seats* |
| 3. Area per Seat | A/# | *6.5 sq ft/seat* | *6.54 sq ft/seat* |
| 4. Floor Area | A | *14,300 sq ft (main floor 11,520)* | *14, 391 sq ft (main floor 10,874)* |
| 5. Volume per Seat | V/# | *220 cu ft/seat* | *224cu ft/seat* |
| 6. Room Volume | V | *484,000 cu ft* | *492,480 cu ft* |
| 7. Room Width | W | *96 ft* | *96 ft* |
| 8. Room Length | L | *120 ft* | *120 ft* |
| 9. Room Height | H | *42 ft* | *42 ft* |
| 10. Room Dimension Ratios | H:W:L | *1 : 2.2 : 2.8* | *1 : 2.2 : 2.8* |
| 11. Stage Opening Dimensions | W X H | *50 ft X 30 ft* | *50 ft X 30 ft* |
| 12. Reverberation Time 500 Hz | $RT_{500}$ | *1.5 sec* | *1.5 sec* |
| 13. Reverberation Time 125 Hz | $RT_{125}$ | *2.25 sec* | *2.52 sec* |
| 14. Constant used in RT Calculation | .049V (.161V) | *23,716* | *24,132* |
| 15. Sabins Needed at 500 Hz | $.049V/RT_{500}$ ($.161V/RT_{500}$) | *15,810 sabins* | *16,088 sabins* |
| 16. Sabins Needed at 125 Hz | $.049V/RT_{125}$ ($.161V/RT_{125}$) | *10,540 sabins* | *10,725 sabins* |

**AIR ABSORPTION CALCULATIONS:**

Relative
Humidity Percentage:

*50* %

| | 1000 | 2000 | 4000 |
|---|---|---|---|
| 4m | *1.17* | *2.93* | *7.28* |
| $\frac{4mV}{1000}$ | *576* | *1,443* | *3,585* |

## ROOM CALCULATIONS STEP TWO

## TAKE-OFF SHEET

JOB NO: ____________

SHEET NO: 1

DATE: 1997

ROOM: Auditorium — BUILDING: School

PROJECT: Sample — CLIENT: ____________

CALC. BY: Irvine — CHECKED BY: Richards

| # | LOCATION | DIMENSIONS | AREA | MATERIALS | | | | | | | | |
|---|---|---|---|---|---|---|---|---|---|---|---|---|
| | | | | Conc | GWB | | open | door | wood | panel | cpt | |
| 1 | Main Floor | per drawings | 10,874 | 10874 | | | | | | | | |
| 2 | Balcony Floor | " | 3,517 | 3517 | | | | | | | | |
| 3 | Ceiling | " | 10,874 | | 10874 | | | | | | | |
| 4 | Under Bal. Ceiling | " | 3517 | | 3517 | | | | | | | |
| 5 | Rear Wall bel. bal. | " | 1,344 | | 1344 | | | | | | | |
| 6 | Rear Wall ab. bal. | " | 1,920 | | 1920 | | | | | | | |
| 7 | Balcony Front | " | 585 | | 585 | | | | | | | |
| 8 | F side to bal. | " | 3,910 | | 3910 | | | | | | | |
| 9 | S wall bel. bal. | " | 404 | | 404 | | | | | | | |
| 10 | S wall ab. bal | " | 740 | | 740 | | | | | | | |
| 11 | Front wall area | " | 1,768 | | 1768 | | | | | | | |
| 12 | | TOTAL SURFACE AREA: | 39,453 | | | | | | | | | |
| 13 | | | | | | | | | | | | |
| 14 | Stage Opening | 20 X 500 | 1500 | | -1500 | | 1500 | | | | | |
| 15 | Doors | 10 X 6 X 7 | 420 | | -420 | | | 420 | | | | |
| 16 | Wainscott | 4 X 340 | 1,360 | | -1360 | | | | 1360 | | | |
| 17 | Diffusor Panels | 12 X 6 X 10 | 720 | | -720 | | | | | 720 | | |
| 18 | Carpet | 3 X 6X 149 | 2,682 | -2682 | | | | | | | 2682 | |
| 19 | | AREA TOTALS: | | 11709 | 21062 | | 1500 | 420 | 1360 | 720 | 2682 | |
| 20 | | | | | | | | | | | | |
| 21 | | TOTAL SURFACE AREA: | 39,453 (Check: Line 12 equals the total of the areas in line 19) | | | | | | | | | |
| 22 | | | | | | | | | | | | |
| 23 | | | | | | | | | | | | |
| 24 | | | | | | | | | | | | |
| 25 | | | | | | | | | | | | |
| 26 | | | | | | | | | | | | |
| 27 | | | | | | | | | | | | |

ACTUAL FLOOR AREA= 14,391 SQ FT

ACTUAL SURFACE AREA= 39,453 SQ FT

ACTUAL ROOM VOLUME= 492,480 CU FT

**ROOM CALCULATIONS STEP THREE**

**REVERBERATION TIME CALCULATIONS**

**JOB NO:** ________

**SHEET NO:** 1

**DATE:** 1995

ROOM: Auditorium

BUILDING: School

PROJECT: Sample

CLIENT: ________

CALC. BY: Irvine

CHECKED BY: Richards

| # | LOC. | MTL | AREA. | α | 125 | α | 250 | α | 500 | α | 1000 | α | 2000 | α | 4000 |
|---|---|---|---|---|---|---|---|---|---|---|---|---|---|---|---|
| 1 | Floor | Conc. | 11709 | .02 | 234 | .02 | 234 | .03 | 351 | .03 | 351 | .03 | 351 | .03 | 351 |
| 2 | Walls/clg | GWB | 21062 | .13 | 2378 | .07 | 1512 | .06 | 1263 | .06 | 1263 | .05 | 1053 | .06 | 1263 |
| 3 | Stage | Opening | 1500 | .25 | 375 | .3 | 450 | .4 | 600 | .45 | 675 | .45 | 675 | .45 | 675 |
| 4 | Doors | Wood | 420 | .15 | 63 | .08 | 34 | .05 | 21 | .04 | 17 | .03 | 13 | .04 | 17 |
| 5 | Wainscott | Wood | 1360 | .13 | 177 | .07 | 95 | .06 | 82 | .06 | 82 | .05 | 68 | .06 | 82 |
| 6 | Panels | GWB | 720 | .13 | 94 | .07 | 50 | .06 | 43 | .06 | 43 | .05 | 36 | .06 | 43 |
| 7 | Aisles | Carpet | 2682 | .05 | 134 | .07 | 187 | .17 | 456 | .33 | 885 | .58 | 1556 | .75 | 2467 |
| 8 | Seating | Occupied | 7845 | .52 | 4079 | .68 | 5335 | .86 | 6747 | .97 | 7609 | .92 | 7217 | .85 | 6668 |
| 9 | | Empty | 3863 | .44 | 1699 | .60 | 2318 | .77 | 2975 | .89 | 3438 | .82 | 3167 | .70 | 2704 |
| 10 | | | Air | | | | | | | | 576 | | 1433 | | 3585 |
| 11 | | | sabins | | 9,593 | | 10,214 | | 12,538 | | 14,943 | | 15,569 | | 17,859 |
| 12 | | | $R_T$ | | 2.52 | | 2.36 | | 1.92 | | 1.61 | | 1.55 | | 1.35 |
| 13 | | | | | | | | | | | | | | | |
| 14 | | sabins design | | | | | | | 16,088 | | | | | | |
| 15 | | sabins short | | | | | | | 3,550 | | | | | | |
| 16 | Ac. tile absorption added | | | .10 | | .24 | | .73 | | .95 | | .86 | | .80 | |
| 17 | less absorption area covered | | | .13 | | .07 | | .06 | | .06 | | .05 | | .06 | |
| 18 | Abs. add | Ac. Tile | 5299 sq ft | 0 | 0 | .17 | 901 | .67 | 3550 | .81 | 4292 | .81 | 4292 | .74 | 3921 |
| 19 | New Total Sabins | | | | 9,593 | | 11,115 | | 16,088 | | 19,235 | | 19,861 | | 21,780 |
| 20 | | | $R_T$ | | 2.52 | | 2.17 | | 1.5 | | 1.25 | | 1.22 | | 1.11 |
| 21 | | | | | | | | | | | | | | | |
| 22 | | | | | | | | | | | | | | | |
| 23 | | | | | | | | | | | | | | | |
| 24 | | | | | | | | | | | | | | | |
| 25 | | | | | | | | | | | | | | | |
| 26 | | | | | | | | | | | | | | | |
| 27 | | | | | | | | | | | | | | | |

**ROOM CALCULATIONS STEP FOUR**

**RECOMMENDED REVERBERATION TIMES**

**JOB NO:** ______

**SHEET NO:** 1

**DATE:** 1995

ROOM: Auditorium BUILDING School

PROJECT Sample CLIENT ______

CALC. BY Irvine CHECKED BY Richards

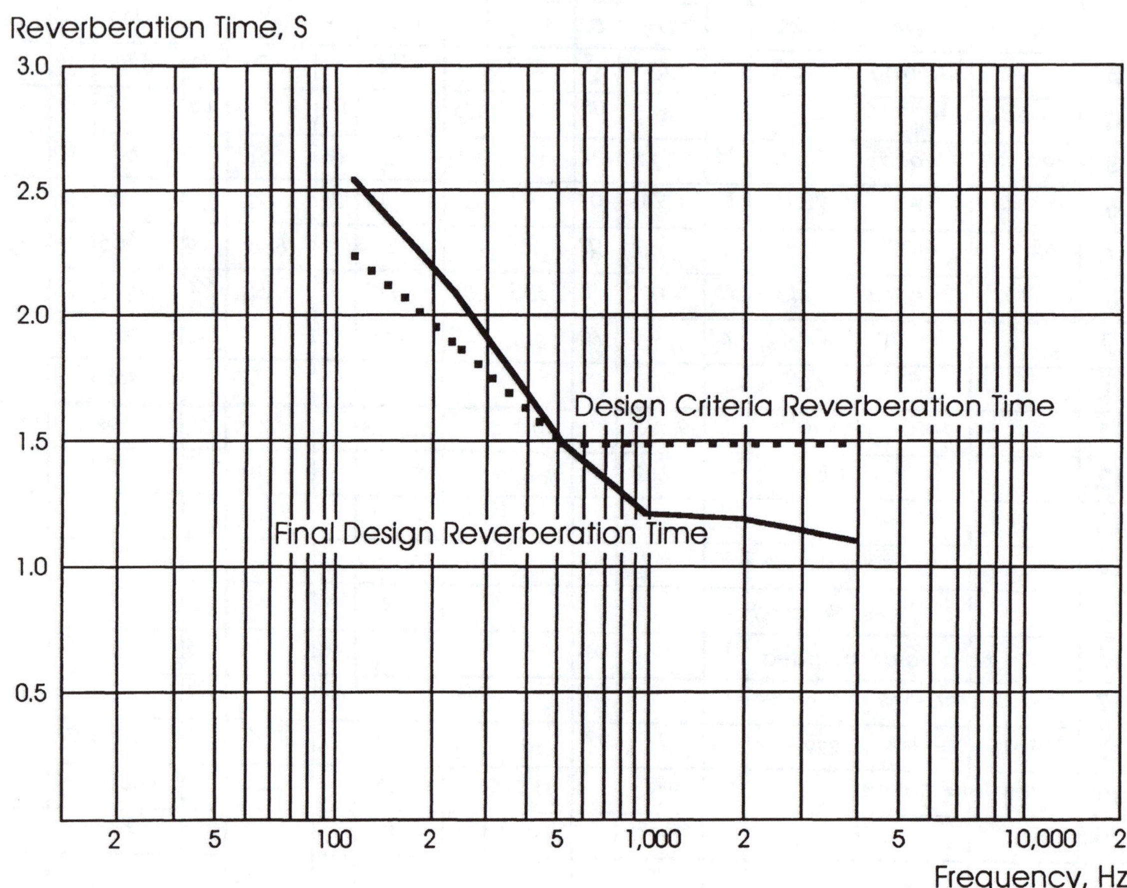

**ROOM CALCULATIONS STEP ONE**

**PROGRAM SHEET**

**JOB NO:** ________
**SHEET NO:** ________
**DATE:** ________

ROOM: ________________ BUILDING ________________

PROJECT ________________ CLIENT ________________

CALC. BY ________________ CHECKED BY ________________

| | | | |
|---|---|---|---|
| 1. Uses | | | |
| | | **Design Criteria** | **Final Design** |
| 2. Number of Seats | **#** | | |
| 3. Area per Seat | **A/#** | | |
| 4. Floor Area | **A** | | |
| 5. Volume per Seat | **V/#** | | |
| 6. Room Volume | **V** | | |
| 7. Room Width | **W** | | |
| 8. Room Length | **L** | | |
| 9. Room Height | **H** | | |
| 10. Room Dimension Ratios | **H:W:L** | | |
| 11. Stage Opening Dimensions | **W X H** | | |
| 12. Reverberation Time 500 hz | $RT_{500}$ | | |
| 13. Reverberation Time 125 hz | $RT_{125}$ | | |
| 14. Constant used in RT Calculation | **.049V (.161V)** | | |
| 15. Sabins Needed at 500 hz | $.049V/RT_{500}$ ($.161V/RT_{500}$) | | |
| 16. Sabins Needed at 125 hz | $.049V/RT_{125}$ ($.161V/RT_{125}$) | | |

**AIR ABSORPTION CALCULATIONS:**

Relative
Humidity Percentage:

______________ %

| | 1000 | 2000 | 4000 |
|---|---|---|---|
| **4m** | | | |
| $\frac{4mV}{1000}$ | | | |

**ROOM CALCULATIONS STEP TWO**

**TAKE-OFF SHEET**

**JOB NO:** ________

**SHEET NO:** ________

**DATE:** ________

ROOM: ______________________ BUILDING ______________________

PROJECT ______________________ CLIENT ______________________

CALC. BY ______________________ CHECKED BY ______________________

| | | | | MATERIALS | | | | | | | | |
|---|---|---|---|---|---|---|---|---|---|---|---|---|
| # | LOCATION | DIMENSIONS | AREA | | | | | | | | | |
| 1 | | | | | | | | | | | | |
| 2 | | | | | | | | | | | | |
| 3 | | | | | | | | | | | | |
| 4 | | | | | | | | | | | | |
| 5 | | | | | | | | | | | | |
| 6 | | | | | | | | | | | | |
| 7 | | | | | | | | | | | | |
| 8 | | | | | | | | | | | | |
| 9 | | | | | | | | | | | | |
| 10 | | | | | | | | | | | | |
| 11 | | | | | | | | | | | | |
| 12 | | | | | | | | | | | | |
| 13 | | | | | | | | | | | | |
| 14 | | | | | | | | | | | | |
| 15 | | | | | | | | | | | | |
| 16 | | | | | | | | | | | | |
| 17 | | | | | | | | | | | | |
| 18 | | | | | | | | | | | | |
| 19 | | | | | | | | | | | | |
| 20 | | | | | | | | | | | | |
| 21 | | | | | | | | | | | | |
| 22 | | | | | | | | | | | | |
| 23 | | | | | | | | | | | | |
| 24 | | | | | | | | | | | | |
| 25 | | | | | | | | | | | | |
| 26 | | | | | | | | | | | | |
| 27 | | | | | | | | | | | | |

ACTUAL FLOOR AREA= ______________________

ACTUAL SURFACE AREA= ______________________

ACTUAL ROOM VOLUME= ______________________

## ROOM CALCULATIONS STEP THREE

## REVERBERATION TIME CALCULATIONS

JOB NO: ________

SHEET NO: ________

DATE: ________

ROOM: ________________ BUILDING ________________

PROJECT ________________ CLIENT ________________

CALC. BY ________________ CHECKED BY ________________

| # | LOC. | MTL | AREA. | α | 125 | α | 250 | α | 500 | α | 1000 | α | 2000 | α | 4000 |
|---|---|---|---|---|---|---|---|---|---|---|---|---|---|---|---|
| 1 | | | | | | | | | | | | | | | |
| 2 | | | | | | | | | | | | | | | |
| 3 | | | | | | | | | | | | | | | |
| 4 | | | | | | | | | | | | | | | |
| 5 | | | | | | | | | | | | | | | |
| 6 | | | | | | | | | | | | | | | |
| 7 | | | | | | | | | | | | | | | |
| 8 | | | | | | | | | | | | | | | |
| 9 | | | | | | | | | | | | | | | |
| 10 | | | | | | | | | | | | | | | |
| 11 | | | | | | | | | | | | | | | |
| 12 | | | | | | | | | | | | | | | |
| 13 | | | | | | | | | | | | | | | |
| 14 | | | | | | | | | | | | | | | |
| 15 | | | | | | | | | | | | | | | |
| 16 | | | | | | | | | | | | | | | |
| 17 | | | | | | | | | | | | | | | |
| 18 | | | | | | | | | | | | | | | |
| 19 | | | | | | | | | | | | | | | |
| 20 | | | | | | | | | | | | | | | |
| 21 | | | | | | | | | | | | | | | |
| 22 | | | | | | | | | | | | | | | |
| 23 | | | | | | | | | | | | | | | |
| 24 | | | | | | | | | | | | | | | |
| 25 | | | | | | | | | | | | | | | |
| 26 | | | | | | | | | | | | | | | |
| 27 | | | | | | | | | | | | | | | |

**ROOM CALCULATIONS STEP FOUR**

**RECOMMENDED REVERBERATION TIMES**

**JOB NO:** ________

**SHEET NO:** ________

**DATE:** ________

ROOM: ____________________ BUILDING ____________________

PROJECT ____________________ CLIENT ____________________

CALC. BY ____________________ CHECKED BY ____________________

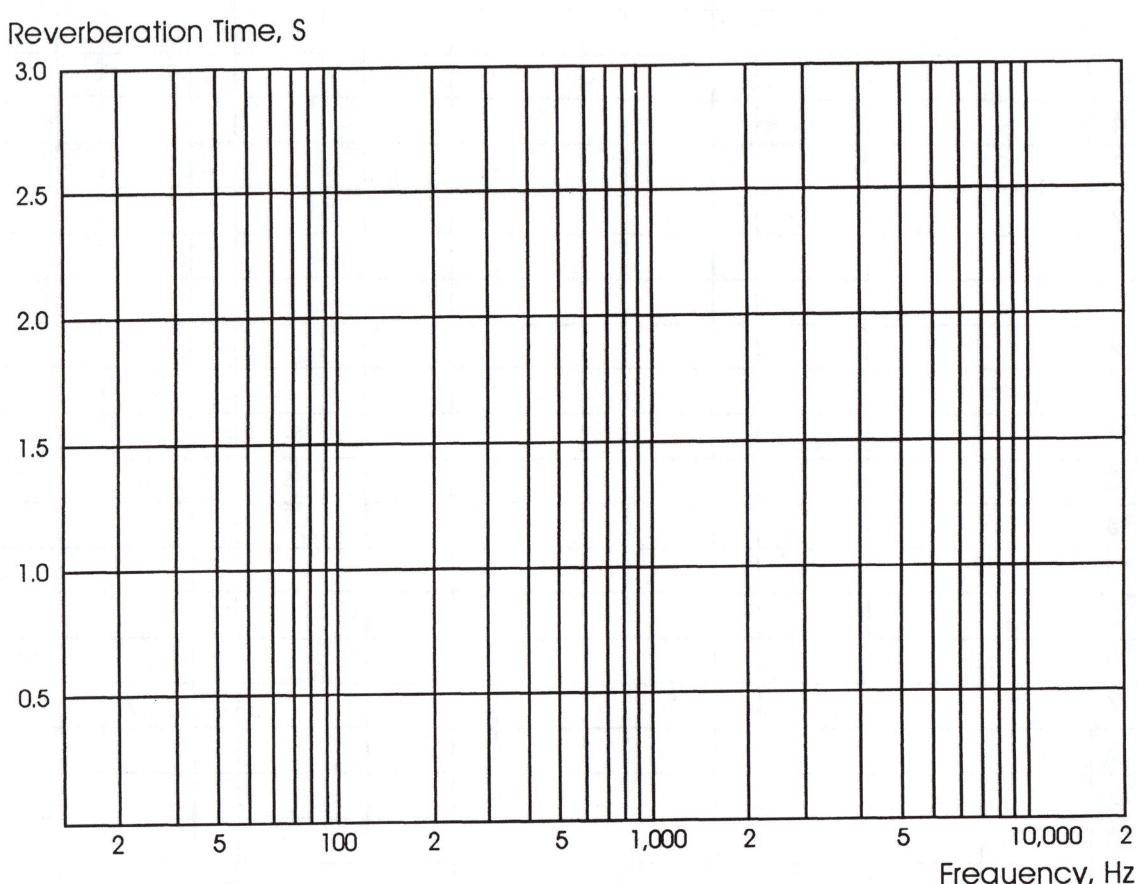

# Chapter 6

# Electroacoustic Systems

Electroacoustic systems serve many purposes in buildings. A primary function is that of sound reinforcement. Effective sound reinforcement is essential when the source is not loud enough for comfortable listening and good intelligibility, when the source is too far away, or when the background noise is too high. A sound system may also be used to assist in a space with acoustical problems, but the preferred method is to first correct the acoustical problems and then install a sound system if required. Spaces requiring sound reinforcement range all the way from conference and board rooms to sports stadiums seating 50,000 or more.

Sound systems are used to reproduce and distribute recorded sound as in a movie theater, for sound effects in a drama theater, and for audio/visual or multimedia presentations.

More specialized sound systems may be used to provide masking to enhance sound privacy in an open office, or may be used to enhance the acoustical environment for traditional live music performances in spaces that lack desirable acoustical qualities.

It is not the intention of this Chapter to enable someone to design and specify sound systems, but rather to inform the building designer of what the systems are used for, basic system requirements, and how they may be properly integrated into building design.

## 6.1 Sound Reinforcement Systems

Sound reinforcement systems consist of three major components:

1. The sound or audio signal source, which may be live, picked up by a microphone or microphones, electronic musical instruments, or it may be prerecorded or computer generated.
2. Signal processing, which may include mixing of several sources, amplification, equalization, volume compression and limiting, audio signal delays and computer control.
3. Loudspeakers, which convert the electrical audio signal to audible sound and distribute this sound to the intended listeners. The selection and placement of the loudspeakers is the single most critical element in the sound system and will generally be a key factor in the success or failure of a sound system to meet user requirements.

There are three basic types of sound systems, relating to the type of loudspeaker arrangement: the central or single-source type, the stereophonic type, and the distributed type. The single-source type utilizes an array or cluster of loudspeakers at a single location designed to deliver sound uniformly to all listeners. Stereo systems are, in effect, two separated single-source systems, since each system, or channel, is independent and must deliver uniform sound to the entire listening area. The distributed type of system utilizes many loudspeakers distributed throughout the listening area, each speaker serving a portion of the area. Many systems utilize an appropriate combination of single-source and distributed speakers. Figures 6-1, 6-2, 6-3 and 6-4 show loudspeaker placement and coverage in single-source, distributed, combination and stereo systems, respectively.

The major objective of a sound reinforcement system is, in effect, to move the listener's ears to a favorable listening position. This means the sound system must produce the required loudness and fidelity of the original sound with an adequate ratio of direct to reverberant sound for every listener. A sound system can be no better than its weakest link, and the weak link more often than not is the selection and placement of the loudspeakers.

Good sound system design starts with the final element of the system, the selection and placement of the loudspeakers, a process greatly enhanced by state-of-the-art computer modeling programs which produce plots of the expected sound coverage throughout the audience area, and can predict the speech intelligibility at various listener locations as well. The design then proceeds towards the front end or input side of the system, first through the power amplifiers that drive the loudspeakers, then the

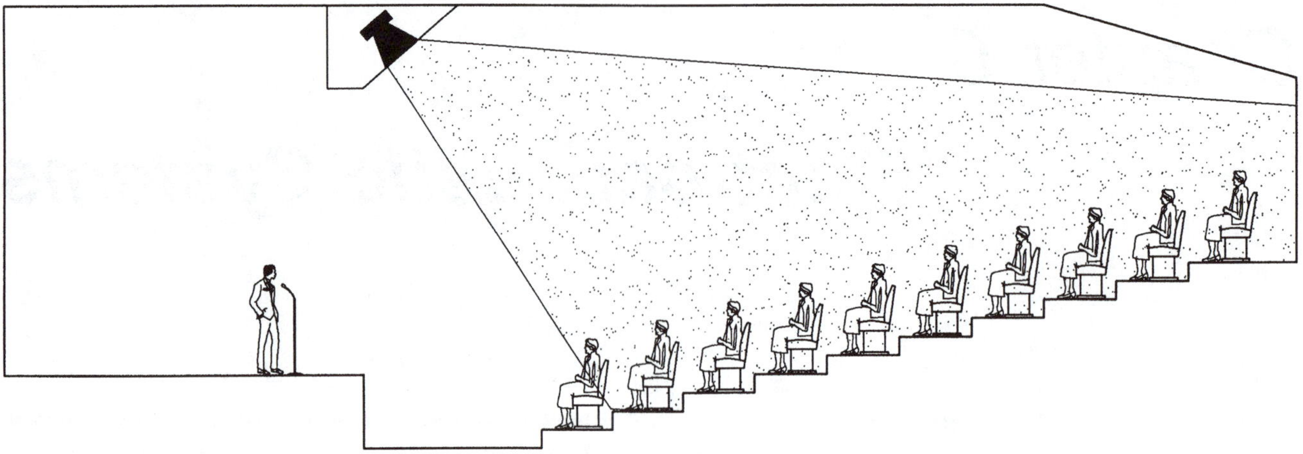

The loudspeaker must be high enough to provide fairly uniform coverage to the audience.

**Figure 6-1.** Single source sound system.

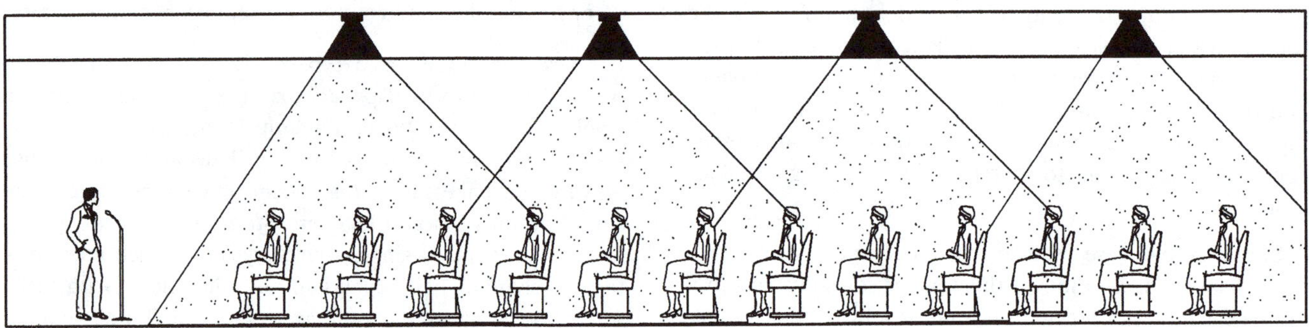

With low ceilings, loudspeakers are distributed to provide uniform coverage to the audience.

**Figure 6-2.** Distributed sound system.

signal processing, and finally the microphones or other sources.

Most sound systems for reinforcement of speech are single channel or monophonic, while other systems, such as for reproducing music, may require multiple channels for stereo, surround sound, or special effects. Traditional stereo systems have a limited listening area since for good stereo effects, all listeners need to have equal opportunity to hear the left and right channels. New system design techniques can provide good stereo imaging for large audiences.

A sound system can perform no better than how it is operated. Certain types of systems can be designed to be essentially "hands free" by utilizing automatic microphone mixers that turn up the gain of a particular microphone only when there is a sound being produced at that microphone. This feature is necessary to minimize undesired pickup of room noise and reverberation, and to maximize system gain before feedback. Such systems also utilize automatic gain controls to even out sound levels. Typical applications are board rooms, council chambers, lecture rooms, courtrooms, meeting rooms, and in houses of worship for simple services.

Sound systems used for reinforcement of live performances require an operator utilizing a mixing board or console. To operate a system properly the operator must be located so that he or she hears what the audience hears. This cannot be accomplished by placing the operator in a sound isolated control booth. Provision should be made to locate the operator somewhere in the seating area.

Probably the most common complaint with sound reinforcement systems is inadequate gain before feedback. Evidence of this is the familiar howling or squealing before the system has adequate gain. Feedback occurs when the sound coming out of the system loudspeakers is picked up by the system microphones and is reamplified. Some of the techniques used to optimize system gain before feedback include the following:

1. Use of directional loudspeakers that will deliver direct sound uniformly throughout the audience area, with a

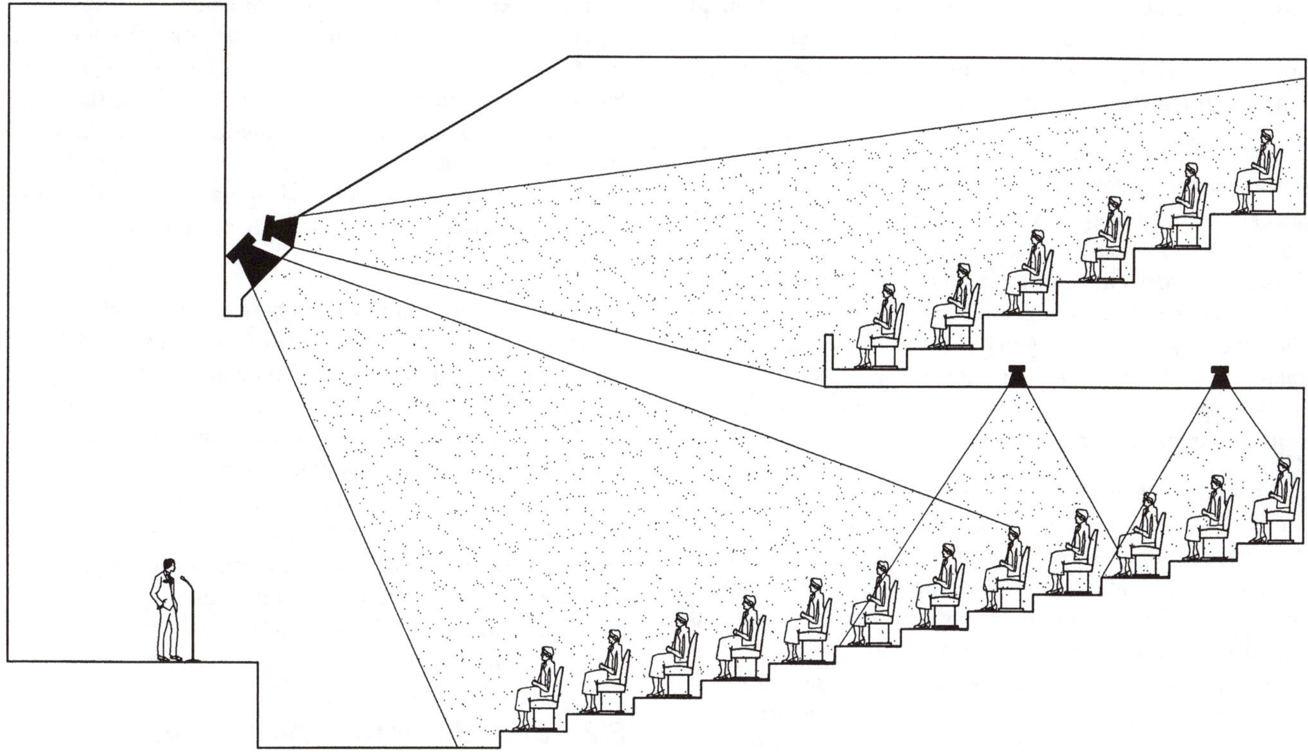

An array of loudspeakers is required at the proscenium for both main floor and balcony coverage. Because the main floor seating under the balcony is not covered by the proscenium loudspeakers, a distributed system, appropriately delayed, is required at this location.

**Figure 6-3.** Combination single source and distributed sound system.

minimum of sound energy directed back into the microphones or at reflective wall or ceiling surfaces.

2. Use of high quality directional microphones that are less sensitive to sound not coming from the person talking.
3. Minimizing the distance from the loudspeakers to the listeners, so the system requires less gain. (The direct sound must arrive at the listener before the reinforced sound.)
4. Minimizing the distance from the sound source to the microphones. With a higher input signal, the system requires less gain.
5. Maximizing the distance from the loudspeakers to the microphones, and by avoiding overlap of coverage patterns.
6. Minimizing the number of microphones that are on at any given time. Each additional microphone that is on and is not required for pickup contributes to system noise.
7. Equalizing the system to have the smoothest possible frequency response.
8. Utilizing state-of-the-art automatic feedback controllers.

Many times a performer will desire "foldback," utilizing stage monitor loudspeakers which can result in higher sound levels at the performers ears than that resulting from the house loudspeakers. Exposure to high sound levels over an extended period of time can cause permanent hearing damage. Performer foldback can be provided by individual in-the-ear monitors which can produce the de-

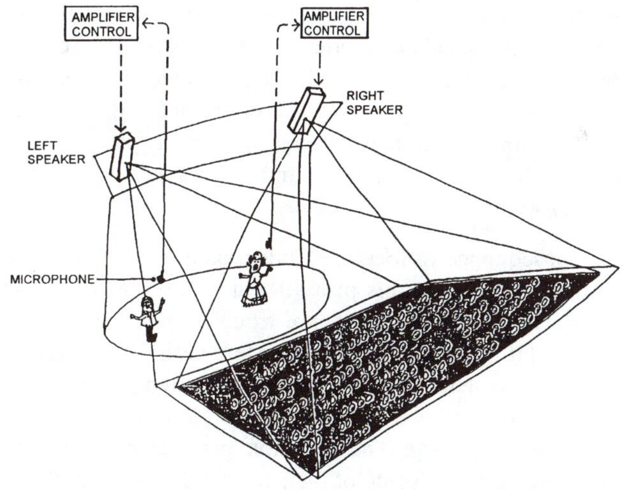


Each loudspeaker must cover the entire audience area.

**Figure 6-4.** Stereo sound reinforcement system.

sired results at much safer sound levels. Introduction of a modest amount of audio signal delay to stage monitor speakers may also permit lower monitor sound levels while still satisfying the performer's foldback requirements.

High powered sound systems can put an endurance strain on the system loudspeakers. Ironically, more loudspeakers are damaged by insufficient amplifier power. In an attempt to achieve the desired sound levels, the amplifiers are overdriven resulting in an effect called "clipping," which can be devastating to loudspeakers. In high performance systems it is good practice to use amplifiers rated at two to four times the continuous power rating of the loudspeakers and make effective use of limiters to prevent the amplifiers from being driven into clipping.

Another consideration with respect to the placement of loudspeakers is that of correct directional information. The most realistic sound reinforcement is that for which the amplified sound appears to come from the originating sound source. In order to accomplish this, the amplified sound must come from the same direction as the original sound. To enhance the effect even further, the amplified sound should arrive at the listener just after the arrival of the sound from the source. If it arrives too late after the original source, there will be a muddying of the sound and in the extreme case will be perceived as an echo.

In many situations, the original sound may be too weak at the listener. However, where the original sound is sufficiently audible, the best practice is to locate the central or single source loudspeakers above the performing area, directed downward at the audience so that the distance from the loudspeakers to the listener is slightly greater than from the performer to the same listener. If circumstances dictate that the loudspeakers be located forward of the performing area, electronic signal delay can be utilized so that the amplified sound arrives just after the original sound.

A consequence of locating loudspeakers above the performing area is that this places them in a prominent visual location. Many architects are troubled by this requirement, and it behooves the sound system designer to advise the architect of this necessity early in design so that the loudspeakers can be effectively integrated into the overall design of the space. The common practice of locating loudspeakers at each side of the performing area is not recommended, as this produces incorrect direction information for many listeners, and makes it more difficult if not impossible to achieve uniformity of coverage and sound quality throughout the audience area.

Electronic signal delay is a very valuable tool which can be used to synchronize the sound from different loudspeakers that may be delivering sound to the same listener, such as seating areas under a balcony overhang where most sound comes from speakers in the ceiling overhead (see Figure 6-3). Even if the sound coming from the source at the front of the hall (live talker or loudspeaker) is much lower in level compared to sound coming from the much closer overhead speakers, if the sound from overhead arrives just after the frontal sound, the perception is that all the sound is coming from the original source. This phenomenon is related to the "Haas effect," discussed in Chapter 4, Section 4.1.

Until recently, sound systems have been assembled from separate components. The advent of audio manipulation in the digital domain has made intelligent "black boxes" that can be programmed to perform the functions of many individual components and be instantly reconfigured to provide other functions. Computer control and monitoring of sound systems is seeing increasing application, particularly in large systems. However, at this writing, a lack of standards for computer control is resulting in incompatibility between various proprietary systems. Even so, digital signal processing and computer control are the wave of the future in sound systems.

## 6.2 Sound Reproduction Systems

While sound reinforcement systems have audio playback capability, certain types of sound systems may serve primarily for sound reproduction. A typical example would be a movie theater. Cinema sound systems were originally all single channel, but now multichannel systems such as THX and Dolby stereo with surround sound are utilized in IMAX and OMNIMAX theaters. Home theater systems are becoming as sophisticated as movie theater systems, but frequently are constrained by limitations of the acoustical environment into which the systems are installed. Even sound reproduction systems now available in many automobiles rivals that of sophisticated home theater systems in spite of a very restrictive listening environment.

Multimedia presentation systems for corporate board rooms require high quality audio reproduction capabilities which must accommodate a variety of program sources such as tape (both analog and digital), laser disks, compact disks and computer-generated sound.

In order to realize the maximum potential of a high quality sound reproduction system, appropriate attention must be paid to the acoustics of the space that encloses the system and the listeners. There must be no distracting intrusion of noise into the listening environment. Other occupied spaces adjacent to the listening room must also be protected from high levels of sound, particularly in the low frequency bass range that state-of-the-art audio systems can generate. The listening room dimension ratios—length to width to height—must be selected so that natural room resonances or modes are appropriately distributed over the frequency range, thus insuring a smooth frequency response (see Chapter 4, Section 4.5).

Discrete sound reflections from hard surfaces in listening rooms can combine with the direct sound from the system loudspeakers to cause undesired frequency response anomalies called comb filtering, or room coloration. An ideal situation for complete freedom from such coloration would be to have the room disappear acoustically and have the ambience generated by the surround tracks of the program material being reproduced. However, such a space would be so acoustically "dead" it would be unpleasant to be in when no sound with recorded ambience was being played. A reasonable compromise is to design a room incorporating both absorptive and diffusing treatments. Appropriate materials are presented in Chapter 2, Section 2.5.2.

## 6.3 Sound Masking Systems

A specialized type of sound system provides sound masking, and is used primarily for providing an unobtrusive, steady background noise in spaces that would otherwise be so quiet that discrete sounds may be distracting or annoying. The most common application is in open offices where the lack of full-height partitions results in insufficient sound privacy between work stations. While acoustical treatments such as sound absorbing space dividers, highly efficient acoustical ceilings and acoustical wall panels all help to reduce sound propagation between work stations, such treatment is still inadequate to achieve optimum acoustical privacy. (See Chapter 5, Section 5.3.2.)

A sound masking system utilizes loudspeakers suspended above the acoustical ceiling to evenly distribute an unobtrusive noise adjusted to a precise frequency spectrum and sound level to make the discrete sound less intrusive. Uniform distribution is very important so that someone moving through the space is not aware of changes in the noise. While by design the masking system is a very effective audio distribution system, it probably should not be used to provide paging announcements or background music because this calls attention to loudspeaker locations, which should remain anonymous.

Masking systems can be effective for closed offices, multifamily dwellings, clinics and hospitals, and other uses where the acoustical separation is inadequate. Masking can be used in spaces adjacent to military briefing rooms to enhance security, and in courtrooms when the judge and attorneys want to confer without the jurors hearing, but don't want to excuse the jury. The judge can switch in masking for the jury area so that they are unable to hear the low level conversation from the bench.

Special design considerations must be applied where there is no suspended ceiling to hide the system loudspeakers, or where the space above the ceiling is a return air plenum. In the latter case, a particular concern is to avoid "hot spots" resulting from the masking noise spilling through unbaffled air return grilles in the ceiling. The solution is to construct a baffle that attenuates the noise coming through the grille without impeding air flow.

It is very important that the noise spectrum produced by the masking system has the specified characteristics and be free of any discernible artifacts or colorations. There must be no information content in the masking noise. It is important that the masking noise have adequate low frequency content lest it have an objectionably harsh characteristic. For these reasons, background music is not a suitable masking signal. An acceptable noise spectrum closely follows an RC-40 contour (see Chapter 1, Figure 1-14), except the slope may be slightly steeper above 2000 Hz. It is important that the low frequency response of the system follow the contour down to about 100 Hz. Such a spectrum will have an A-weighted sound level of about 47 dBA. The level should be within the range of 46 to 48 dBA.

The sound level produced by a masking system is very critical. If it is too low the system will be ineffective, and if too high the masking noise becomes objectionable. The window of acceptable masking noise level may be only a few decibels wide. A masking system which is improperly designed, installed or operated can be worse than no masking at all. However, a good masking system can be a very cost-effective solution to a sound privacy problem.

In large spaces or buildings having several areas to receive masking, the system will utilize centralized equipment consisting of a noise generator, equalizers to set the noise spectrum and amplifiers to distribute the signal to loudspeakers above the suspended ceiling. The system may be zoned to facilitate independent adjustment of sound levels in different areas. If some areas served by the system have acoustical ceilings of a different type than other areas, each area must be served by its own equalizer and amplifier to adjust for the different sound transmission characteristics of the ceiling materials.

There are differing design philosophies regarding masking systems. Some designers advocate utilizing several noise generators feeding alternate loudspeakers in a given area. The contention is that this makes it more difficult to tell where the noise is coming from, and if there is a failure of a component in a certain noise channel, the area will not be totally devoid of masking noise. The effect can be rather remarkable if the masking suddenly is gone.

Other designers advocate automatic variation in the masking noise level with the time of day or level of activity in the area. The danger here is that any changes in system characteristics may call attention to the system itself, while the objective is to have the system "disappear" from the acoustical environment of the area being served. Variable level masking systems are not recommended.

## 6.4 Electroacoustic Enhancement Systems

The concept of using a sound system to change the perceived acoustical characteristics of a space is not new. Over a period of several decades many systems have been developed and implemented with varying degrees of success. Many systems have been very costly and sophisticated, as well as difficult to operate and maintain.

There has been a prevailing attitude that sound enhancement systems are artificial and therefore unacceptable regardless of how well they may perform. The advent of advanced digital signal processing as well as the development of better system components, particularly loudspeakers, is bringing this technology to a point where it will be economically viable and acceptable to the most critical listeners and performers.

Essentially such a system picks up the sound of the performers, processes it to add discrete delayed sound reflections and reverberation to achieve the desired acoustical environment, and reintroduces it to the space where it blends with the natural sound produced by the performers. Loudspeakers are placed with great care on wall and ceiling surfaces to simulate reflections and add reverberation. Since the signal is processed in the digital domain, it is a simple matter to vary the characteristics of the processed signal to emulate any desired acoustical environment. There is no feasible method by which passive acoustical adjustments within the space could achieve a comparable range of acoustical conditions. If it doesn't sound right, the signal parameters can be changed at the push of a button or the stroke of a key.

## 6.5 Design Coordination and Finalizing the Sound System

The most successful sound systems require full coordination with the architect, electrical engineer and theater consultant if there is one on the design team. Because loudspeaker placement is critical and often times in visually prominent locations, the architect should be so advised to permit full integration in the design. If loudspeakers are to be concealed, the enclosure must satisfy both visual and acoustical requirements. For background sound masking systems, the type of acoustical ceiling is critical to the design requirements of the system.

For theaters, design of the production intercom and house monitor/paging systems is usually a joint effort of the theater consultant and sound system designer.

The electrical engineer's documents will incorporate the sound system "rough-in" requirements, such as conduit size and runs, terminal boxes and power requirements. This information must be provided by the sound system designer. Pulling the cable through the conduit can be done by either the electrical or sound system subcontractor; this needs to be established prior to bidding,

The full potential of a sound system can be realized only if the system is properly adjusted and balanced at the completion of installation, but not before all room interior finishes are in place, particularly acoustical absorbents. This process will include exact aiming and setting of power levels of the system loudspeakers, accurate setting of audio signal delays, equalization of the system frequency response and adjustment of signal levels within the electronic portion of the system, a process referred to as gain staging. The effective testing and adjustment of a sound system requires specialized instrumentation and skills. Generally a team effort of the design consultant and contractor will yield the best results, and it is recommended that both parties be required to participate by specification.

It is very important to provide both written and hands-on operating instructions to the people responsible for running the system. Written operating instructions should be like a cookbook recipe that any literate person can follow and get the system working properly, even if they have never seen it before.

Equally important is as-built documentation of the system as well as the manufacturer's service manuals for the major system components. This information should be complete enough so that a competent technician who is unfamiliar with the system can troubleshoot and service the system effectively. The system should not be accepted and final payment withheld until the operations and maintenance manuals have been reviewed and approved by the design consultant.

# Chapter 7

# Mechanical Systems Noise and Vibration Control

Building mechanical systems contain many potential sources of noise and vibration. System design, equipment selection, installation methods and operating parameters of all components must be considered to avoid problems. Attempting to correct for poor design or incorrect installation after construction can be very expensive, and may not even be feasible to correct from an engineering standpoint.

An excellent source of information for the control of noise and vibration in mechanical systems is contained in publications issued by the American Society of Heating, Refrigerating and Air-Conditioning Engineers (ASHRAE). Chapter 7 of the 1993 Fundamentals Handbook, and Chapter 42 of the 1991 Applications Manual are the most recent versions as this Handbook is being prepared.

The ASHRAE manuals address all factors affecting acoustical performance, including appropriate noise and vibration criteria, equipment selection, system design, appropriate operating parameters and installation procedures. The reader is encouraged to consult these manuals for specific design information, performance data for noise and vibration generation of mechanical equipment, and acoustical characteristics of noise and vibration reduction materials applied to mechanical systems. The publications are periodically updated to account for information available on new mechanical equipment, and to present the results of ongoing research conducted by ASHRAE.

Figure 7-1 shows the frequency range of types of mechanical (and associated electrical) equipment. It may be possible to gain some knowledge of the source of noise just by listening to it. If the noise has a "hissy" quality, it may be from regeneration in the diffuser, while a lower frequency rumble may be due to noise from the fan or turbulence in air ducts. Like solving acoustical problems in architectural systems, frequency is an important consideration in the type and effectiveness of controls for both noise and vibration in mechanical systems.

## 7.1 Mechanical System Design Criteria

Because mechanical systems may produce both noise and vibration, it is necessary to assign appropriate criteria before beginning design of a system. Noise criteria are specified using either the RC or NC family of curves, or dBA sound levels, for designated uses. These noise descriptors were defined in Chapter 1, Sections 1.6 and 1.7, and the design criteria are presented in Appendix A.4. These criteria have evolved over a period of many years, and while they are intended primarily for mechanical system noise, they are often used for setting limits for other sources of noise such as traffic and machinery.

Design criteria for vibration associated with mechanical equipment or other vibrating surfaces have been developed in terms of machine "roughness" and human "feelability." Vibration criteria are discussed in Section 7.5.1.

The process of setting design criteria comes from a comprehensive understanding of the owner's or developer's intended use or program for the project. It is important that the owner is made aware of the cost and construction difficulties associated with a desired design objective. For example, designing an auditorium to achieve an RC-15 design curve will be considerably more expensive that meeting an RC-20 curve, while the latter would be acceptable for most types of programs. Mechanical penthouses for high-rise buildings are preferred by many mechanical engineers, but could involve costly floating floors to avoid problems in noise-sensitive spaces immediately below. Design-build projects may set structural limitations on space available for air distribution ducts, thus requiring high air velocities before space uses are known, with the possibility of noisy ducts in noise-sensitive areas.

## 7.2 Location of Mechanical Equipment

Common sense dictates that mechanical equipment should not be located close to noise-sensitive activities. Probably the most difficult design problem occurs when noisy and/or vibrating equipment is located above a critical space. Such a location requires extremely careful detailing for vibration isolation as well as appropriately designed supporting construction to retard transmission of airborne noise and structural vibration. The dynamic response of the supporting structure is often a very difficult engineering problem to solve (see Section 7.5.4).

Self-contained air handling units can be inexpensively located on roofs if they feed and return air to spaces directly below. The noise problem occurs because there is no opportunity to install noise reduction in the very short duct runs. Indeed, the return air is sometimes achieved simply by punching a hole in the roof below the unit, and drawing air from a plenum space above a suspended ceiling. Also, such units are frequently placed at mid-span on a frame roof structure, the worst location in terms of potential vibration problems. If possible, roof top equipment should be placed over supporting columns or walls, and not over noise-sensitive spaces.

Another problem created by having air-handling equipment too close to critical spaces is insufficient room for long, low velocity duct runs. If the distance from the discharge of a fan to the outlet diffuser is short, it may be difficult or even impossible to reduce the noise to acceptable levels. Large duct penetrations in the mechanical room wall can create a noise leakage path if the duct immediately enters a noise-sensitive space.

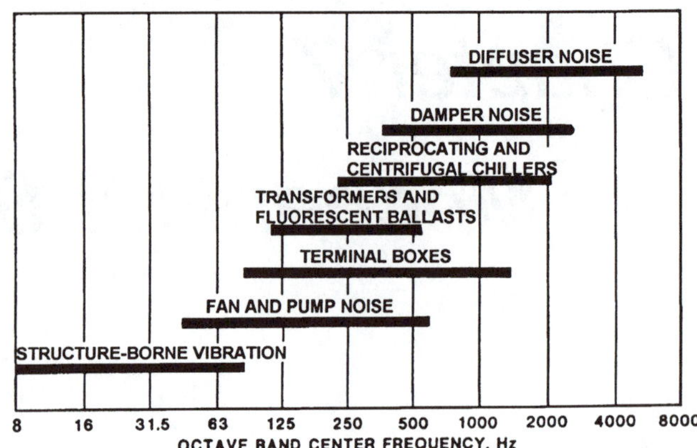


Frequencies at which various types of mechanical and electrical equipment generally dominates sound spectra.

**Figure 7-1.** Frequency range of mechanical and electrical equipment.

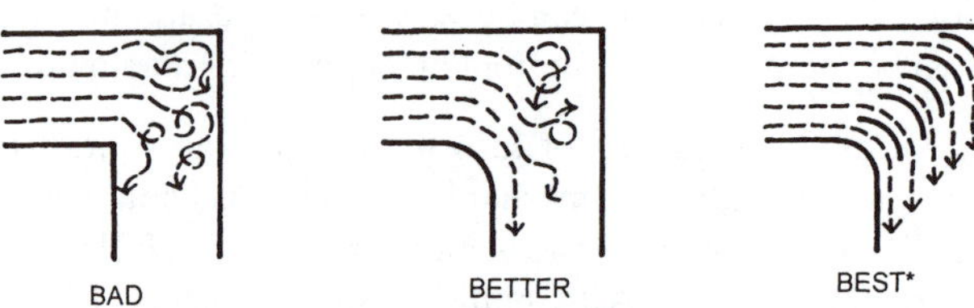


*Except at high air velocities when turning vanes cause noise regeneration.

**Figure 7-2.** Design objective should be to achieve and maintain laminar flow throughout the entire duct system.

## 7.3 Air System Noise Sources

There are a number of potential noise sources in air distribution systems, including

1. Noisy fans or blowers
2. Duct design which causes turbulence
3. Air velocities that are too high
4. Protuberances in the air stream
5. Noisy dampers or diffusers

Each component has the potential for adding noise to the system, and each has unique noise characteristics, to be discussed in the following sections.

### 7.3.1 Fans and Blowers—Sound Power

Because fans must move air, they create turbulence and hence are noisy. Some types of fans are noisier than others (and often less efficient as well). In critical applications, selection of a quiet fan may be required to reduce costly noise reduction treatments in the supply and return systems, or there may be insufficient space to install such treatment in the first place. Plug fans are among the quietest available compared to other types with comparable air moving capacities. Among the worst offenders are multizone air handling units, that may have discharge duct configurations following a tortuous path.

Fan noise is usually rated as sound power levels (see Section 1.2) in octave bands with center frequencies from 63 to 4000 Hz. Sound power level data are available from manufacturers for several fan operating conditions, or may be calculated by methods described in the ASHRAE Handbooks referred to at the beginning of this Chapter. For large fans, low frequency noise is usually the most troublesome, and the most difficult to attenuate (see Section 7.4).

### 7.3.2 Duct Geometry—Air Velocity

Air ducts should be designed to achieve and maintain laminar flow of the air through them, which means a uniform velocity profile at any location along the air stream. Duct geometries that create air turbulence are noise sources, and reduce the efficiency of the system by increasing static pressure losses. Figures 7-2 and 7-3 show good and bad design for a 90° elbow and branch take-off in a duct system. Figure 7-4 shows the effects of proper and improper feed to an outlet, and misalignment in connecting an outlet to a duct using flexible duct.

The air discharge from a centrifugal fan does not have a uniform velocity profile, and it is important not to have turns or insert duct silencers, or impose a transition in duct shape within several duct diameters of the discharge. Good and bad design of a fan discharge into ductwork is shown in Figure 7-5. Discharging into an absorbent-lined plenum is a good way to avoid noise generation and achieve significant noise reduction, particularly at the low frequencies.

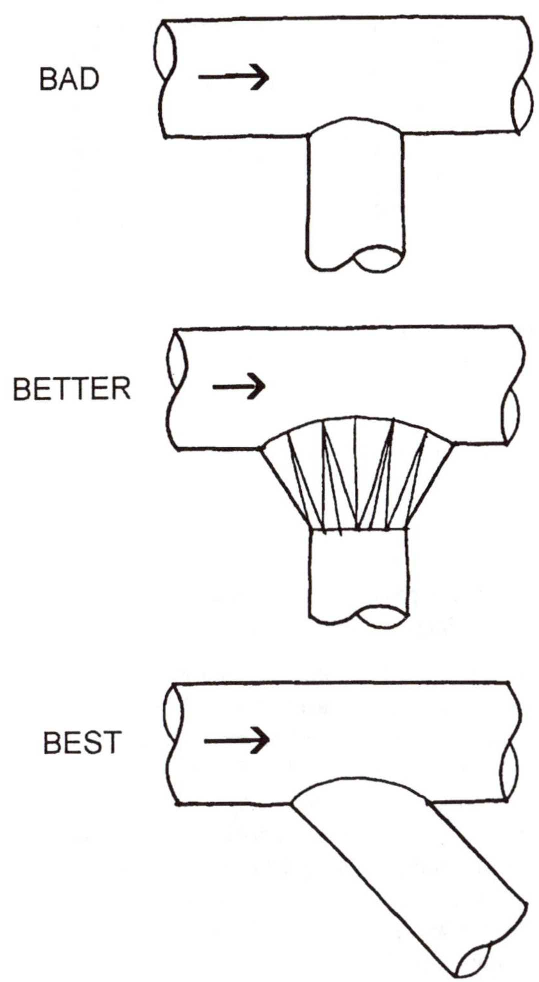


Static pressure losses are minimized by properly designed take-offs.

**Figure 7-3.** Duct configurations for high velocity take-offs.

Branch take-offs from main supply ducts should be done with radiused turns, or with 45° elbows. Avoid 90° take-offs, or if necessary, use turning vanes. Caution: Turning vanes, like other in-duct devices, can generate noise if air velocities are too high. The guiding principle should be to avoid abrupt changes in air direction or velocity.

Excessive air velocities are frequently a cause of noise in air ducts. In Table 7-1 are shown the approximate air velocities for main, branch and runout ducts that must not be exceeded to meet the RC criteria for the spaces being served (see Appendix A.4 for RC design criteria).

Because low air velocities require large air ducts, it is important that the design architect for a building allow sufficient space for duct runs. Design-build projects are particularly at risk because floor-to-floor spacing may be established before the air distribution system is designed, and there may be inadequate space for low velocity duct runs in critical areas.

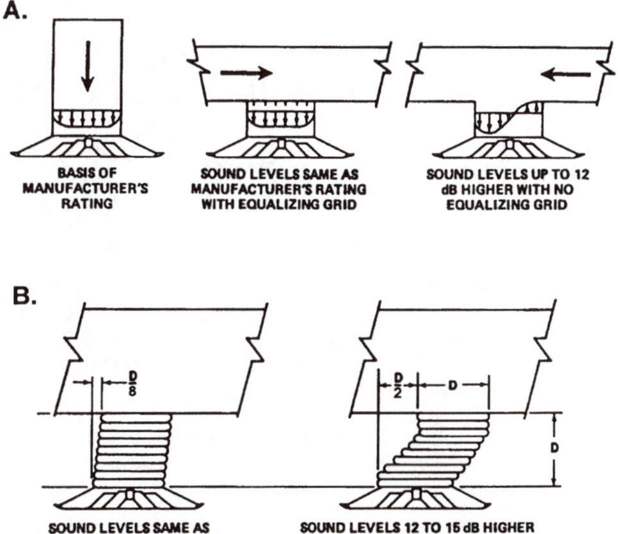


**Figure 7-4.** A. Proper and improper airflow conditions to an outlet. B. Effect of proper and improper alignment of flexible duct connector.

### 7.3.3 Terminal Devices

Because terminal devices such as fan-powered mixing boxes, dampers and diffusers are located very close to or within occupied spaces, the noise they generate is especially important. Some background noise in an office can provide masking and improve acoustical privacy, but the noise must have a neutral spectrum and not be so loud as to be annoying. The fact is that mechanical noise is not reliable for producing a satisfactory masking signal because it can vary with time and is difficult to accurately predict (see Sections 5.3.2 and 6.3).

Excessive air velocities through diffusers is a common source of noise, and may be compounded by attempting to reduce the volume velocity with a damper located directly behind the diffuser. Dampers should be located well up stream of the diffuser so the noise it produces can be attenuated before reaching the room. A well designed air distribution system utilizing techniques such as static regain can minimize the use of dampers to regulate air flow.

Calculation of noise from diffusers is usually done for the person closest to the diffuser, and involves the noise characteristics of the diffuser for a range of air velocities (manufacturers can provide noise ratings), distance to the receiver, and the room absorptive characteristics. Usually the calculation considers both the direct and reverberant fields (see Section 4.6.2). If the receiver location is midway between two diffusers, a separate calculation should be performed for each diffuser and the levels added (see Section 1.2.2). Acceptable background noise levels can be specified as RC- or NC-curves, or as A-weighted values. Rec-

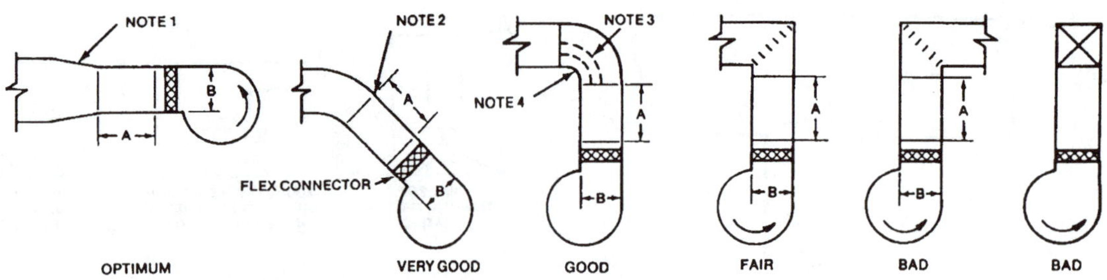


NOTES:
1. Slopes of 1 in 7 preferred. Slopes of 1 in 4 permitted below 2000 ft/min (10 m/s).
2. Dimension A should be at least 1.5 times B, where B is the largest discharge duct dimension.
3. Rugged truning vanes should extend the full radius of the elbow.
4. Minimum of 6″ (150 mm) radius.

**Figure 7-5.** Various outlet configurations and their possible rumble conditions.

**TABLE 7-1.** Maximum Permissible Air Velocities in Ducts

| | Velocities in ft/min (m/min) | | |
|---|---|---|---|
| RC-curve | Main | Branch | Runout |
| RC-20 | 885 (270) | 690 (210) | 395 (120) |
| RC-25 | 985 (300) | 885 (270) | 490 (149) |
| RC-30 | 1280 (390) | 1080 (329) | 640 (195) |
| RC-35 | 1475 (450) | 1180 (360) | 785 (239) |
| RC-40 | 1770 (539) | 1380 (421) | 985 (300) |
| RC-45 | 2065 (629) | 1575 (480) | 1180 (360) |

**TABLE 7-2.** Single-Pass Sound Transmission Loss of Acoustical Materials

| | Sound Transmission Loss, dB, at Frequency, Hz | | | | | | |
|---|---|---|---|---|---|---|---|
| Material | 63 | 125 | 250 | 500 | 1000 | 2000 | 4000 |
| 1″ (25 mm) glass fiber clg panels | 5 | 6 | 6 | 4 | 6 | 10 | 12 |
| Same, foil-backed | 12 | 13 | 13 | 22 | 14 | 27 | 31 |
| 5/8″(16 mm) mineral fiber panel | 11 | 13 | 15 | 20 | 24 | 32 | 32 |
| 5/8″ (16 mm) gypsum board | 11 | 16 | 22 | 28 | 32 | 29 | 31 |

ommended levels for different use categories are given in Appendix A.4.

Fan-powered mixing boxes are often located above suspended ceilings in occupied rooms, and can produce excessive noise for someone located below the unit. The attenuation of the suspended ceiling depends on the type of material, and is least for glass fiber ceiling panels. A foil backing on the panels can provide some additional attenuation. Mineral fiber panels tend to have more attenuation, and a solid gypsum board ceiling higher values yet. Typical single-pass sound attenuation data for several types of ceilings are given in Table 7-2.

The data given in Table 7-2 assume the mixing box is several feet above the suspended ceiling. If the box is just a few inches above the ceiling, the attenuation values will be less, particularly at the low frequencies. If the attenuation provided by the ceiling is inadequate, and the box cannot be relocated to a less critical area, the noise can be further reduced by placing a layer of gypsum board on the ceiling directly below the box. Batt insulation can also be used, but it will provide very little additional attenuation at the low frequencies.

## *7.4 Air System Noise Control*

Noise control is best achieved in air distribution systems by selecting quiet components, and designing for laminar flow by avoiding abrupt changes in air flow (velocity or direction) and excessive velocities. Other means of noise control may also be required and are discussed in the following sections.

In recent years concern has been expressed about the use of glass fiber or other porous materials in air distribution systems, because such materials may pollute the air stream or harbor harmful micro-organisms. Such concerns would be of particular interest to hospitals, medical clinics and clean rooms. A position statement on this matter has been issued by the North American Insulation Manufacturers Association (NAIMA) which minimizes the risks associated with glass fiber insulations. In addition, the Association provides useful guidelines on proper work practices for the handling and installation of glass fiber materials. NAIMA can be reached at 44 Canal Center Place, Suite 310, Alexandria, VA, 22314, Tel: (703) 684-0084. At the same time, new products are being developed for customers desiring to avoid using glass fiber materials, and active noise cancellation offers another approach for those concerned.

### 7.4.1 Lined and Braced Ducts

Absorbent linings inside of air ducts can add considerable attenuation to noise within the duct. The linings typically consist of glass fiber insulation blanket or board in thicknesses of 1/2″ (13 mm) and 1″ (25 mm). The boards are treated with a porous black neoprene coating to prevent erosion of the glass fibers in the air stream.

Absorbent linings are particularly effective at bends in the duct, where the noise must either reflect off the duct wall or diffract to change direction. The lining of long, straight duct runs becomes less and less effective as the duct length increases, particularly at the higher frequencies.

Unlined ducts will provide a small amount of attenuation, mainly at the low frequencies, due to sound transmission through the duct walls. However, transmission through duct walls portends another problem. If the duct contains high noise levels, such as close to a fan, while passing through a noise-sensitive area, this through-the-wall transmission or "break-out" noise can be excessive. The duct will have to be enclosed, be constructed of heavier gauge metal, or have additional noise reduction added between the fan and transmitting duct.

Large, rectangular sheet metal air ducts are prone to vibrating or "oil-canning" if the air is pulsating due to some instability in the system. This condition can be minimized by good system design, but it may be prudent to stiffen large sheet metal panels with angle braces. Circular ducts are not subject to this problem.

### 7.4.2 Duct Silencers

Duct silencers (sound traps) are in-duct noise attenuators designed to maximize noise reduction and minimize static pressure loss. Typically they are 3, 5 or 7 ft in length (0.91, 1.52 or 2.13 m) and contain streamlined, perforated dividers filled with absorbent. Acoustical performance data are available from the manufacturers for the various models, lengths and air flow volumes. Noise attenuation is lowest at the low frequencies, and silencer selection is often based on low frequency performance.

Duct silencers can become noise sources (sometimes termed "self-noise") if not properly installed or if air velocities are too high. The inlet to a silencer should have a uniform air velocity profile, requiring that inlet ducts connected to the silencer be straight for several duct diameters with no branch connections. If a transition at the inlet (or outlet) of the silencer is required to minimize static pressure loss, it should be gradual to avoid turbulence. Note that silencers may be required on both the inlet and outlet of fans.

In mechanical rooms, the duct silencers should be located close to the room wall, thus minimizing duct exposure to room noise on the quiet side of the silencer. If the duct is leading directly into a noise-sensitive space (to be avoided if possible), additional protection for the ducts and silencers on the mechanical room side may be required. A gypsum board enclosure is an effective treatment for this purpose. The enclosure must not contact the duct or silencer, and the space between should be packed with sound control blanket.

### 7.4.3 Plenums and Enclosures

An effective means of achieving noise reduction at fan inlets and discharges is to provide a plenum lined with an acoustical absorbent. A plenum is no more than a large enclosure, usually constructed of sheet metal panels with the inner wall perforated and absorbent behind.

The air inside the plenum has a uniformly positive or negative pressure, thus resulting in near laminar air flow at penetrating ducts. The plenum also serves as an "energy sink," effecting a considerable reduction in sound energy. The advantage over duct silencers is that there is very little static pressure loss. The disadvantage is that it requires a lot of space.

### 7.4.4 Active Sound Cancellation

There is considerable interest in active sound cancellation for installation in air ducts. The ducts provide a confined source and propagation path, a condition favoring the use of this technology. At the beginning of Section 7.4 is a brief discussion of developing concerns about the use of glass fibers and porous materials in air distribution systems. Noise cancellation avoids these concerns, and it has already been demonstrated that the method can achieve significant noise reduction, particularly at the troublesome lower frequencies.

An active noise cancellation system consists of a sensing microphone located in a duct just upstream of the system loudspeakers (see below). The microphone output is fed into a signal processing unit which shifts the phase 180° and then energizes loudspeakers placed in the duct wall. Noise cancellation occurs when the two signals combine, where the phase difference is 180°. Perfect cancellation does not occur because the noise characteristics at the sensing microphone and loudspeaker locations are not identical over the duct crossections. A second downstream microphone senses noise transmitted beyond the cancellation zone and feeds back to the controller data to optimize the system. Companies offering noise cancellation systems are listed in Appendix A.12.

## 7.5 Vibration Isolation

Mechanical equipment that has rotating or reciprocating components will have unbalanced forces associated with its operation that can cause the equipment to vibrate. If the equipment is not isolated from the supporting structure, the vibration will be transmitted to the structure where it can cause objectionable noise, an unpleasant sen-

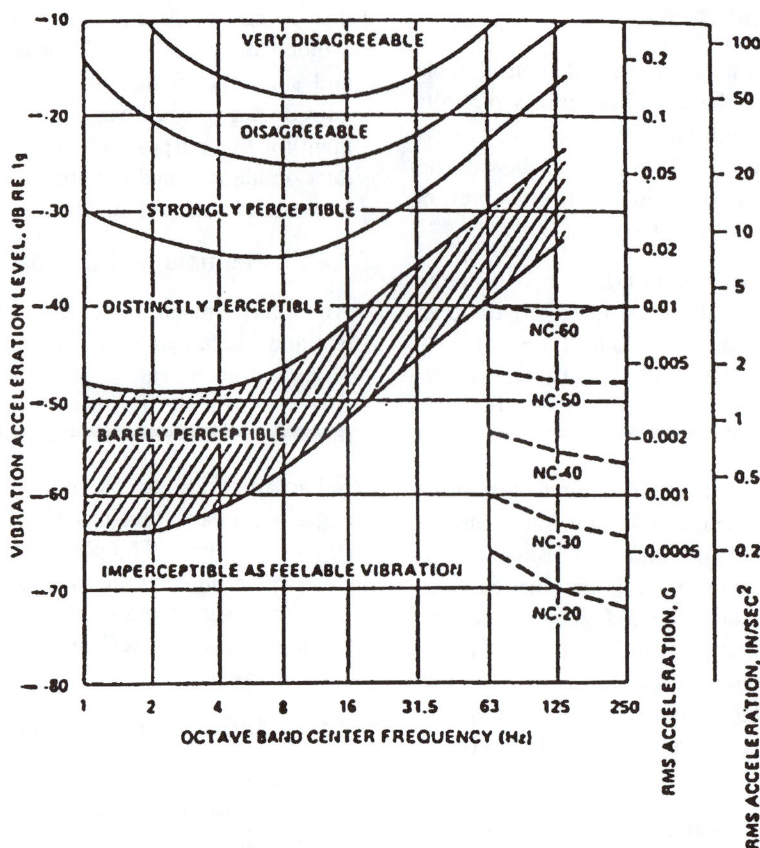


People are most sensitive to vibration at frequencies around 10 Hz. See text for an explanation of the dashed NC curves.

**Figure 7-6.** Approximate sensitivity and response of people to feelable vibration.

sation of motion, or disturb imaging instruments or manufacturing processes that are vibration sensitive. Other causes of vibration in mechanical systems are air turbulence in fans and ducts, and turbulence in plumbing from valves or other restrictions to flow.

The purpose of vibration isolators is to interrupt the transfer of vibrational energy from source to building structure. The subject of the following sections is to explain the limits of acceptable structural vibration, how isolators work, the types that are available, and applications for the various types.

The vibration parameters usually referred to in discussions of vibration levels or criteria are displacement, velocity and acceleration. Displacement is the total distance a vibrating surface moves from the rest position (termed peak-to-peak), and is measured in some unit of length such as inches or millimeters. Velocity is the time rate of change of displacement, measured in in/sec or mm/sec. Acceleration is the time rate of change of velocity, measured in in/sec$^2$ or mm/sec$^2$. Vibration measurements are usually made with accelerometers, the output of which is proportional to acceleration. The other two parameters of displacement and velocity can be obtained by signal processing. The parameter appropriate for a given application depends on which correlates best with the observed reaction to the vibration. A discussion of how vibration measurements are made is contained in Appendix A.10.

### 7.5.1 Vibration Isolation Design Criteria

Vibration criteria have been developed in terms of human perception and mechanical equipment performance. Additionally, manufacturers of vibration sensitive instruments or manufacturing processes provide maximum permissible vibration levels for the installation of their product. Meeting the criteria is likely to be a challenging design problem, requiring the architect, mechanical and structural engineers, and instrument manufacturers to work in concert to insure success of the design. Another very common source of structural vibration is from footfalls, and is almost always present in areas where vibration is a potential problem (see Chapter 3, Sections 3.5 and 3.6).

Figure 7-6 shows the sensitivity of people to feelable vibration as function of the vibration acceleration level. It is interesting to note that human sensitivity to "disagree-

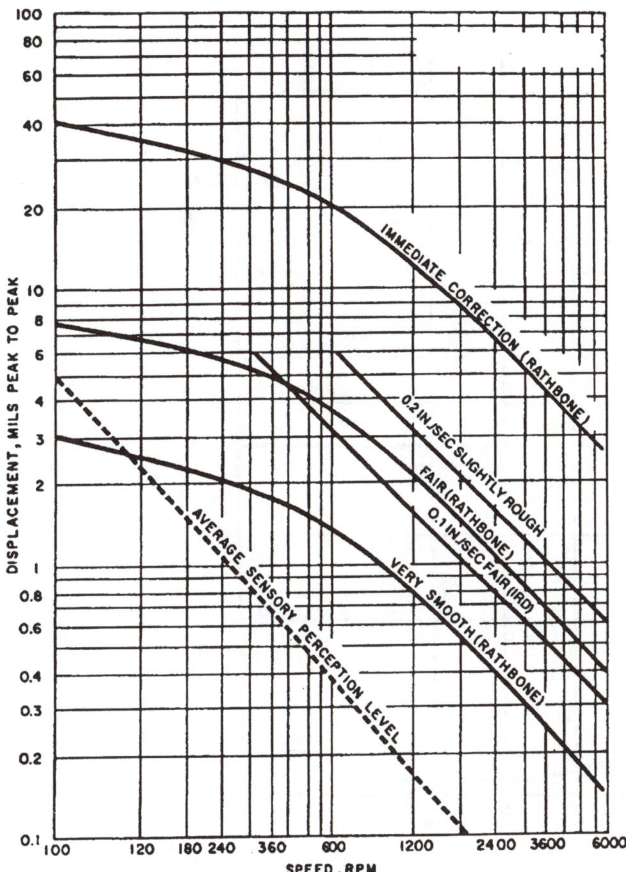


For a given displacement, machine vibration problems become more critical with increasing RPM

**Figure 7-7.** Machine vibration criteria.

able" vibration is greatest in the region of 10 Hz. It is also interesting to note that vibration can result in clearly audible sound at acceleration levels well below the threshold of "feelability." At the right side of Figure 7-6 are plotted the low frequency end of the NC noise criterion curves. These curves show the approximate octave band sound pressure levels (refer to Figure 1-13 in Section 1.7) that will be radiated from the indicated acceleration levels of vibrating large surfaces such as walls, floors or ceilings.

Another important consideration for the effects of vibration is possible damage to machinery. Figure 7-7 shows plots of roughness criteria for machinery vibration as a function of displacement. A method has been developed in recent years for detecting incipient machinery failure prior to the actual breakdown. This method, termed "machine health," involves periodic vibration measurements at machine supports. If an increasing trend in vibration at a particular frequency is detected for subsequent measurements, and the frequency can be identified with a particular machine component, replacement of the component can be scheduled at a convenient time. This method is obviously preferred to allowing the component to fail with the possibility of catastrophic results.

### 7.5.2 Methods of Vibration Isolation

Vibration isolators are springs that are partially deflected when a static load is applied. The resistance imposed by the spring is the spring constant, and is measured in lb/in (kg/m). A spring so loaded has a *natural* or resonant frequency. If a dynamic load is applied to a spring having a frequency (termed *disturbing* frequency) that is near or below the natural frequency of the spring, more force will be transmitted to the supporting structure than is applied to the spring, and the isolator becomes, in fact, a vibration amplifier. Obviously, this is a condition to be avoided. If the disturbing frequency is above the natural frequency, less force is transmitted to the structure than is applied to the spring.

The amount of energy transmitted by an isolator, or transmissibility (T), is proportional to the square of the ratio of disturbing frequency (fd) to natural frequency (fn), and is given by

$$T = \frac{1}{1 - (fd/fn)^2}$$

Figure 7-8 shows the vibration transmissibility as a function of the disturbing frequency divided by the natural frequency. Notice that in the equation above when the ratio fd/fn = 1, T goes to infinity. In real systems, however, energy losses in the system, called damping, limit the value of T. Indeed, damping is an important means of restricting the amount of vibration. Even so, a coincidence of disturbing and natural frequencies is to be avoided in any vibration isolation system.

The percent efficiency of an isolator is equal to the quantity 100 (1 − T). For example, if the applied force to an isolator is 100 lb (45.4 kg), and the transmitted force is 5 lb (2.27 kg), T = 0.05, and the percent efficiency is 100 (1 − 0.05) = 95%.

Figure 7-9 shows the transmissibility (efficiency) of an isolator for various combinations of static deflection and disturbing frequencies. For example, if a fan operating at 1200 RPM (fd = 20 Hz) requires an efficiency of 98%, the isolator must have a static deflection of 1.25″ (32 mm), with a natural frequency fn = 2.8 Hz (read from the resonance line at the top of the diagonal lines in Figure 7-9). As a general rule, isolation efficiencies greater than 95% requires that fn be approximately 1/10 of fd.

The isolation efficiencies shown in Figure 7-9 assume a system of minimal damping, single degree of freedom, and a rigid supporting structure. For flexible structures, resulting in a multiple degree of freedom system, the efficiency will be less (see Section 7.5.4).

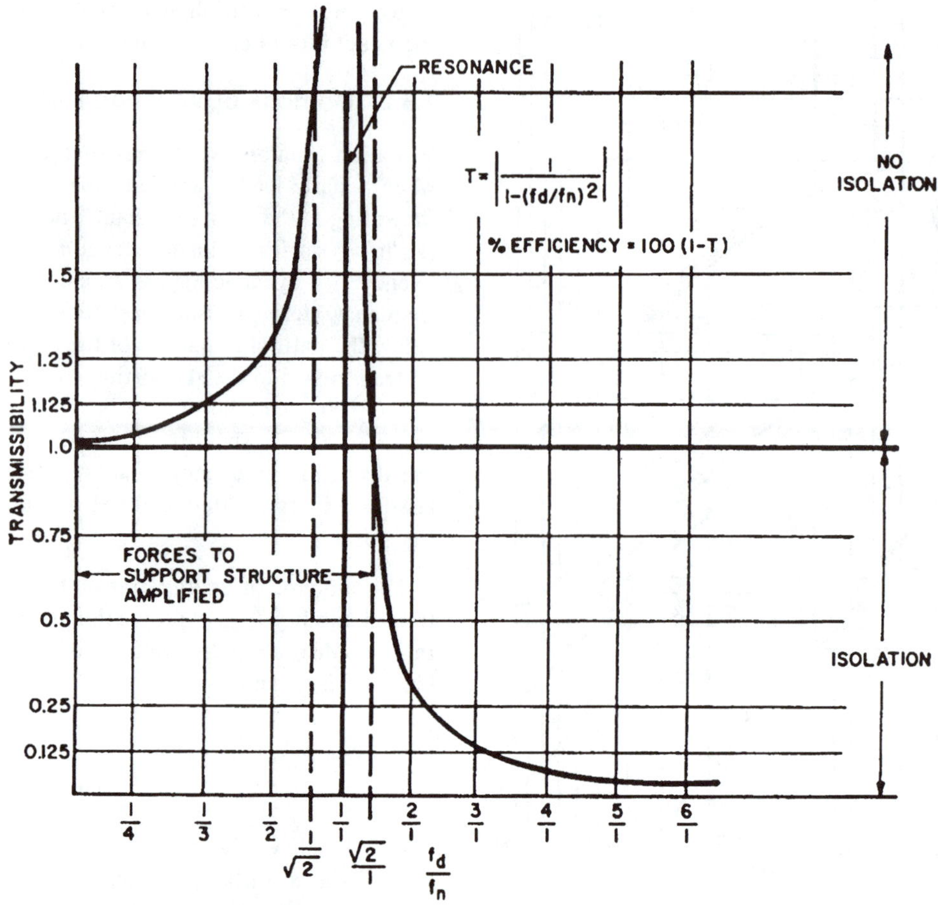


Amplification of the force applied to an isolator occurs near resonance—a condition to be avoided.

**Figure 7-8.** Vibration transmissibility as a function of fd/fn.

The are several basic types of vibration isolators, each operating in a preferred range of static deflections. The various types are pictured in Figure 7-10, and are described below, with typical applications for each type.

*Neoprene.* Isolators made of neoprene are designed to operate in either compression or shear. Neoprene is more efficient in shear than in compression. Various configurations are available, such as sheets, waffle pads, cylinders and more complicated shapes. They are used in applications where a static deflection of up to about 1/4″ (6.4 mm) is required, such as small mechanical equipment operating at over 2000 RPM. Neoprene has high internal damping, thus reducing the magnitude of transmitted vibration at frequencies in the range of positive efficiencies.

*Glass Fiber Blocks.* Medium density glass fiber blocks are used in a few applications, such as the resilient support of floating floors, and suspension systems for pipes and ceilings. Static deflections of up to 1/2″ (13 mm) are typical. If the blocks are to be used in an application where water or other liquids may be present, the blocks are encased in a neoprene jacket.

*Steel Springs.* Deflections of up to 6″ (152 mm) can be obtained with steel springs, and are used where low frequency isolation is required, such as large fans and reciprocating compressors. Because the springs can transmit high frequency vibration, they quite often incorporate a neoprene isolating element as well. Springs with large static deflections tend to be unstable, thus the design must incorporate some means of keeping the spring vertical during times of excessive vibration, such as during start-up or seismic events. The ratio of height to diameter is an important consideration for spring stability.

*Inflatable.* This type of isolator is capable of very large static deflections, and is used for test chambers and tables holding vibration-sensitive instruments. They are inherently unstable, however, and some type of lateral restraining system is mandatory.

*Active Vibration Cancellation.* There is a new type of

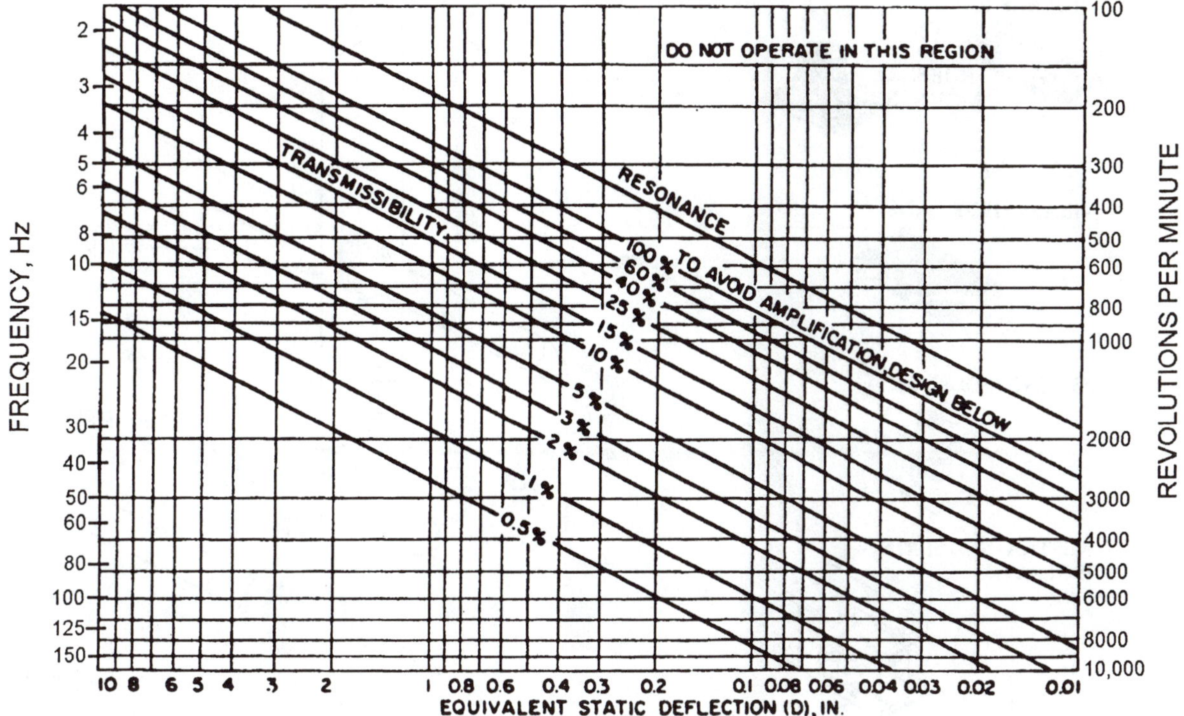


Typical isolation efficiencies are about 98%, corresponding to a transmissibility of 2%.

**Figure 7-9.** Transmissibility of a vibration isolation system as a function of isolator static deflection and forcing frequency.

vibration "isolator" currently under development. The technique involves developing an opposing force to cancel the vibration of the system. It will probably consist of a sensor, a processing unit, a power amplifier and a force generator to be mounted on the system. The processor reverses the signal to provide a force 180° out of phase with the driving force to secure a cancellation of the vibration. This is somewhat similar to noise cancellation systems that are available for reducing fan noise in air ducts (see Section 7.4.4). At this time limits on applications for the methodology are unknown.

Because most mechanical equipment has different weights at the mounting locations, it may be necessary to select isolators with different spring constants to maintain the same static deflection at each mount. If this is not done, there may be some instability during operation because of the inconsistent natural frequencies.

In some applications it is desirable to mount the equipment on an inertia base, usually a heavy concrete base weighing about twice the equipment weight. These bases do not absorb energy, but they will reduce the amplitude of vibration and provide greater stability by lowering the center of gravity for large isolator deflections.

Mechanical and electrical connections to isolated equipment must incorporate an isolation attachment to prevent vibration from short circuiting the isolators. Pipe connections to fans, pumps, chillers etc. should be made with twin sphere neoprene isolators. Woven metal wire connectors are not effective isolators. Duct connections should incorporate a fabric band or "flexible sleeve" for an isolation break. Electrical connections should incorporate flexible conduit with two 90° bends to allow the equipment to move in any direction without the vibration being transmitted.

Pipes carrying water or other liquids may be a source of vibration even if isolated from the equipment to which they are attached. Pipes are not very efficient radiators of sound because of their small areas, particularly the smaller diameter pipes, but once the vibration is transmitted to a larger wall or ceiling surface, noise can result. It is necessary to isolate the pipe from the building structure to the maximum extent possible. Figure 7-11 illustrates a method for isolating small pipes from structural elements. For hanging, larger pipes may require spring isolators, while neoprene hangers are satisfactory for smaller pipes. Where the pipes penetrate a wall or ceiling, an oversize hole should be used, caulked to prevent airborne sound leakage.

Suspended ceilings must not be hung from air duct supports. Any vibration in the duct will be carried to the ceiling system and radiated as noise.

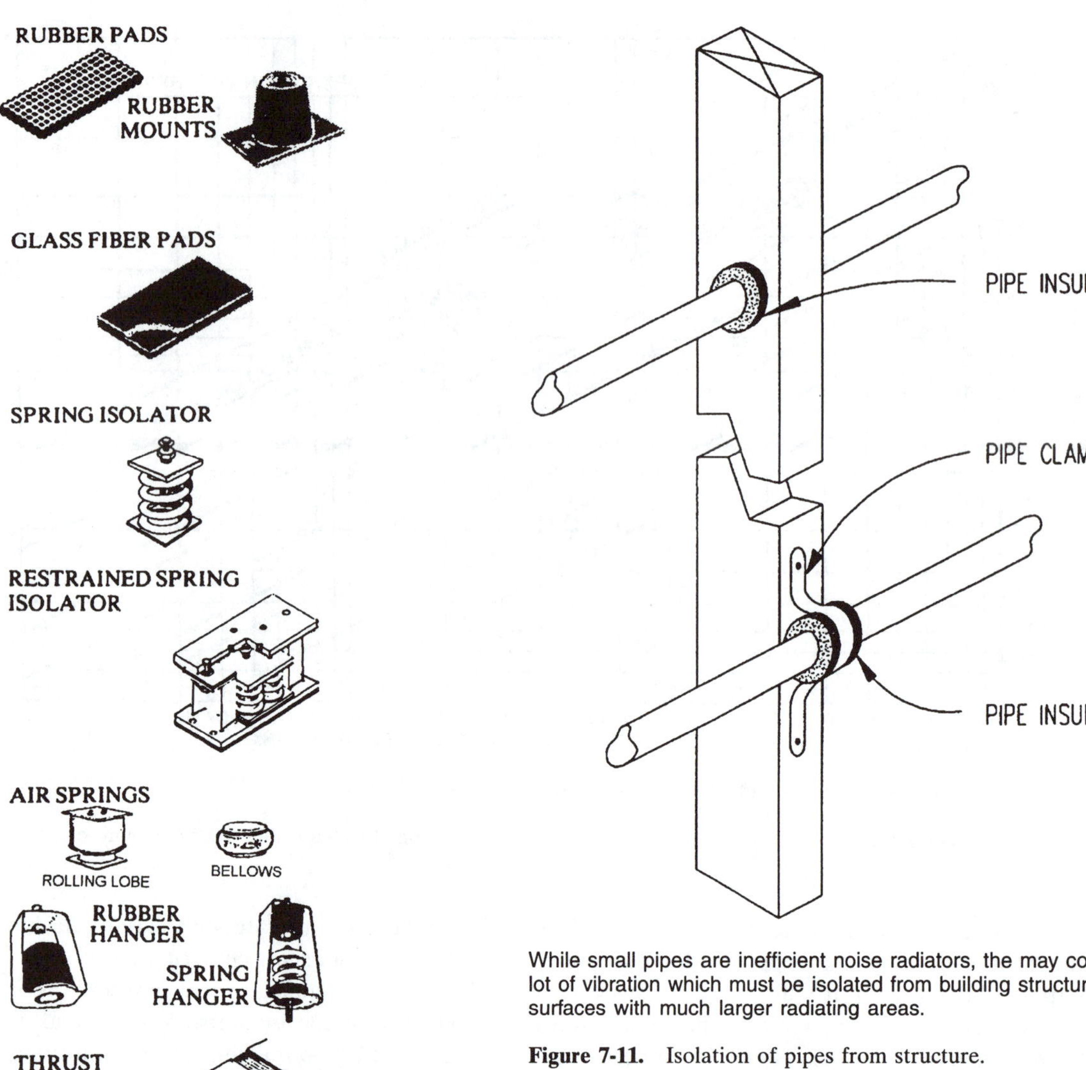


**Figure 7-10.** Typical vibration isolation devices.

While small pipes are inefficient noise radiators, the may contain a lot of vibration which must be isolated from building structures and surfaces with much larger radiating areas.

**Figure 7-11.** Isolation of pipes from structure.

## 7.5.3 Seismic Restraints

All major pieces of mechanical equipment supported on or suspended with vibration isolators must be restrained during seismic events, to prevent damage to the equipment or its surroundings. Restraining devices under normal conditions have no effect on the function of the isolator, but restrict the motion of the isolated equipment if subjected to excessive lateral or vertical forces, or frequencies coincident with the resonant frequency of the isolator.

Some isolator types are contained in housings that restrain the motion of the equipment it is supporting. Another type of restrainer is a "snubber," which mounts adjacent to the isolator, but not quite in contact with it. Whatever the type, it must meet applicable codes that consider the seismic zone of the area, where the equipment is located in the building, and if the equipment is used for life safety.

## 7.5.4 Multiple Degrees of Freedom

The isolation efficiencies specified in Figure 7-9 assume that the isolator is supported on a foundation that does not move. If the foundation is a concrete slab on grade or a massive concrete structure above grade, that assumption may be valid. This condition is considered to be a single degree of freedom system. However, many times mechanical equipment is supported on relatively light weight or flexible structural systems, such as mid-span on steel trusses.

If the supporting structure deflects significantly when the weight of mechanical equipment is imposed, a new resonant system is created with a unique set of resonant frequencies. The isolators supporting the equipment, therefore, will not behave in the same way as with a single degree of freedom system. A multiple degree of freedom system has been created, and the isolation efficiency of such a system is much more difficult to accurately predict.

It is generally accepted practice to avoid multiple degree of freedom systems. As discussed in Section 7.2, mechanical equipment should be located on grade, or as close as possible over supporting columns or bearing walls. In mechanical penthouses where a floating concrete slab may be used for airborne sound isolation, major items of mechanical equipment should never be installed on the floating slab, but rather on piers supported on the base structure. The floating slab must be isolated from the piers in the same manner as the perimeter isolation break.

## 7.6 Mechanical Equipment Out-of-Doors

Many times it is advantageous to locate mechanical equipment out-of-doors. Chillers and cooling towers are so located because they must dissipate heat, and fans and air handling units are placed on roofs to avoid the cost of a mechanical penthouse. These locations are not only potential sources of noise and vibration in the buildings they serve, but may cause community noise problems as well.

Chillers and cooling towers are quite noisy, and may have pure tone components in the low to midfrequency range that make the noise more annoying. A barrier surrounding the equipment can add some noise reduction, but the barrier must not be so close that circulation of cooling air is restricted. If the equipment is located next to a wall, reflections off the wall will reduce the effectiveness of the barrier. Application of an absorbent suitable for exterior use can minimize the reflection. If adjacent buildings are high rise, the upper floors may be looking over the barrier directly to the equipment.

Roof-mounted air handling units often times have air ducts penetrating the roof directly below the unit. The return air may be no more than a hole in the roof, looking directly up to the fan inlet. Some means of noise reduction must be employed if the fan is located over a noise-sensitive space. There is usually no room for a return air plenum, but a baffle suspended below the opening with an absorbent on top is better than nothing. Supply air is, of course, ducted, and there is more of an opportunity to install noise reduction in the form of lined ducts or duct silencers.

Vibration is another potential problem of roof-mounted equipment. Mid-span locations are particularly vulnerable; refer to Section 7.5.2 for appropriate methods of incorporating the required vibration isolation.

Roof-mounted equipment can also be a community noise problem (see Chapter 8, Section 8.8). Many communities have ordinances specifying the maximum permissible noise from such equipment at property lines. If the specified levels are exceeded, the installation of barriers may add sufficient noise reduction to be in compliance. As mentioned earlier, make sure the barriers do not restrict the flow of cooling air. If the equipment operates at night, lower nighttime noise limits would apply.

## 7.7 Elevators—Machine Rooms

Elevators can cause noise problems if the shaft is adjacent to a noise-sensitive space. Even if the shafts are constructed of concrete or concrete blocks, noise can be created by the direct attachment of the car guide rails. It is possible to incorporate vibration isolation in the attachment of the rails to the shaft wall, but elevator codes impose strict limits on acceptable maximum deflection of the guide rails. In critical situations, double wall construction is recommended.

Possible sources of noise for high-rise drum and cable elevator systems are the motor/gear reduction assembly, and activation of the solenoid-operated drum brake, characterized by a low frequency impulsive "thump." The motor/gear box/drum mechanism can be installed on a concrete inertia base, and the entire assembly supported with vibration isolators. The isolator loading at the drum end of the inertia base will be considerably heavier than at the motor/gear box end, requiring different spring constants to obtain the same deflection. Also, check with the elevator manufacturer for the maximum permissible isolator deflection.

The most likely source of objectionable noise associated with hydraulic elevators is the characteristic whine of the motor/pump assembly. The machine room should be located in an area not close to noise-sensitive spaces. The motor/pump assembly should set on an inertia base supported with vibration isolators, and the hydraulic lines from pump to piston should be isolated from the building structure.

# Chapter 8

# Community Noise

The most common sources of community noise that architects and builders must consider in building design are surface vehicular traffic, airplanes, railroads and some industries. Such noise can adversely affect building occupants unless special design precautions are taken, or they may even prohibit entirely any noise-sensitive development.

The project itself may cause an unacceptable source of community noise. The noise generated by mechanical equipment located outside a building, such as chillers, cooling towers, air handling units and heat pumps, can produce unacceptable levels at adjacent land uses. Many utilities projects may be in this same category, such as electrical substations, waste water treatment plants, recycling stations and power generating facilities.

Special precaution must be taken to consider the frequency content of many of the noise sources listed above. They often contain strong low frequency components, and use of STC ratings for selecting facade building elements will not adequately consider these low frequencies. (See Chapter 3, Section 3.2.1, for an explanation of STC.) Walls, windows and doors typically have a lower TL rating at the low frequencies (compared to the higher frequencies; see Section 3.3), where these noise sources may have high levels. In such circumstances, the STC ratings for facade elements should be derated from 3 to 5 points. Even then, the low frequency content should be evaluated separately to insure compatibility with design criteria.

Noise sources with particularly strong low frequency components are diesel train engines, asphalt plants, and power generating plants or natural gas pumping stations powered by diesel or turbine engines.

## 8.1 Noise Regulations

Regulations pertaining to community noise usually have two basic purposes. The first is to specify noise limits that a business or resident may not exceed at adjacent properties. Such limits are often specified by state, county or municipal noise ordinances or regulations.

The second purpose is to establish acceptable land uses, often including minimum construction requirements, for noise-sensitive projects that are planned for areas subjected to high noise levels, such as near airports and highways. Regulations of this type usually appear in zoning codes and requirements for project financing, such as those imposed by the Federal Department of Housing and Urban Development for multifamily housing.

### 8.1.1 Community Noise Limits

While there are many approaches as to how community noise regulations are drafted, the more effective ones have several features in common:

1. The location at which the regulation applies, usually the property line or within some distance of an occupied building.
2. The noise descriptor that must be used, usually an A-weighted level expressed as a percentile or an average, such as Leq. Occasionally, octave band or one-third octave band levels, or NC-curves, are specified (see Chapter 1, Sections 1.3.2, 1.1.1 and 1.7).
3. The type of meter to be used and a method for conducting the measurements.
4. Specification of noise levels that cannot be exceeded. Uses or zoning of the source and receiving properties are usually a consideration, such as residential, commercial or industrial.
5. Maximum permissible increase of a new noise source above existing noise.
6. A penalty for noise limits at night (often 10 dBA less), and for noise generated on weekends, particularly Sundays.
7. Special considerations for unusual noise characteristics, such as impulsive, pure tones or strong low frequency components.
8. Provisions for penalties, variances and exemptions. Exemptions often include carriers engaged in interstate transportation, emergency vehicles and construction noise, although the latter category often has limits that are somewhat less restrictive except at night.

**TABLE 8-1.** Land Use Compatibility With Yearly Day-Night Average Sound Levels

| Land Use | Yearly Day-Night Average Sound Level (Ldn) | | | | |
|---|---|---|---|---|---|
| | Below 60 | 60–65 | 65–70 | 70–75 | 75–80 |
| Residential | Y | Y(30) | Y(35) | N | N |
| Transient Lodgings | Y | Y | Y(30) | Y(35) | Y(40) |
| Schools | Y | Y(30) | Y(35) | N | N |
| Hospitals, Churches | Y | Y(30) | Y(35) | Y(40) | N |
| Auditoriums | (Special consideration for any location) | | | | |
| Offices | Y | Y | Y | Y(30) | Y(35) |
| Retail | Y | Y | Y | Y(30) | Y(35) |
| Manufacturing | Y | Y | Y | Y | Y(30) |
| Outdoor Sports (with Sound Reinforcement) | Y | Y | Y | Y | N |

NOTES: Y=YES, N=NO, (XX) is the required noise reduction, dBA. For all Y designations without a noise reduction rating, normal construction is assumed to achieve a noise reduction of at least 25 dBA. For all noise reduction ratings of 30 or more, open windows are not permitted; mechanical ventilation is required.

Some municipal ordinances have nuisance provisions for noise sources such as barking dogs, loud music, public address systems, etc. Ordinances of this type usually do not specify an objective noise limit and are, therefore, difficult to enforce.

### 8.1.2 Building Requirements

It is not uncommon for government agencies to require that buildings planned for noisy areas meet minimum sound reduction performances. Local zoning codes are frequently vehicles for implementing such requirements, following guidelines established by federal or state agencies. Examples of such requirements are shown in Tables 8-1 and 8-2, Section 8.2.1.

The building designer must verify that the design will meet the code requirements. This process requires that the noise exposure be defined, and that the exterior construction be adequate to achieve the required noise reduction (see Section 8.2.2). Methods for designing building facades to achieve specified noise reductions are discussed in Chapter 3, Section 3.4.

## *8.2 Rating Methods*

The documentation of community noise has two basic purposes. The first is for land use planning, to accommodate the process of selecting appropriate uses for the degree of noise exposure. The second is for building design, so that the exterior construction will prevent the intrusion of high noise levels.

### 8.2.1 Land Use Compatibility

The most common noise descriptor for land use planning in use today is the day-night sound level, Ldn. The Ldn is a single-number, 24-hour descriptor obtained by averaging the 24 A-weighted hourly $L_{eqs}$ (see Chapter 1, Sections 1.3.2 and 1.6), but with 10 dBA added to each $L_{eq}$ for the 9 hours between 10 p.m. and 7 a.m. This penalty is added because of the presumed greater sensitivity to noise during the nighttime hours. The Ldn is used in noise guidelines and regulations promulgated by the U.S. Departments of Transportation and Housing and Urban Development, and the Environmental Protection Agency. Figure 1-17 in Chapter 1 shows the correlation between communities highly annoyed and Ldn.

Table 8-1 is derived from FAA and HUD documents, in which limitations are placed on various land uses according to the Ldn zone in which they are located. Notice that some uses are permitted in high noise zones, provided the construction meets certain noise reduction requirements. In some cases, ratings are upgraded from those presented in the referenced documents. Many municipal building departments will require proof that such requirements are met before issuing a building permit.

Because the Ldn descriptor is frequently referred to in meetings and hearings attended by nontechnical people, it is often thought of as the *maximum* noise level permitted at a site. This is not the case; the actual noise level can exceed the Ldn value, since Ldn is a type of average. Many different temporal noise scenarios can have the same Ldn value. For example, a few very noisy events (such as jet aircraft at close range) can have the same Ldn rating as a steadier noise at a lower level (such as near a freeway with heavy traffic).

Other rating methods which may be encountered in land use planning documents are defined below. Approximate numerical equivalents to Ldn are given in Table 8-2. (See also Chapter 1, Section 1.3.2.)

*Community Noise Equivalent Level (CNEL).* This descriptor is unique to the State of California, and is determined in the same manner as Ldn, except for a 3 dBA penalty for the 3 hours from 7 p.m. to 10 p.m. As a practical matter, the two descriptors are virtually identical for most noise environments.

*Noise Exposure Forecast (NEF).* A predecessor to Ldn, this descriptor still appears in some older planning documents. It is based on Perceived Noise Level data (see below) rather than A-weighted data.

*Composite Noise Rating (CNR).* This descriptor predates both Ldn and NEF, but may appear in land use planning documents that have not been updated since the 1960s. Like NEF, it is based on Perceived Noise Level data (see below).

*Perceived Noise Level (PNL).* This is a complicated noise descriptor requiring a frequency analysis of the noise, and a calculation procedure closely following the method for determining loudness level (see Chapter 1, Section 1.5.1, and Appendix A.8). At the present time it is used for jet aircraft noise certification procedures in accordance with Federal Air Regulation (FAR) Part 36.

**$L_x$.** This descriptor, or percentile, specifies that sound level which is exceeded "X" percent of the time for a specified time period. For example, the $L_{50}$ is that sound level exceeded 50% of the time, the $L_{10}$ 10% of the time, and so forth. The $L_1$ or $L_{0.1}$ is usually close to the $L_{max}$, and the $L_{99}$ close to the $L_{min}$. The $L_{eq}$ is usually between the $L_{50}$ and the $L_{10}$. For a steady sound, the $L_1$ and $L_{99}$ will be close together. The further apart they are, the more the sound fluctuates.

### 8.2.2 Building Design

Design of the exterior construction of a building to prevent noise intrusion requires knowledge of the site noise characteristics. This can be accomplished by measuring the noise (always the preferred method) or by calculating it based on an appropriate noise model. While octave band data are preferred for design, many times only A-weighted noise levels are available. In such cases it may be possible to assign octave band levels from knowledge of the source characteristics, adjusted to the value of the A-weighted level. Octave band levels for a number of common noise sources are presented throughout this book, including the following sections of this chapter.

Most community noise sources are not constant in either level or frequency characteristics with time. A decision must be made as to the appropriate noise descriptor to use for design purposes. For example, if NO noise intrusions are acceptable, such as for a performing arts theater or recording studio, it would be appropriate to use the maximum noise level, providing it occurs with some degree of regularity, such as aircraft noise. For other uses, where occasional low level intrusions can be tolerated, such as classrooms or living quarters (other than sleeping rooms), the Leq would probably be the appropriate descriptor for design purposes. See also the end of Sections 8.4.2 and 8.4.3.

The noise reduction that must be achieved by a building's exterior construction is the difference between the site noise levels and the permissible levels in building spaces with outside walls of roofs. The procedures for performing calculations of this type are presented in Chapter 3, Section 3.4.

**TABLE 8-2.** Comparison of Noise Rating Methodologies

| Ldn | CNE | LNEF | CNR | Noise Exposure Rating Recomm. Noise Control |
|---|---|---|---|---|
| <55 | <55 | <20 | <90 | Minimal exposure, normally requires no special consideration. |
| 55–65 | 55–65 | 20–30 | 90–100 | Moderate exposure, land use controls should be considered. |
| 65–75 | 65–75 | 30–40 | 100–110 | Significant exposure, land use controls recommended. |
| >75 | >75 | >40 | >110 | Severe exposure, land use controls mandatory. |

## 8.3 Highway Noise

Noise from vehicles on streets, highways and freeways probably affects more people than any other source of community noise. Building projects located within several hundred feet of heavy traffic should consider the noise exposure in the design of the exterior construction. Many noise-sensitive uses favor locations close to heavily traveled roadways because of business exposure and ease of access. Examples are motels, office buildings, theaters, and even hospitals. Residential housing developments near freeways are occurring with greater frequency as prime development sites with more suitable noise environments are becoming scarce, and quick access to the freeway is considered a benefit by many people.

### 8.3.1 Highway Noise Generation and Prediction

The primary source of noise from traffic on city streets, with speeds below about 40 MPH (64.4 KPH) is from the engine and exhaust. A typical octave band frequency spectrum for downtown, start-and-stop traffic in a hilly area is shown in Figure 8-1. Also shown in Figure 8-1 is a spectrum for high speed freeway traffic, where the predominant source is tire noise. Notice the relatively higher noise levels at low frequencies for the downtown traffic. Reflections off adjacent buildings also causes higher noise levels for downtown streets.

There are a number of factors affecting the noise generated by surface traffic. The more important of these are as follows:

*Mix of Vehicles.* Heavy trucks and busses generate much higher noise levels than passenger cars. A typical octave band frequency spectrum for each is shown in Figure 8-2, for the vehicles at a distance of 50 ft (15 m) and traveling at a speed of 50 MPH (80 KPH). If the percentage of heavy vehicles in the total traffic mix exceeds about 2% to 3%, those vehicles will produce more acoustic energy than the remaining 97 to 98% passenger cars.

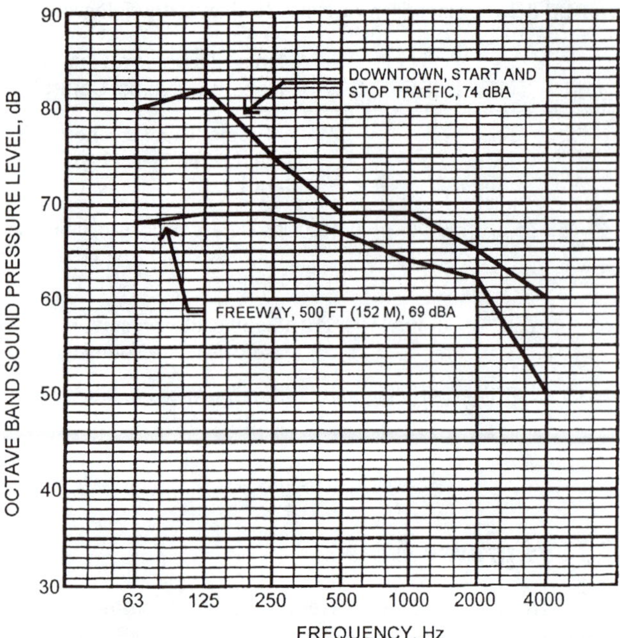


Start and stop traffic—particularly with trucks—generates high noise levels at the low frequencies. Freeway noise traffic, with freely flowing, high speed vehicles, has a flatter spectrum.

**Figure 8-1.** Typical traffic spectra.

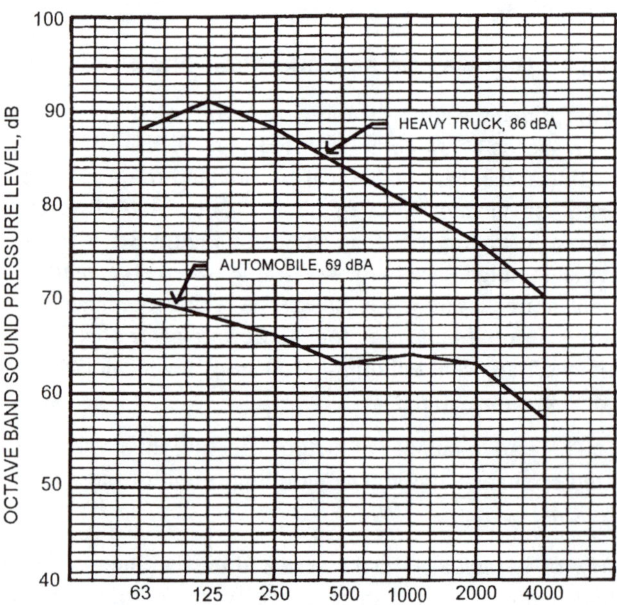


Heavy trucks generate relatively more noise at the low frequencies, compared to automobiles. If the vehicle speed is increased or decreased by 50%, the noise for heavy trucks changes by about 3.5 dBA, and for automoblies about 5.5 dBA.

**Figure 8-2.** Typical vehicle spectra, distance 50 ft (15.2 m), 50 mph (80 kph)

*Number of Vehicles.* Each time the traffic volume is doubled, the average noise level increases by 3 dB (the acoustic energy generated is doubled; see Chapter 1, Sections 1.2.1 and 1.2.2). How the noise level varies with time is also dependent on traffic volume. Large fluctuations in noise level are produced by light traffic volumes, while heavy traffic generates a fairly steady noise level. See Chapter 1, Figure 1-7.

*Traffic Speed.* Noise generation increases with speed for both passenger cars and heavy vehicles, but the effect is more pronounced for cars. Trucks tend to maintain a fairly constant engine RPM at all speeds, resulting in a smaller change in noise level with speed. Noise generation as affected by vehicle speed is explained in Figure 8-2.

*Surface Characteristics.* Rough or worn surfaces generate more tire noise than smooth surfaces. The difference may be as much as 5 or 6 dBA. However, smooth surfaces may have unsafe traction characteristics, and will become rougher following years of heavy traffic. Wet surfaces will produce noise levels about 5 dB higher than for dry pavement, but only at frequencies above 2000 Hertz.

Because of the prevalence of traffic noise in our urban environment, accurate prediction of the noise under a variety of conditions is necessary for effective building design. In addition to the factors affecting vehicular noise generation, characteristics of the propagation path must also be considered. The more important of these are as follows:

*Distance.* Individual vehicles act as point sources, and the noise diminishes at a rate of 6 dB per doubling of distance (see Chapter 1, Section 1.4.1). If the concern is maximum noise levels from individual vehicles, the point source model should be used. Heavy traffic produces a continuous line source, and the rate of diminution is 3 dB per doubling distance as shown in Figure 8-3, *provided* the solid angle of highway exposure remains the same. If the solid angle becomes narrower with increasing distance, such as would be the case if highway exposure is limited by intervening hills or buildings, the highway approaches the point source condition. Because this transition is a fairly common occurrence, the rate of noise decrease is often set at 4.5 dB per doubling distance. (See also Ground Cover following.)

An outcome of the conditions described above is that the noise from individual vehicles is more pronounced close to the highway, while at greater distances the noise from individual vehicles merges with noise from many other vehicles, and the noise becomes steadier with time. A steady noise is generally considered to be less intrusive than a fluctuating noise. It may be more appropriate, therefore, to select a noise descriptor favoring maximum levels close to a highway, and an average level, such as the Leq, at greater distances.

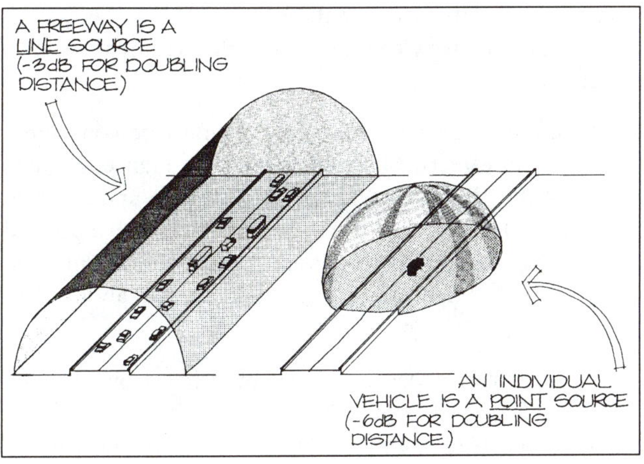


**Figure 8-3.** Sound radiation from point and line sources.

**TABLE 8-3.** Absorption of Sound in Air, dB/300 ft (91.4 m).

| Temperature | Relative Humidity% | Frequency, Hz 1000 | 2000 | 4000 |
|---|---|---|---|---|
| 86°F (30°C) | 10 | 0.87 | 2.9 | 9.6 |
| | 50 | 0.70 | 1.2 | 2.5 |
| | 90 | 0.73 | 1.4 | 2.4 |
| 68°F (20°C) | 10 | 1.4 | 4.5 | 10.9 |
| | 50 | 0.47 | 0.99 | 2.9 |
| | 90 | 0.53 | 0.91 | 2.0 |
| 50°F (10°C) | 10 | 2.2 | 4.2 | 5.7 |
| | 50 | 0.43 | 1.3 | 4.7 |
| | 90 | 0.35 | 0.81 | 2.6 |
| 32°F (0°C) | 10 | 1.4 | 1.7 | 1.9 |
| | 50 | 0.68 | 2.4 | 7.1 |
| | 90 | 0.37 | 1.2 | 4.3 |

*Barriers.* A hill, building or solid wall blocking the path of sound propagation will attenuate the noise. The acoustical characteristics of barriers are discussed in the following section on Controlling Highway Noise.

A row of houses or similar size buildings will cause between 5 and 10 dBA of shielding, depending on the spacing between them. A second row of houses will cause some additional shielding, but only about one-half as much as the first row.

*Ground Cover and Trees.* For sound propagation paths within about 5 ft (1.5 m) the ground, the type of ground cover can have a pronounced effect on attenuation values. The values vary significantly as a function of type of ground cover, frequency and distance above the ground. The effects of these variables are so complicated that they preclude the use of any simplified "rule of thumb" for predicting reliable ground attenuation values. However, subtracting 4.5 dB per doubling of distance not only compensates for a diminishing solid angle of highway exposure (see the preceding discussion on "Distance"), but also makes some allowance for ground attenuation.

Trees and bushes have very little effect on sound propagation, unless the foliage is very dense and the distance through it is 100 ft (30.5 m) or more. For A-weighted sound levels, a deduction of 3 dBA may be subtracted per 100 ft (30.5 m) of distance. Remember that deciduous trees lose their foliage in winter.

*Air Absorption.* At frequencies above 1000 Hz., air absorption becomes an important factor in predicting noise levels at distances of more than about 100 ft (30.5 m). Table 8-3 gives the absorption of sound per 300 ft (91.4 m) for frequencies of 1000 Hz and above. Air absorption is not significant at lower frequencies.

*Weather Conditions.* Wind and temperature gradients will change the direction of sound propagation through the air. This phenomenon is called *refraction*. Because of obstacles along the ground, wind velocities usually increase with increasing altitude. This causes the sound rays to bend down toward the surface downwind of the source, as shown in Figure 8-4. In this case the sound may pass over obstacles that normally would provide some shielding. In an upwind direction, the sound rays bend upward, leaving a shadow zone at some distance from the source.

Sound travels faster through warmer air. With a typical daytime temperature gradient, where the air temperature cools with increasing altitude, the sound rays are bent upward, causing shadow zones at some distance from the source. Downward bending occurs with a temperature inversion or "stable atmosphere," characterized by warmer air above a cool surface layer. Temperature inversions often occur over water during the evening hours, accounting for the ability of someone on one side of a small lake to hear conversation from the other side. Temperature inversions are frequently accompanied by fog, because the surface layer of air has cooled below the dew point.

Cloud layers are often thought of as sound reflectors, causing sound propagation over much longer distances than in clear weather. Although the bottom of a cloud layer does indicate a change in temperature, hence a change in the direction of sound propagation, clouds occur at altitudes much too high to affect surface sound propagation.

The Federal Highway Administration (FHWA) has developed a computer program for predicting A-weighted sound levels generated by highway facilities. This program, named STAMINA, uses data on traffic and propagation path characteristics for input. The output consists of A-weighted sound levels at locations of interest along the highway. The output data may also be incorporated into Geographical Information System (GIS) type programs

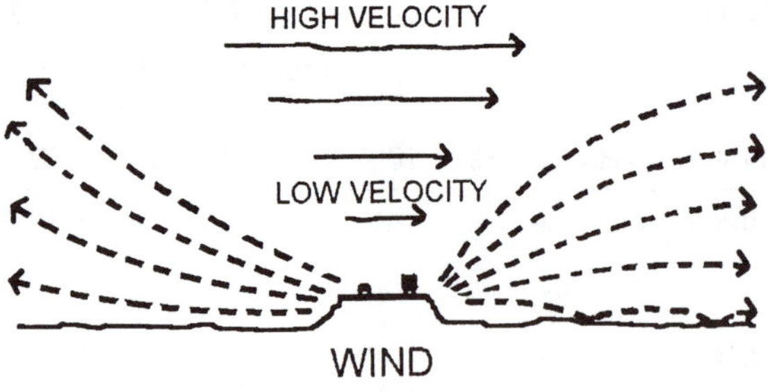


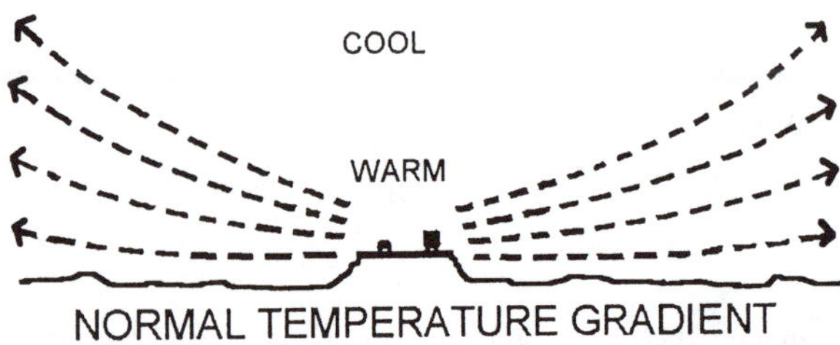


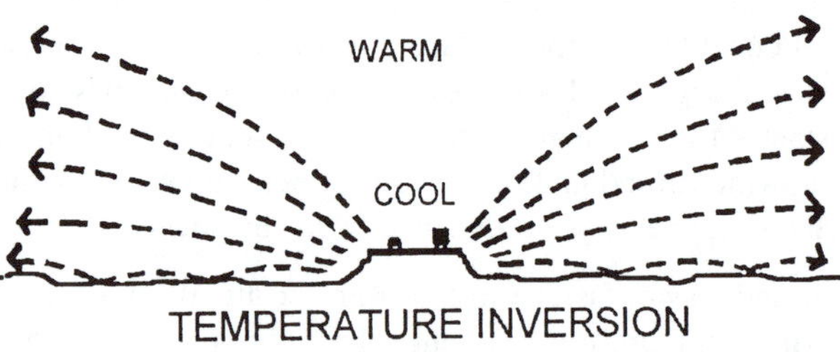


Wind velocity and temperature gradients affect the speed of sound at different altitudes, causing the sound to refract, or change direction.

**Figure 8-4.** Weather effects on sound propagation.

for developing noise level contours based on land topography. The FHWA is currently developing a new computer program called TRAFFIC NOISE MODEL (TNM) to replace STAMINA. This program should be available in 1996.

Traffic data for STAMINA calculations are usually provided by highway traffic engineers. Experience shows that these estimates are occasionally too low for both traffic volume and speed. If the data seem unrealistic, it may be prudent to request that the projections be reviewed to insure accurate noise predictions.

### 8.3.2 Controlling Highway Noise

*Barriers.* By far the most common means of reducing highway noise is through the use of barriers, which block the direct sound transmission path between source and receiver. Barriers may take the form of walls, heavy solid fences, earth berms, the sides of a highway in a cut section, or any combination of these features. One design approach for a highway barrier in a residential area is shown in Figure 8-5.

There are a number of factors that must be considered to design an effective barrier. It must be high enough to interrupt the "line of sound" between source and receiver. If the sound just grazes over the top, it will have a small amount of attenuation (less than 5 dB). The attenuation increases with height up to a practical limit of about 15 to 20 dB.

Barrier performance is limited by diffraction over the top (see Chapter 1, Figure 1-10). To prevent a significant contribution from sound transmitted *through* the barrier, it must weigh at least 4 lb/sq ft (19.5 kg/sq m), and must not have any gaps or cracks, including along the base. Since the top of the barrier now becomes the sound source for the receiver, the propagation path may be 5 ft (1.5 m) or more above the ground, thus negating losses from ground cover that were present before the barrier, and the net affect of the barrier is less than expected.

A barrier is most effective close to either the source or receiver, and is least effective half way between. A problem often encountered with barriers close to a receiver is loss of view. Transparent barriers have been tried in a few instances, but suitable materials that do not crack or yellow with age are yet to be developed.

The noise reduction provided by a barrier is frequency-dependent, a consequence of diffraction phenomena. The performance is poorest at the low frequencies and improves as the frequency increases. A procedure for determining the noise reduction provided by barriers of specified height is shown in Appendix A.7. A barrier providing less than about 5 dBA of attenuation is not cost-effective. Most state departments of transportation require that highway noise barriers achieve at least 8–10 dBA attenuation (recalling that a 10 dB reduction in level is judged to be about one-half as loud—see Chapter 1, Section 1.5.1).

*A precaution:* Barriers must extend well beyond the last building to be protected, to control sound coming around the end of the walls. An alternative is to return the wall along the side of the end properties. As a rule of thumb, no more than 10% of the highway should be exposed beyond the ends of the barrier.

While barriers may be feasible for ground activities and the first story of buildings, it is usually not practical for barriers to protect upper floors where windows may have to be opened for ventilation. If the windows are to remain closed, ventilation will have to be provided by other means, usually a forced-air system with no penetrations on the side exposed to highway noise.

A problem that is frequently overlooked by designers

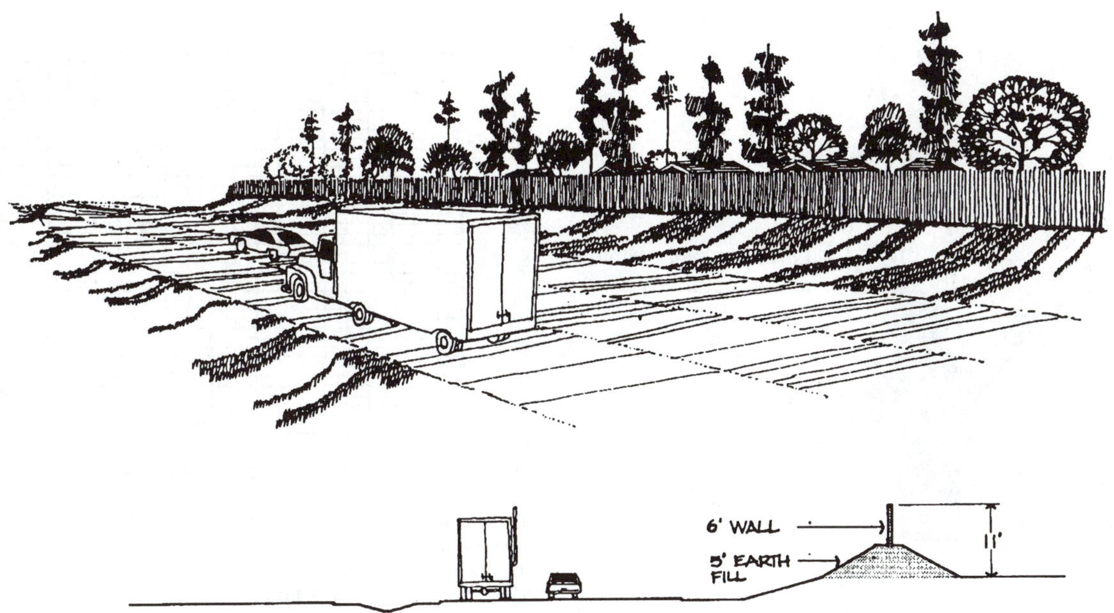


There must not be any cracks in the wall, including along the base.

**Figure 8-5.** Noise barrier for a residential area.

when placing barrier walls on both sides of a highway is that sound will be reflected from the far wall over the near wall and reduce its effectiveness considerably. In this situation, the tops of the walls should be sloped outward at an angle that will reflect the sound upward, high enough to avoid striking buildings on the other side. Alternatives are to use barriers with an absorptive surface, available from several manufacturers, or a construction consisting of interlocking concrete units which creates a sloping surface to reflect the sound with an upward component.

*Lids.* If the highway is completely covered, the noise is virtually eliminated except near the portals, where the receiver is exposed to traffic beyond the lid. Lids, however, are very expensive, require forced air ventilation if more than several hundred feet (about 100 m) in length, may require lighting, and represent safety hazards that may exclude vehicles carrying flammable liquids or explosives.

*Surface Characteristics.* A number of studies have been conducted on the noise produced by different surface materials. Some are slightly less noisy than others when new, but the rank-ordering changes with wear. Traction and safety requirements will usually dictate surface materials.

*Traffic Restrictions.* Moderate noise reductions are possible by limiting traffic speeds. For example, a 5 dBA reduction if achievable by reducing the speed from 55 MPH to 45 MPH (88.5 to 72.4 KPH). Reductions are also possible by restricting truck traffic, at least during the more noise-sensitive nighttime hours.

*Landscaping.* Trees and shrubs are frequently considered for control of noise from traffic. While such landscaping may make a good visual screen, it must be very dense and 100 ft (30.5 m) or more in depth to be at all effective. The plantings must also be a type that does not lose its leaves in the winter, losing what little benefit it does provide. It is very seldom that this method is a satisfactory solution for controlling traffic noise.

## 8.4 Aircraft Noise

Aircraft noise exposure near airports typically is confined to long "fingers" centered on the approach and departure paths, broadening out somewhat at the airport boundaries because of jet engine run-ups, taxiing, and takeoff and reverse thrust operations. Figure 8-6 shows the typical shape of noise contours around a large airport, each contour a designated Ldn value. Aircraft noise places a severe limitation on acceptable land uses within these zones, as shown in Table 8-1. The architect will be required to consider noise exposure for the design of buildings within the 65 Ldn contour.

In recent years there has been increased interest in extending the noise impact zone further out to the 55 Ldn contour. Such a mandate would greatly increase the area in which precautions must be taken in land use planning and building design. It is important that building designers and planners understand aircraft noise impacts, to ensure noise compatible development in airport environs. Most

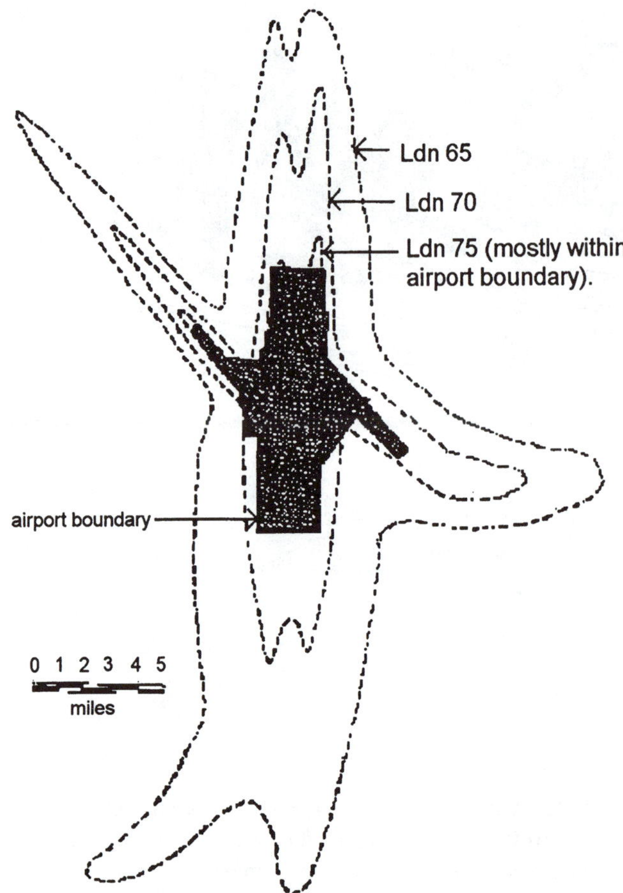


Refer to Table 8-1 for land use recommendations within the Ldn contours.

**Figure 8-6.** Typical Ldn contours for a large airport.

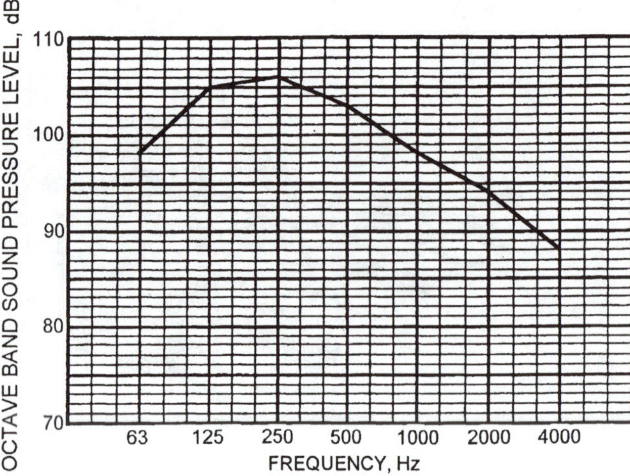


The distance to the runway was 1000 ft (305 m), but the maximum level (which is shown) occured when the jet had passed beyond the microphone location. Actual distance to the jet was closer to 1200 ft (366 m). Noise from stage 3 aircraft will be substantially less.

**Figure 8-7.** Typical spectrum of a jet departure (Stage 2 Boeing 727, 104 dBA).

major airports in the United States attempting to implement much needed expansion are being challenged by area residents complaining about noise. Such problems can be avoided, or at least minimized, with appropriate land use planning and building design, coupled with aircraft operations sensitive to noise impacts.

### 8.4.1 Sources of Aircraft Noise

The major source of noise around airports is departing jet aircraft. For a stationary listener on the ground, the spectrum changes as the aircraft approaches, passes overhead, and departs. The spectrum is predominantly mid- and high frequency on the approach, then shifts to lower frequencies as the aircraft departs. This frequency change is caused by the *Doppler* effect, where the frequency of a moving noise source is shifted up when the source is moving towards the listener, and down as it moves away.

When designing a building exposed to jet aircraft noise, the departing spectrum should be used for selecting facade elements. That is because the low frequency noise components will be the controlling factor in the design. Also, the highest noise levels radiated from a jet engine are at an angle of about 45° from the aircraft axis, toward the rear. A typical spectrum of a stage 2 (see below) departing jet aircraft at a distance of 1000 ft (305 m) is shown in Figure 8-7.

Part of the Federal Aviation Administration's (FAA) aircraft certification process involves a fairly complicated procedure for assessing noise. The aircraft are placed into one of three noise categories, designated stage 1, stage 2 and stage 3. Stage 1 aircraft are the oldest and noisiest (707, DC-8), and are no longer permitted to operate at most U.S. airports. The second generation stage 2 aircraft are about 10 dBA quieter than the stage 1, and include such aircraft as the DC-9, DC-10, 727, 737, 747 and L1011. Stage 3 aircraft, such as the 757, 767, and 777 are another order of magnitude quieter than the stage 2. All military aircraft are exempt from noise regulations.

Stage 2 aircraft are gradually being replaced by stage 3 aircraft, or are being retrofitted with noise reduction packages to bring them into compliance with stage 3 performance requirements. The FAA has mandated that only stage 3 aircraft may be in operation by the year 2000. Total noise exposure will not decrease by as much as might be expected, however, because of the anticipated increase in the number of operations.

Noise in the immediate vicinity of airports can be generated by engine maintenance run-ups, taxiing, the takeoff roll, and reverse thrust during landing. The latter two categories contribute to what is called "sideline noise," and occurs at airport boundaries parallel to the runway(s).

### 8.4.2 Aircraft Noise Prediction

The Federal Aviation Administration has developed a computer program called the Integrated Noise Model (INM) for predicting noise exposure around airports. The program generates equal noise contours using the Ldn descriptor (see Figure 8-6). Input data includes the average number of daily operations, type of aircraft, destination (which affects the rate of climb-out), flight path, landing and departure profiles, weather conditions and runway usage. Program updates will include land topography for predicting the sideline noise components.

The Ldn contours are useful for land use planning such as making appropriate zoning assignments, as discussed in Section 8.2.1. Caution should be exercised when interpreting contours in the 55 to 65 Ldn range, because at this distance from the airport flight paths may vary substantially from those assumed in the program. It is prudent to check with local residents about actual departure conditions.

Ldn contours should be available for all major airports, prepared in accordance with the FAA's part 150 noise abatement program. For building design purposes, however, it is better to use octave band data, particularly because of the strong low frequency components. Such data will not be as readily available as Ldn data, and will probably require the services of an acoustical consultant. A typical jet aircraft spectrum shown in Figure 8-7 can be used if properly adjusted for distances other than 1000 ft (305 m).

One other noise descriptor often used to describe aircraft noise is the Single Event Level (SEL, except SENEL in California). The SEL is the A-weighted level of a single aircraft event. The noise level is sampled each second for the time period that the level is within 10 dBA of the maximum level (usually 20 to 40 seconds), and all levels are summed together. The SEL, then, is the A-weighted level of an event as though it lasted only one second.

### 8.4.3 Aircraft Noise Mitigation

Airports have very little control over aircraft flight trajectories once the aircraft is airborne. While the FAA has designated departure paths and power cutbacks for noise abatement purposes, such procedures are limited by safety and economic considerations. Airport operators may, however, restrict night operations, particularly for the noisier stage 2 aircraft.

Acoustically treated blast fences are often used to control engine run-up noise in a designated maintenance area. Sideline noise may be reduced by barriers, though once the aircraft is airborne, the barrier is no longer effective.

It is imperative that the design of any building housing noise-sensitive uses in the vicinity of an airport consider the noise exposure. The relevant conditions of noise exposure are the maximum noise spectrum and level, and the duration of the noise at or near the maximum level. Typical aircraft noise events will last from about 20 seconds close to the airport, to 45 seconds further out. Momentary interruptions from noise may be permissible in an office building, for example, and even in school classrooms within certain limits, but aircraft should not be heard at anytime in studios, performing arts theaters and other critical uses.

Sleep interference from aircraft noise is an important consideration in residences, hotels and motels. Figure 1-16 in Chapter 1 shows the effect of an aircraft noise event on sleep. If it is decided that a 10% probability of being awakened is acceptable (corresponding to a maximum noise level of about 50 dBA), and the maximum level outside is known, the required facade noise reduction is the difference.

## *8.5 Electrical Substations*

Electric power distribution facilities are often placed in residential areas. The large transformers can be a source of objectionable "hum," and may be equipped with noisy cooling fans as well. The noise of a typical transformer consists of a series of single frequencies (or pure tones) with a fundamental frequency of 120 Hz, and higher harmonics at each multiple of 120. A typical noise spectrum of a large power transformer is shown in Figure 8-8. The cooling fan produces more of a broad-band noise, although the blade passage frequency may be predominant.

Barriers may be utilized to control transformer noise, but a number of precautions must be observed. The top of the transformer should be used as the source height. Large power transformers are over 3 meters in height. The barrier walls must not be so close to the transformer as to restrict maintenance or the circulation of cooling air. The interior of the barrier enclosure (if three- or four-sided) should be absorptive, using a material suitable for exterior application (see the description of SOUNDBLOCKS™ in Chapter 2, Section 2.5.2). Finally, the design goal should be 5 dBA less than the regulation limit because of the objectionable nature of the transformer hum.

## *8.6 Race Tracks*

Facilities for automotive racing can be extremely noisy. Some racing vehicles can produce sound levels exceeding 110 dBA at a distance of 50 ft (15.2 m). Some track operators have placed noise restrictions on racing vehicles if the track is located in the vicinity of noise sensitive areas. Even then, complaints about noise may come from 1 or 2 miles (1.6 or 3.4 km) away, particularly if weather conditions favor noise propagation (see Section 8.3.1, "Weather Conditions").

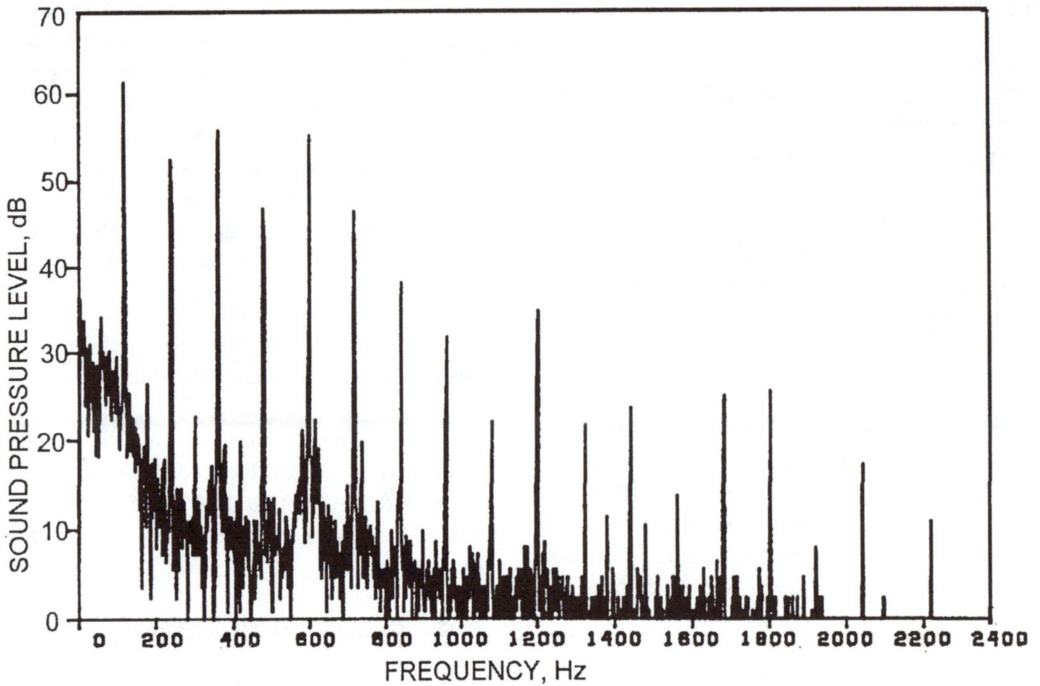


The pure tone frequency components at multiples of 120 Hz are clearly evident.

**Figure 8-8.** Typical transformer noise spectrum. 1 Hz Bandwidth, distance 20 ft (6.1 m), 56 dBA

Barriers can help to attenuate noise radiation away from the track, but they must be quite high if attenuation values of 15 dBA or more are required. Grandstands or other buildings can be effective barriers if properly located. Because car acceleration coming out of turns is particularly noisy, increasing the radius of the curves can help to reduce noise.

## 8.7 Railroad Noise

Noise from railroads is caused primarily by the diesel engines and secondarily from steel wheels on steel rails, particularly with cars that have wheels with flat spots. Typical engine and track noise spectra are shown in Figure 8-9.

There is very little that can be done to mitigate railroad noise. Because of the strong low frequency components from the diesel engines, barriers are not very effective. In addition, there is no practical means for reducing train whistle signaling for crossings.

While track noise can be decreased by using continuously welded track, solutions for railroad noise must come from adjacent property developers. Building design must consider the exposure from both noise and vibration, paying particular attention to the strong low frequency components.

There are special noise problems associated with railroad switchyards. The diesel switch engines are constantly starting up and slowing down, and the reactive-type engine exhaust mufflers actually reinforce the noise at certain engine RPMs, creating very strong low frequency components.

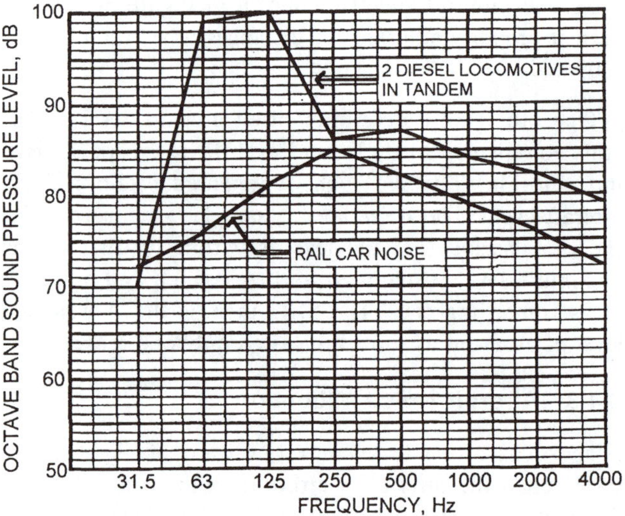


The strong low frequency components of the diesel locomotives are clearly evident in this figure.

**Figure 8-9.** Typical train noise levels. Distance: 200 ft (61 m). Speed: 40 mph (64 kph)

Another noise problem is caused by "humping," where

rail cars are pushed over a rise in the tracks, then shunted to different tracks to make up trains. After passing over the hump, the cars are slowed down by a "retarder," a third rail which rises up and presses against the car wheel. This causes a very loud and annoying high-pitched squeal (frequencies above 1000 Hz.). Following the retarder operation, the car continues on to couple with the train being made up. This causes an impact noise which can be very loud if the car was not slowed sufficiently by the retarder.

Trackside barriers are not used for reducing retarder and car impact noise. Yard workers sometimes ride on cars by hanging on to the sides, and may come in contact with the barrier. Rail companies will not use such barriers because of safety considerations.

Again, the responsibility for reducing rail yard noise impacts rests with the developer of adjacent properties, and the same considerations apply as for railroads.

## 8.8 Outside Mechanical Equipment

One of the most common complaints about outside mechanical equipment is noise coming from a neighbor's air conditioning or heat pump unit. For convenience, these units are frequently mounted on the side of a home and, invariably, it seems, opposite the neighbor's bedroom window. Enclosures are usually unacceptable as they interfere with the circulation of cooling air. A practical solution is to relocate the unit at the rear of the house away from a neighbor's window and, if necessary, put up a barrier wall on the side toward the neighbor.

Outside and rooftop mechanical equipment are often installed at buildings of all types. Typical equipment includes air handling units, compressors, chillers, cooling towers and trash compactors. If residential areas are nearby, the noise from these units may exceed allowable levels. *Caution*: If the equipment can operate at any hour, nighttime noise limits would apply.

Barriers are an effective way of reducing this type of noise, whether the equipment is rooftop or on the ground. Care must be exercised in locating the barrier so as not to interfere with air circulation or servicing requirements. Effective barrier design is presented in Section 8.3.2, and Appendix A.7 includes a procedure for calculating barrier attenuation.

Another potential problem from rooftop equipment is vibration, which can cause noise in critical spaces below. Proper methods of vibration isolation of mechanical equipment are presented in Chapter 7, Section 7.5.

## 8.9 Service Station Car Washes

Service stations are frequently located in residential areas and associated car washes can cause serious noise problems. The noise of the pumps and drying fans is high and the units are usually operated with open doors at both ends.

It is not practical to reduce the noise at the source, so barriers are an effective means for reducing noise radiated by the open ends and walls of car washes. If there is much opposition to the car wash, it may be advisable to include a safety factor of at least 5 dB below the applicable noise ordinance. People can be particularly sensitive to a noise source which they have initially opposed, or consider unnecessary. A detailed discussion of barrier design is given in Section 8.3.2, and Appendix A.7 includes a procedure for calculating barrier attenuation.

Even if the walls of the car wash are masonry or concrete, noise radiated from the roof of a car wash could be a problem. An effective method of reducing this source of noise is to extend the side walls above the edge of the roof.

High power vacuum cleaners are usually associated with car washes to clean the inside of cars. These are usually very noisy and need to be located where they will not cause noise problems to adjacent properties. A wall, partial enclosure or the car wash building between the unit and residences may be beneficial, but there must be room for proper air circulation and servicing. On the other hand, if the unit is placed near or facing a wall, the reflected sound path may be a problem.

While operation of most drive-through car washes in residential areas is limited to the daylight hours, the increasingly popular self-service car washes operate at all hours. Many noise ordinances are more restrictive at night and will be the limiting factor for design.

Residential structures of more than one story may pose problems that make it impractical to install car washes, as the necessary barrier heights may be impractical.

## 8.10 Mining Activities

Mining operations are often located in rural areas where background noise levels are low and residents are accustomed to a quiet environment. The operation will probably involve haul trucks and front-end loaders, while many will also have rock crushers, screens, washers and asphalt or concrete plants.

Predicting noise levels from mining operations at adjacent locations can be quite involved because of moving sources, rough terrain and variable ground cover. Such predictions usually require the services of an acoustical consultant.

Control of noise from mining activities begins with ensuring that all equipment is properly maintained and adequately muffled. Special noise mitigation packages are available for some of the equipment, such as high perform-

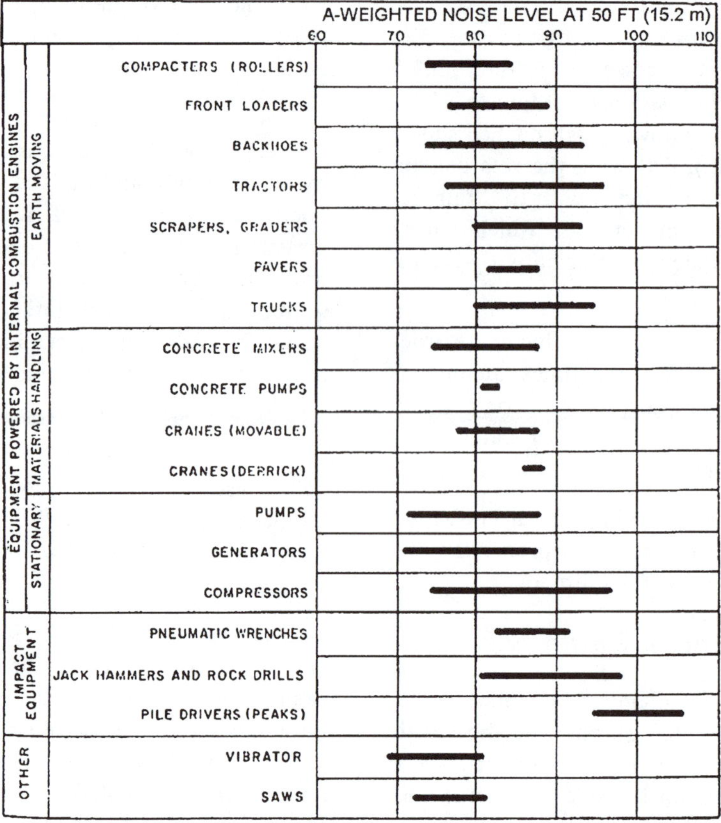


Noise at the upper limits is probably from poorly maintained equipment. At most construction sites, several machines will be operating simultaneously. Caution: Noise at the operator for hand-held equipment will be substantially higher than the levels shown, which are for a distance of 50 ft (15.2 m).

**Figure 8-10.** Noise levels of construction equipment.

ance mufflers for loaders, shrouds or barriers for rock crushers, and rubberized surfaces for screens. It may be possible to locate haul roads away from noise-sensitive areas, and restrict hours of operation to avoid nights and weekends. Frequent grading of haul roads to avoid pot holes helps to reduce noise from banging tailgates or other loose equipment.

Barriers can be very effective in attenuating noise from both fixed and moving equipment. Excess material is usually plentiful at mining sites and can be used to build earth berms. Design of barriers is covered extensively in Section 8.3.2, and Appendix A.7 includes a procedure for calculating barrier attenuation.

## 8.11 Construction Noise

At one time or another almost everyone will be exposed to construction noise. Fortunately such noise is usually temporary, although it can be very annoying and bring with it dust, fumes and vibration.

Noise regulations usually recognize the temporary nature of construction noise, and impose less restrictive limits than for other types of environmental noise. It is not uncommon, however, to restrict construction at night and on Sundays.

Noise from construction equipment varies widely in both frequency and level depending on the type of equipment. Predicting construction noise at distant locations is further complicated by equipment that moves around the construction site and use factors that may vary from infrequent operation to continuous. The range of A-weighted noise levels produced by various types of equipment at a distance of 50 ft (15.2 m) is shown in Figure 8-10.

Methods for controlling construction noise are similar to those used for mining activities as discussed in the preceding section. Compliance with noise regulations will be the responsibility of the general contractor. A common problem is communicating noise mitigation practices to the various subcontractors working at the site.

# Chapter 9

# Industrial Noise Control

The control of noise generated by industrial activity is concerned with preventing permanent damage to hearing, with ensuring reliable speech communication and audibility of other signals, and with preventing unacceptable noise levels in the adjacent community. There are state and federal noise regulations for the prevention of hearing damage. Community noise is usually regulated at the state and/or local level and is discussed in detail in Chapter 8. Interference with speech communication must be controlled so that instructions and warning messages are clearly understood. The effects of noise on speech are discussed in Chapter 1, Section 1.8.

Vibration caused by machinery can be a major concern. Many industrial processes are controlled by computers, which can be sensitive to vibration with the possibility of malfunction. The noise environment of the computer operators must also be considered. Excessive vibration can accelerate machine wear, leading to premature failure. Methods of vibration isolation are presented in Chapter 7.

The control of industrial noise is specific to the type of machine or operation, and it is beyond the scope of this handbook to recommend treatments for individual noise sources. This Chapter will present the principles and basic methods of achieving noise control in industrial environments.

## 9.1 Damage to Hearing—OSHA Criteria

Permanent damage to hearing caused by excessive noise has received considerable attention during the past 30 years and continues to be studied with ongoing research programs. Federal regulations are promulgated by the Department of Labor through the Occupational Safety and Health Administration (OSHA), and by similarly structured state agencies and agencies in other countries.

Noise levels which exceed 85 to 90 dBA can result in permanent loss of hearing if the 8-hour daily exposure continues for a period of years. OSHA regulations specify maximum permissible exposure times without hearing protection for various noise levels, as shown in Table 9-1:

**TABLE 9-1.** OSHA Permissible Exposure Times Without Hearing Protection

| Noise Level, dBA | Permissible Exposure Time, hours |
|---|---|
| 90 | 8 |
| 95 | 4 |
| 100 | 2 |
| 105 | 1 |
| 110 | 0.5 |
| 115 | 0.25 |

Even though the base noise level specified in Table 9-1 is 90 dBA, OSHA requires that a hearing conservation program be instigated if the level exceeds 85 dBA. The program consists of periodic checks of worker noise exposure and hearing acuity. If the noise exposure exceeds 90 dBA, the regulations require "engineering controls," stating that the noise exposure of the worker be reduced by actually reducing the noise level, or by moving the worker to a quieter location for part of the work day. If these methods are insufficient to reduce worker exposure to safe levels, hearing protectors are required.

As shown in Table 9-1, OSHA decreases permissible exposure time by 1/2 for every 5 dBA increase in noise level, starting at 90 dBA for 8 hours. This 5 dBA increment is termed the "exchange rate." As this handbook is being prepared, the National Institute for Occupational Safety and Health (NIOSH) is preparing a draft noise exposure criteria document recommending an 85 dBA, 8 hour base line exposure, using a 3 dBA exchange rate instead of the present 5 dBA rate. This effect of such a change in both base level and exchange rate is shown in Table 9-2.

The Institute of Noise Control Engineering (INCE) has reviewed the NIOSH draft document and prepared comments as follows from a letter to NIOSH from Dr. David M. Yeager, 1996 INCE/USA President.

> This is a much more difficult criterion to meet, significantly affecting the approach to noise control engineering. Under the draft criterion, many more workers will be exposed to unacceptable noise, requiring engineering control. Industry will be faced with the compounding efforts to implement engineering controls

**TABLE 9-2.** Comparison of A-weighted Noise Levels and Exchange Rates

| | Permissible Exposure Time, hours | | | | | |
|---|---|---|---|---|---|---|
| Agency | 8.0 | 4.0 | 2.0 | 1.0 | 0.5 | 0.25 |
| Current OSHA Regulation | 90 | 95 | 100 | 105 | 110 | 115 |
| NIOSH Proposal | 85 | 88 | 91 | 94 | 97 | 100 |

in more areas of its plants, plus more extensive controls to meet the draft criterion. The medical community may deem that the draft 85 dBA/3 dBA criterion is appropriate for the health of workers; NIOSH should understand that meeting the criterion will be a significant financial burden to industry.

## 9.2 Interference with Speech or Warning Signals

Clear understanding of speech is important in most industrial operations. Workers may be required to verbally communicate while operating on, or working near noisy machines. Use of the telephone may be required, and paging messages, warning signals or instructions must be clearly heard at all worker locations. As mentioned earlier, the effects of noise on speech are presented in Chapter 1, Section 1.8.

In most industrial environments a sound reinforcement system will be required for paging and delivering warning instructions or signals. A properly designed system must have an adequate number of loudspeakers with the audio output concentrated in the frequency range appropriate for the understanding of speech, or match the frequency content of the warning signal. The loudspeaker spacing must be determined so that the signal is not lost in the background noise.

If the shop where the loudspeakers are to be installed is highly reverberant, the installation of acoustical absorbents can be beneficial. Added absorption will reduce the reverberant noise level, and will reduce the number of loudspeakers required by permitting a larger spacing between them. Hanging baffles is an efficient acoustical treatment that is frequently used in industrial shops. Sound absorption data on baffles is presented in Chapter 2, Section 2.5.3. If absorption cannot be added, the directivity pattern of the loudspeakers should concentrate the sound where it needs to be heard, and not allow sound to spill over into areas where the signal is not required. The extra acoustic energy simply adds to the reverberant noise level, thus reducing the signal-to-noise ratio of the paging or warning systems.

If workers are required to wear hearing protection (earmuffs or plugs), the signal reduction caused by such devices must be considered in determining proper sound levels for the paging or warning systems. The decibel attenuation of protective hearing devices is available from the manufacturers. Fortunately, hearing devices reduce the noise level by about the same amount as they reduce the signal level, so the ratio between the two levels remains the same.

Reliability of telephone usage can be improved by locating the instrument in a small partial enclosure, the inside surfaces of which are highly absorbent. Also, this is an application for active noise cancellation, since the area where the noise reduction is needed is very well defined; refer to Chapter 2, Section 2.8

## 9.3 Noise Control Techniques

Different aspects of the subject of noise control are treated throughout this handbook, and the basic principles presented apply to industrial noise control as well. A systematic approach to such problems starts with treating the source, then the transmission path, and finally the receiver. Even though it is usually along the transmission path where most of the noise reduction will be achieved, the source and receiver location noise characteristics must be well understood, in order to define the amount of noise reduction required.

In existing industrial facilities, a noise survey can provide the necessary data on source and receiver location characteristics. The basic measurement techniques are discussed in Appendix A.10, though a qualified acoustical consultant should be employed for such a survey, particularly for obtaining data on a large, complicated noise source. For new facilities in the design phase, computer programs have been developed that plot noise contours throughout the building, based on the location and noise radiation characteristics of major noise-producing machines, and on the acoustical characteristics of the building.

### 9.3.1 Noise Control at the Source

The first consideration for reducing noise from a machine is to treat the source itself, which means the treatment is applied directly to the machine. Obviously this is a machine-specific task, and must be done with full consideration of impacts any noise reduction treatments may have on safety, efficient operation, maintenance, cost and warranties. It is advisable that the manufacturer be consulted on any treatments to be applied to or near a machine.

A machine that has large noise radiating surfaces, such as sheet metal panel enclosures, may be treated by installing suitable absorptive materials on the inside surfaces, and by applying damping materials to the surfaces to reduce vibration. The use of vibration isolators between the machine and the panels can also reduce the amount of vibration transmitted to the panels. *Caution*: Panels or baffles attached directly to a machine for noise shielding must

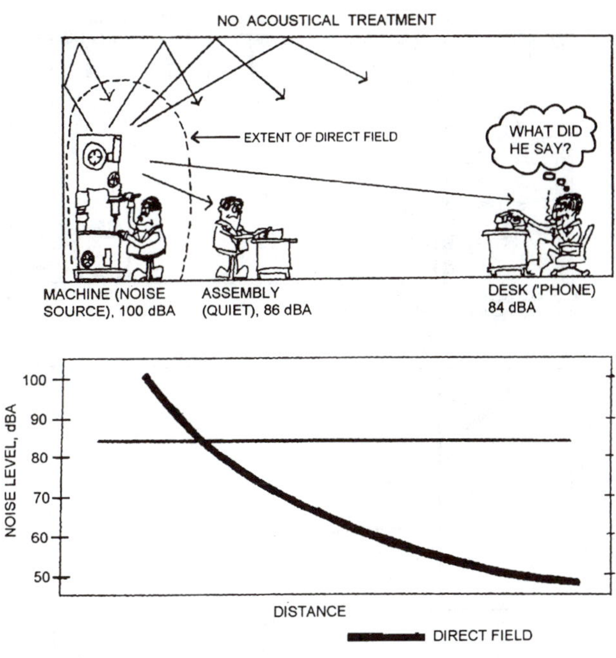


**Figure 9-1a.** Noise control through the use of absorbents, barriers and ear protection.

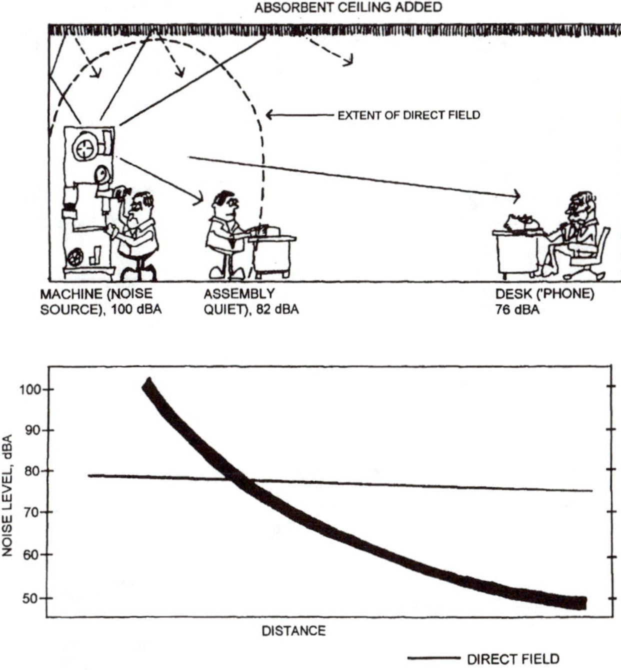


**Figure 9-1b.** Noise control through the use of absorbents, barriers and ear protection.

incorporate vibration isolation at the points of attachment. If this is not done, the panels may become noise radiators and defeat their intended purpose. Manufacturers of vibration isolation mounting devices are listed in Appendix A.12.

### 9.3.2 Noise Control along the Transmission Path

In order to correctly assess the benefits of transmission path noise reduction treatments, a clear understanding of direct and reverberant sound fields is necessary. This subject was presented in Chapter 4, Section 4.6.2, and will be further treated in this Section.

Figures 9-1a through 9-1e depict a progressive reduction in noise at three work stations with the application of acoustical treatments. The top cartoon in each figure shows a worker at a noisy machine, a second worker at a nearby assembly area, and a third worker at a distant desk using the telephone. Below each cartoon is a graph showing the effect of the various treatments on the direct and reverberant noise fields in the shop as a function of distance from the source. Each figure is discussed in the following text.

*Figure 9-1a.* The shop has all hard surfaces. The operator of the machine is in the direct field, the limits of which are within the dashed line in the cartoon, and represented by the heavy line in the lower graph. The noise level is 100 dBA at this location close to the machine. Outside of the direct field area, the reverberant field is dominant, and is uniform throughout the shop at a level of 84 dBA. The assembly worker is at a location where the reverberant field is dominant, but with a small contribution by the direct field, resulting in a noise level of 86 dBA. The worker at the desk using the telephone is in the 84 dBA reverberant field, with no significant direct field component.

An evaluation of the noise exposures represented in Figure 9-1a places the machine operator in danger of permanent hearing damage, and the assembly worker just over the threshold for being subject to a hearing conservation program according to OSHA regulations. Furthermore, 86 dBA is uncomfortably loud and some noise reduction would provide a better working environment. At the desk, 84 dBA will cause considerable difficulty in using the telephone.

*Figure 9-1b.* A highly absorbent acoustical ceiling has been added to the shop, lowering the reverberant field to a range of 76 to 78 dBA. The reverberant field is no longer uniform throughout the shop, because the absorbent ceiling causes a slight reduction with increasing distance from the source. The direct field is unaffected by the ceiling treatment, but the *cubic volume* now dominated by the direct field has increased and includes the assembly worker. The noise exposure at the machine operator remains unchanged, while the exposure at the assembly worker has been lowered below the hearing conservation program

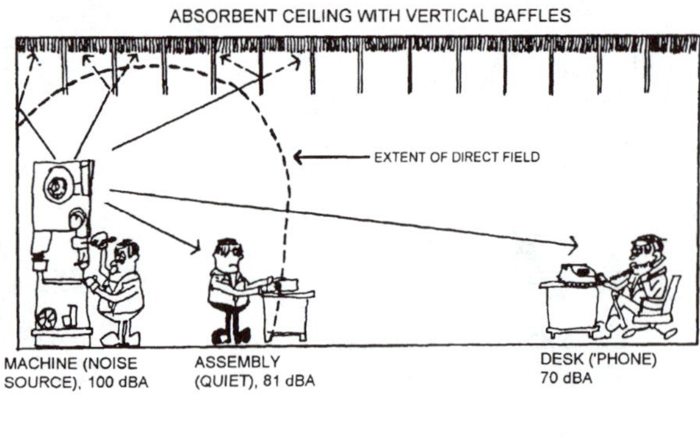


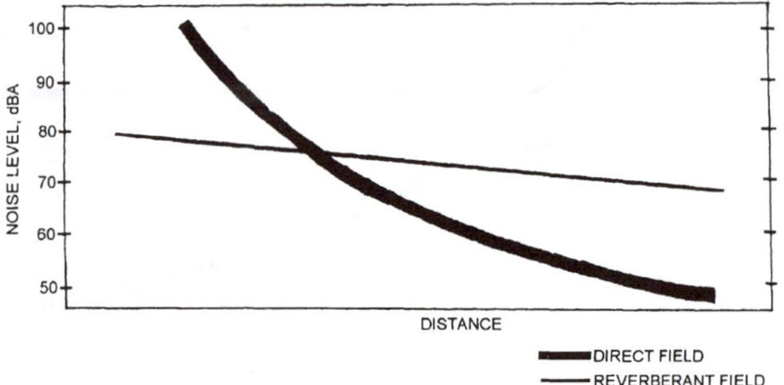


**Figure 9-1c.** Noise control through the use of absorbents, barriers and ear protection.

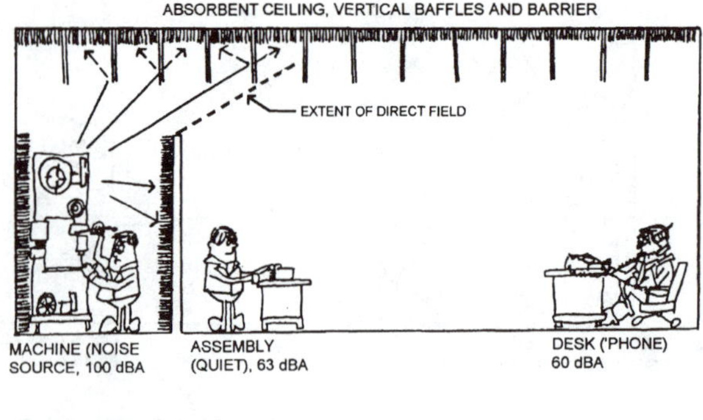


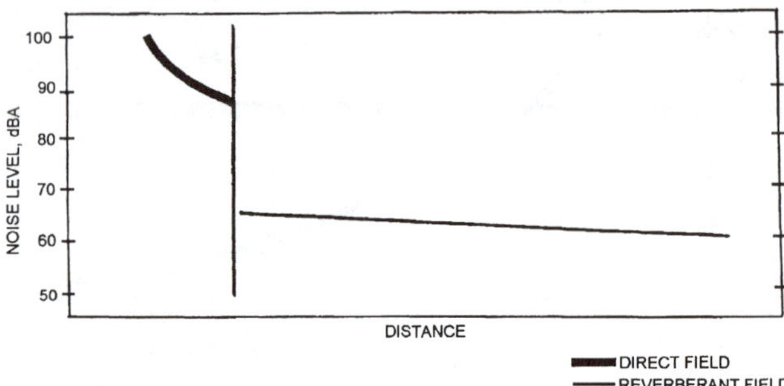


**Figure 9-1d.** Noise control through the use of absorbents, barriers and ear protection.

threshold. The noise level at the desk, while significantly lower, is still too high for telephone usage.

*Figure 9-1c.* Vertically hanging absorbent baffles have now been added to the ceiling, resulting in a more rapid decrease in the reverberant field level with distance. Only the desk location benefits from this treatment; the other two worker locations are in the direct field, which is unaffected by shop surface treatments.

*Figure 9-1d.* A vertical barrier has been erected between the machine operator and the assembly worker. The barrier interrupts the direct field affecting the assembly worker, resulting in a substantial reduction in noise. Also, highly absorbent acoustical treatment has been added to the machine side of the barrier and the wall surface in back of the machine. This treatment prevents a buildup of noise within the partial enclosure, which could actually *increase* the exposure of the machine operator. Additionally, by placing the treatment close to the noise source, the amount of acoustic energy escaping into the shop is reduced, causing a lowering of the reverberant field level. This benefits both the assembly worker and the desk location.

*Figure 9-1e.* For the final step in the treatment schedule, earmuffs have been provided for the machine operator, and a second barrier has been erected close to the desk location. Notice this barrier is not nearly as effective as the machine barrier because it is not close to the source. The earmuffs, by providing 20 dBA of attenuation, provide the machine operator with sufficient protection to avoid permanent hearing damage. The assembly worker is at a rela-

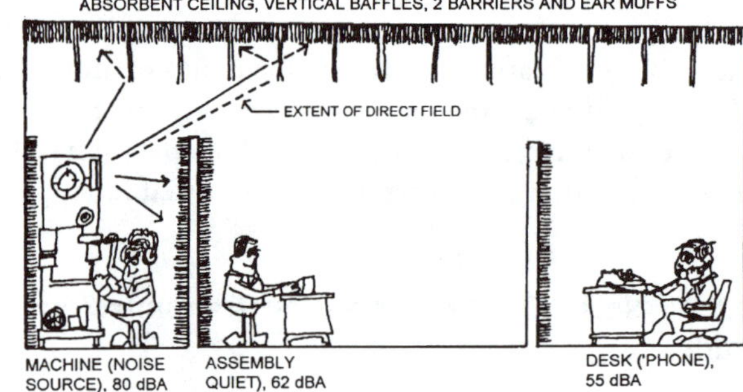


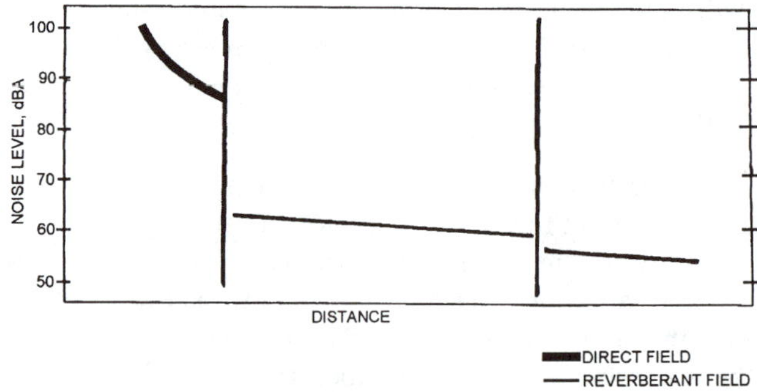


**Figure 9-1e.** Noise control through the use of absorbents, barriers and ear protection.

tively quiet 62 dBA, and the telephone user should experience no difficulty at a level of 55 dBA.

An alternative treatment for the machine operator that might be possible is active noise cancellation. In essence, a microphone would sample the noise field at an appropriate location, probably near the operator's head. The signal from the microphone would undergo a phase shift, then feed loudspeakers such that the reproduced sound would be exactly out of phase with the direct sound arriving at the operator's ears. This noise cancellation technique is relatively new but may be quite effective in applications such as this. The process probably would have little effect on the noise field in other parts of the shop.

As demonstrated in Figure 9-1, barriers or partial enclosures around noisy machines can be quite effective. A more complete enclosure is practical where the operator does not have to be present in the immediate vicinity. The noise source may be enclosed with ports to feed material through, such as a punch press. Heavy vinyl curtains on a track may be placed around a noise source to permit easy access. These curtains may have an absorbent lining to reduce the noise level in the enclosure. It is also important to have an absorptive ceiling above the curtain to reduce reflected noise off the ceiling. In Appendix A.12 are listed manufacturers of these products, numbers 18 and 37.

Caution must be exercised in erecting barriers or enclosures around machines where air circulation is needed for cooling. Supplementary blowers and ducting may be necessary to provide sufficient air. Without these precautions, the machine may exceed its operating temperature, and warranties may no longer be binding.

Noise reduction of the reverberant field can be achieved with absorptive materials applied to the ceilings and walls. The noise reduction (NR) attainable can be determined by

$$NR = 10 \log \frac{A1 + A2}{A2}$$

where A1 = original absorption, sabins (U. S. or metric)

A2 = added absorption, sabins (U. S. or metric)

Since the total absorption in the shop is known, the reverberation time (RT, or $T_{60}$) may be calculated using the Sabine equation (see Section 4.6.3):

$$RT\,(U.S.) = \frac{0.049 \times \text{Volume}}{\bar{\alpha} \times \text{Area}},$$

$$(\text{metric}) = \frac{0.161 \times \text{Volume}}{\bar{\alpha} \times \text{Area}}$$

The RT is useful in evaluating the shop for speech intelligibility. Also, the reduction in loudness may be calculated using a method described in Appendix A.8. An example calculation of these measures follows.

A shop constructed of concrete walls and ceiling measures 100 ft (30.5 m) × 200 ft (61.0 m) in plan with a ceiling height of 20 ft (6.1 m). The absorption for the concrete floor, walls and ceiling combined with an assumed absorption for all machines and other shop contents gives an *average* sound absorption coefficient, $\bar{\alpha}$, at 1000 Hz for all shop surfaces of 0.04. The RT is calculated as follows:

U. S. units:

Ceiling area: 100 ft × 200 ft = 20,000 sq ft

Floor area: 100 ft × 200 ft = 20,000 sq ft

Wall area: 600 ft × 20 ft = 12,000 sq ft

Total area: 20,000 sq ft + 20,000 sq ft + 12,000 sq ft = 52,000 sq ft

Volume: 100 ft × 200 ft × 20 ft = 400,000 cu ft

Reverberation time: $\frac{0.049 \times 400{,}000}{0.04 \times 52{,}000} = 9.4$ sec

Metric units:

Ceiling area: 30.5 m × 61.0 m = 1,860 sq m

Floor area: 30.5 m × 61.0 m = 1860 sq m

Wall area: 183 m × 6.10 m = 1116 sq m

Total area: 1860 sq m + 1860 sq m + 1116 sq m = 4836 sq m

Volume: 30.5 m × 61.0 m × 6.10 m = 11,349 cu m

Reverberation time: $\frac{0.161 \times 11{,}349}{0.04 \times 4837} = 9.4$ sec

Assume the shop is treated with acoustical baffles suspended in 25 rows spaced 4 ft (1.22 m) apart. With 50 baffles per row, the total number of baffles is 1250. The absorption per baffle at 1000 Hz is 6 U.S. sabins (0.56 metric sabins).
The total absorption is:

U. S. units:

Absorption of baffles: 6.0 × 1250 = 7500 sabins

Original absorption of shop: 0.04 × 52,000 sq ft = 2080 sabins

Total absorption: 7500 + 2080 = 9,580 sabins

Reduction in noise level: $10 \log \frac{(2{,}080 + 7500)}{2080} = 6.6$ dB

Metric units:

Absorption of baffles: 0.56 × 1250 = 700 sabins

Original absorption of shop: 0.04 × 4837 sq m = 193 sabins

Total absorption: 700 + 193 = 893 sabins

Reduction in noise level: $10 \log \frac{193 + 700}{193} = 6.7$ dB

If the original reverberant noise level in the shop was 90 dB at 1000 Hz, the added absorption will reduce it 6.6 dB to a level of 83.4 dB. A quick way to find the percentage loudness reduction at 1000 Hz is to consult the band loudness indexes at 1000 Hz for sound levels of 90 db and 83 dB in Appendix A.8, Table A-8-1. The sone values for 90 dB and 83 dB are 32.9 and 20, respectively.
The percentage loudness reduction is determined as follows:

$$\text{Loudness reduction} = 100\ \frac{(32.9 - 20)}{(32.9)} = 39.2\%$$

The procedure for calculating the total loudness reduction for all frequencies is presented in Appendix A.8.

The new reverberation time is calculated as follows:

U. S. units:

$$\text{RT} = \frac{0.049 \times 400{,}000}{9580} = 2.0\ \text{sec}$$

Metric units:

$$\text{RT} = \frac{0.161 \times 11{,}349}{893} = 2.0\ \text{sec}$$

These calculations have been made for a frequency of 1000 Hz only. To determine the change in the A-weighted (dBA) noise level, these calculations would have to be repeated for each frequency band (octave or 1/3 octave), apply the A-weighting adjustment (see Appendix A.4), then total the adjusted levels following the method for adding decibels as described in Chapter 1, Section 1.2.2. To determine the effects of reverberation on speech communication, average the noise levels at the 3 Speech Interference Level frequencies of 500, 1000 and 2000 Hz, and assess the impact using Figure 1-15 in Chapter 1.

Prefabricated enclosures designed to provide a quiet environment in a noisy area are available from several manufacturers. These rooms are usually portable or relocatable and provide air ventilation. In some cases there may be vibration-sensitive equipment in the room such as computers, and isolation of the room from floor vibration may be necessary. Refer to Appendix A.12 for the manufacturers with products for industrial noise control, numbers 1, 18, 33 & 37.

### 9.3.3 Noise Control at the Receiver

A worker that is exposed to high noise levels that are not amenable to mitigation must participate in a hearing conservation program involving periodic testing of hearing acuity. Such a program may be developed in-house, or may utilize the services of an outside industrial hygienist. For in-house testing, booths constructed in a similar manner as those intended for quiet areas in noisy locations are available from the same manufacturers.

Other options involve relocating the worker part of the day to a quieter area. In many cases the use of hearing protection will be required, such as earplugs or earmuffs. There are several types of ear plugs with varying degrees of efficiency and comfort. Earmuffs may be used in place of earplugs, both for greater comfort and efficiency. However, noise reduction is always the first option, thus freeing the worker from the burden of providing self-protection.

# Appendix

## A.1 U.S. to Metric Conversion Factors

Conversion factors listed below are only for quantities used in the handbook.

| To Convert — | To — | Multiply By — |
|---|---|---|
| inches (in) | millimeters (mm) | 25.40 |
| feet (ft) | meters (m) | 0.3048 |
| miles (mi) | kilometers (km) | 1.6093 |
| square feet (sq ft) | square meters (sq m) | 0.0929 |
| cubic feet (cu ft) | cubic meters (cu m) | 0.0283 |
| pounds (lb) | kilograms (kg) | 0.4536 |
| pounds/sq foot (lb/sq ft) | kilograms/sq meter (kg/sq m) | 4.8824 |
| pounds/cu foot (lb/cu ft) | kilograms/cu meter (kg/cu m) | 16.0185 |
| degrees Fahrenheit (°F) | degrees centigrade (°C) | $\frac{5\,(°F - 32)}{9}$ |

Note: Physical sound characteristics, such as sound pressure and sound power, are always expressed in metric units.

A useful reference is the *Metric Guide for Federal Construction*, published by the National Institute of Building Sciences, 1201 L Street N.W., Washington, D.C. 20005. Call (202) 289-7800 for ordering information.

## A.2 Frequency Bands and Weightings

This list contains preferred frequencies for use in acoustics, and the relative response of sound level meter frequency weightings meeting ANSI specification: S1.4. See Chapter 1, Sections 1.1.1 and 1.6.

| | | Relative Response, dB | | |
|---|---|---|---|---|
| Octave Band Centers, Hz | One-third Octave Band Centers, Hz | A | B | C |
| | 12.5 | −63.4 | −33.2 | −11.2 |
| 16 | 16 | −56.7 | −28.5 | −8.5 |
| | 20 | −50.5 | −24.2 | −6.2 |
| | 25 | −44.7 | −20.4 | −4.4 |
| 31.5 | 31.5 | −39.4 | −17.1 | −3.0 |
| | 40 | −34.6 | −14.2 | −2.0 |
| | 50 | −30.2 | −11.6 | −1.3 |
| 63 | 63 | −26.2 | −9.3 | −0.8 |
| | 80 | −22.5 | −7.4 | −0.5 |
| | 100 | −19.1 | −5.6 | −0.3 |
| 125 | 125 | −16.1 | −4.2 | −0.2 |
| | 160 | −13.4 | −3.0 | −0.1 |
| | 200 | −10.9 | −2.0 | 0 |
| 250 | 250 | −8.6 | −1.3 | 0 |
| | 315 | −6.6 | −0.8 | 0 |
| | 400 | −4.8 | −0.5 | 0 |
| 500 | 500 | −3.2 | −0.3 | 0 |
| | 630 | −1.9 | −0.1 | 0 |
| | 800 | −0.8 | 0 | 0 |
| 1000 | 1000 | 0 | 0 | 0 |

*(Continued on next page)*

| Octave Band Centers, Hz | One-third Octave Band Centers, Hz | Relative Response, dB A | B | C |
|---|---|---|---|---|
| | 1250 | +0.6 | 0 | 0 |
| | 1600 | +1.0 | 0 | −0.1 |
| 2000 | 2000 | +1.2 | −0.1 | −0.2 |
| | 2500 | +1.3 | −0.2 | −0.3 |
| | 3150 | +1.2 | −0.4 | −0.5 |
| 4000 | 4000 | +1.0 | −0.7 | −0.8 |
| | 5000 | +0.5 | −1.2 | −1.3 |
| | 6300 | −0.1 | −1.9 | −2.0 |
| 8000 | 8000 | −1.1 | −2.9 | −3.0 |
| | 10000 | −2.5 | −4.3 | −4.4 |
| | 12500 | −4.3 | −6.1 | −6.2 |
| 16000 | 16000 | −6.6 | −8.4 | −8.5 |
| | 20000 | −9.3 | −11.1 | −11.2 |

## A.3 Weights of Common Building Materials

| Material | Density lbs/cu ft | Density kg/cu m |
|---|---|---|
| Concrete, calcareous gravel aggregate | 167 | 2675 |
| Concrete, sand & gravel aggregate | 150 | 2403 |
| Concrete, slag aggregate | 110 | 1762 |
| Concrete, cinder aggregate | 93 | 1490 |
| Concrete, shale aggregate | 80 | 1281 |
| Asphalt shingles (fire retardant) | 71 | 1137 |
| Linoleum | 86 | 1378 |
| Brick | 139 | 2227 |
| Glass | 144 | 2307 |
| Sand, wet | 125 | 2002 |
| Sand, dry | 100 | 1602 |
| Gypsum board | 48 | 769 |
| Gypsum board, type X | 50 | 801 |
| Glass fiber batt insulation | 1 | 16 |
| Rock wool insulation | 3 | 48 |
| Glass fiber insul. board, low density | 2.5 | 40 |
| Glass fiber insul. bd., med. density | 6 | 96 |
| Glass fiber insul. bd., high density | 10 | 160 |
| Magnesium | 122 | 1954 |
| Aluminum | 165 | 2643 |
| Zinc | 415 | 6648 |
| Brass | 534 | 8554 |
| Copper | 556 | 8906 |
| Lead | 710 | 11373 |
| Sheet steel | 489 | 7833 |
| Gypsum plaster, 68% sand | 109 | 1746 |
| Gypsum plaster, 15% vermiculite | 51 | 817 |
| Insulation board, cellulose fiber | 20 | 320 |
| Insulation board, mineral fiber | 24 | 384 |
| Plywood, douglas fir | 36 | 577 |
| Wood, douglas fir | 38 | 609 |
| Hardboard, medium | 67 | 1073 |
| Hardboard, tempered | 96 | 1538 |

## A.4 Recommended RC, NC and A-Weighted Criteria*

Criteria for Acceptable HVAC Noise Levels in Unoccupied Rooms (See Chapter 1, Section 1.7, and Chapter 7, Section 7.1.)

| Occupancy | Preferred | Alternate | dBA |
|---|---|---|---|
| Private Residences | RC 25–30 | NC 25–30 | 30–35 |
| Apartments | RC 30–35 | NC 30–35 | 35–40 |
| Hotels/Motels | | | |
| Individual rooms or suites | RC 30–35 | NC 30–35 | 35–40 |
| Meeting/banquet rooms | RC 30–35 | NC 30–35 | 35–40 |
| Halls, corridors, lobbies | RC 35–40 | NC 35–40 | 40–45 |
| Service/support areas | RC 40–45 | NC 40–45 | 45–50 |
| Offices | | | |
| Executive | RC 25–30 | NC 25–30 | 30–35 |
| Conference rooms | RC 25–30 | NC 25–30 | 30–35 |
| Private | RC 30–35 | NC 30–35 | 35–40 |
| Open-plan areas | RC 35–40 | NC 35–40 | 40–45 |
| Business machines/computers | RC 40–45 | NC 40–45 | 45–50 |
| Public circulation | RC 40–45 | NC 40–45 | 45–50 |
| Hospitals and clinics | | | |
| Private Rooms | RC 25–30 | NC 25–30 | 30–35 |
| Wards | RC 30–35 | NC 30–35 | 35–40 |
| Operating rooms | RC 25–30 | NC 25–30 | 30–35 |
| Laboratories | RC 35–40 | NC 35–40 | 40–45 |
| Corridors | RC 30–35 | NC 30–35 | 35–40 |
| Public Areas | RC 35–40 | NC 35–40 | 40–45 |
| Churches | RC 30–35 | NC 30–35 | 35–40 |
| Schools | | | |
| Lecture and classrooms | RC 25–30 | NC 25–30 | 30–35 |
| Open-plan classrooms | RC 35–40 | NC 35–40 | 40–45 |
| Libraries | RC 35–40 | NC 35–40 | 40–45 |
| Courtrooms | RC 30–35 | NC 30–35 | 35–40 |
| Multipurpose auditoriums | RC 20–25 | NC 20–25 | 25–30 |
| Movie theaters | RC 30–35 | NC 30–35 | 35–40 |
| Restaurants | RC 35–40 | NC 35–40 | 40–45 |
| Concert and recital halls | RC 15–20 | NC 15–20 | 20–25 |
| Recording studios | RC 15–20 | NC 15–20 | 20–25 |
| TV studios | RC 20–25 | NC 20–25 | 25–30 |

*Reprinted (slightly modified) by permission of the American Society of Heating, Refrigeration and Air-Conditioning Engineers, Atlanta, Georgia, from the 1991 *ASHRAE Handbook—Applications*.

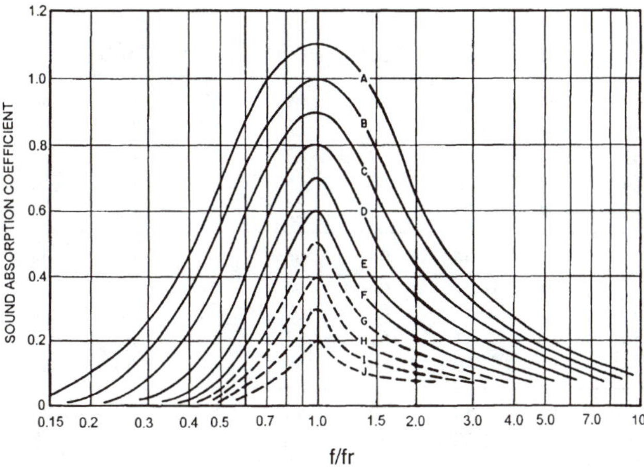


**Figure A.5-1.** Sound absorption coefficients for resonant plywood panels.

## *A.5 Sound Absorption of Solid Panels*

Wood, hardboard or plastic panels attached to a surface with an airspace behind can exhibit significant sound absorption at low frequencies. Such panel types and typical applications are discussed in several sections of Chapter 2, and Chapter 5, Section 5.2.4.

The factors that affect the sound absorption characteristics of solid panels are thickness, density and stiffness of the panel, edge conditions of the panel installation, distance between the panel and back surface, and the presence of absorptive material in the enclosed space. A given panel installation will have a resonant frequency at which the maximum absorption occurs.

Figures A.5-1 and A.5-2 show how the resonant frequency and absorption versus frequency characteristics of plywood panels can be approximated. Following are instructions for use of the figures:

1. From Figure A.5-1, select the desired curve of absorption vs frequency and note its letter designation. The peak of the curve will coincide with the resonant frequency $f_r$ determined in the next step.
2. Decide on the resonant frequency desired for the panel treatment. Remember that the purpose of the treatment is to provide low frequency absorption in a room lacking such absorption from other acoustical treatments. For example, this is likely to occur in an instrumental music rehearsal room, or to avoid "boominess" in a room intended primarily for speech.
3. Referring to Figure A.5-2, locate the intersection of the 2 diagonal lines corresponding to the chosen resonant frequency $f_r$ and the letter designation of the chosen absorption curve.
4. Find the required thickness of the plywood panel by moving vertically to the top of the graph. Select the closest standard plywood thickness. To use this graph for other panel materials, use the surface density scale at the bottom of the graph.
5. Find the required average depth D of the airspace in back of panel by moving horizontally to the left of the graph.
6. Determine whether or not an absorbent material is required in the enclosed airspace behind the panel. If the selected absorption curve is G, H, I or J (indicated by the dashed lines in the 2 graphs), the airspace should be left completely empty.

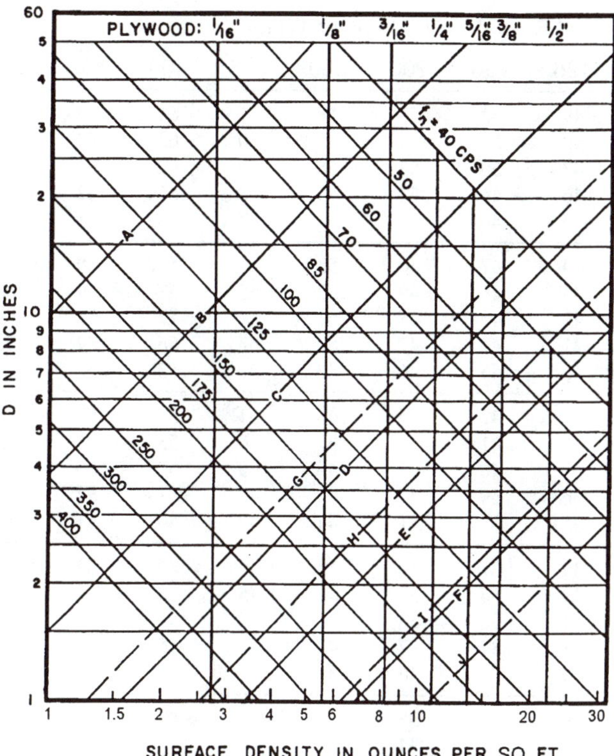


**Figure A.5-2.** Design curves for resonant plywood panels.

There are several precautions that should be observed in using these graphs. The panels should not be braced on their back, but left to vibrate freely like a drum head. Panel supports should be no closer that 16″ (406 mm) for plywood up to 3/8″ (9.5 mm) thick, and 24″ (610 mm) for 1/2″ (13 mm) plywood. There is some evidence that absorption can be increased by installing the panels using a resilient gasket all around the edges.

If a wood finish is desired over some other material such as concrete or masonry, and no increase in low frequency absorption is desired (such as in a concert hall), apply the wood directly to the backing material with no airspace. If that procedure is not possible, use thicker plywood such as 3/4″ (19 mm) and brace the back with randomly spaced strips cut from the same plywood stock and securely fasten to the panel.

This methodology is an approximation which assumes the physical characteristics of common plywoods with a density of about 35 lbs/cu ft (561 kg/cu m). If heavier ply-

**TABLE A.6-1.** STC Contour Values

| 125 | 160 | 200 | 250 | 315 | 400 | **500** | 630 | 800 | 1000 | 1250 | 1600 | 2000 | 2500 | 3150 | 4000 |
|---|---|---|---|---|---|---|---|---|---|---|---|---|---|---|---|
| 49 | 52 | 55 | 58 | 61 | 64 | **65** | 66 | 67 | 68 | 69 | 69 | 69 | 69 | 69 | 69 |
| 48 | 51 | 54 | 57 | 60 | 63 | **64** | 65 | 66 | 67 | 68 | 68 | 68 | 68 | 68 | 68 |
| 47 | 50 | 53 | 56 | 59 | 62 | **63** | 64 | 65 | 66 | 67 | 67 | 67 | 67 | 67 | 67 |
| 46 | 49 | 52 | 55 | 58 | 61 | **62** | 63 | 64 | 65 | 66 | 66 | 66 | 66 | 66 | 66 |
| 45 | 48 | 51 | 54 | 57 | 60 | **61** | 62 | 63 | 64 | 65 | 65 | 65 | 65 | 65 | 65 |
| 44 | 47 | 50 | 53 | 56 | 59 | **60** | 61 | 62 | 63 | 64 | 64 | 64 | 64 | 64 | 64 |
| 43 | 46 | 49 | 52 | 55 | 58 | **59** | 60 | 61 | 62 | 63 | 63 | 63 | 63 | 63 | 63 |
| 42 | 45 | 48 | 51 | 54 | 57 | **58** | 59 | 60 | 61 | 62 | 62 | 62 | 62 | 62 | 62 |
| 41 | 44 | 47 | 50 | 53 | 56 | **57** | 58 | 59 | 60 | 61 | 61 | 61 | 61 | 61 | 61 |
| 40 | 43 | 46 | 49 | 52 | 55 | **56** | 57 | 58 | 59 | 60 | 60 | 60 | 60 | 60 | 60 |
| 39 | 42 | 45 | 48 | 51 | 54 | **55** | 56 | 57 | 58 | 59 | 59 | 59 | 59 | 59 | 59 |
| 38 | 41 | 44 | 47 | 50 | 53 | **54** | 55 | 56 | 57 | 58 | 58 | 58 | 58 | 58 | 58 |
| 37 | 40 | 43 | 46 | 49 | 52 | **53** | 54 | 55 | 56 | 57 | 57 | 57 | 57 | 57 | 57 |
| 36 | 39 | 42 | 45 | 48 | 51 | **52** | 53 | 54 | 55 | 56 | 56 | 56 | 56 | 56 | 56 |
| 35 | 38 | 41 | 44 | 47 | 50 | **51** | 52 | 53 | 54 | 55 | 55 | 55 | 55 | 55 | 55 |
| 34 | 37 | 40 | 43 | 46 | 49 | **50** | 51 | 52 | 53 | 54 | 54 | 54 | 54 | 54 | 54 |
| 33 | 36 | 39 | 42 | 45 | 48 | **49** | 50 | 51 | 52 | 53 | 53 | 53 | 53 | 53 | 53 |
| 32 | 35 | 38 | 41 | 44 | 47 | **48** | 49 | 50 | 51 | 52 | 52 | 52 | 52 | 52 | 52 |
| 31 | 34 | 37 | 40 | 43 | 46 | **47** | 48 | 49 | 50 | 51 | 51 | 51 | 51 | 51 | 51 |
| 30 | 33 | 36 | 39 | 42 | 45 | **46** | 47 | 48 | 49 | 50 | 50 | 50 | 50 | 50 | 50 |
| 29 | 32 | 35 | 38 | 41 | 44 | **45** | 46 | 47 | 48 | 49 | 49 | 49 | 49 | 49 | 49 |
| 28 | 31 | 34 | 37 | 40 | 43 | **44** | 45 | 46 | 47 | 48 | 48 | 48 | 48 | 48 | 48 |
| 27 | 30 | 33 | 36 | 39 | 42 | **43** | 44 | 45 | 46 | 47 | 47 | 47 | 47 | 47 | 47 |
| 26 | 29 | 32 | 35 | 38 | 41 | **42** | 43 | 44 | 45 | 46 | 46 | 46 | 46 | 46 | 46 |
| 25 | 28 | 31 | 34 | 37 | 40 | **41** | 42 | 43 | 44 | 45 | 45 | 45 | 45 | 45 | 45 |
| 24 | 27 | 30 | 33 | 36 | 39 | **40** | 41 | 42 | 43 | 44 | 44 | 44 | 44 | 44 | 44 |
| 23 | 26 | 29 | 32 | 35 | 38 | **39** | 40 | 41 | 42 | 43 | 43 | 43 | 43 | 43 | 43 |
| 22 | 25 | 28 | 31 | 34 | 37 | **38** | 39 | 40 | 41 | 42 | 42 | 42 | 42 | 42 | 42 |
| 21 | 24 | 27 | 30 | 33 | 36 | **37** | 38 | 39 | 40 | 41 | 41 | 41 | 41 | 41 | 41 |
| 20 | 23 | 26 | 29 | 32 | 35 | **36** | 37 | 38 | 39 | 40 | 40 | 40 | 40 | 40 | 40 |
| 19 | 22 | 25 | 28 | 31 | 34 | **35** | 36 | 37 | 38 | 39 | 39 | 39 | 39 | 39 | 39 |
| 18 | 21 | 24 | 27 | 30 | 33 | **34** | 35 | 36 | 37 | 38 | 38 | 38 | 38 | 38 | 38 |
| 17 | 20 | 23 | 26 | 29 | 32 | **33** | 34 | 35 | 36 | 37 | 37 | 37 | 37 | 37 | 37 |
| 16 | 19 | 22 | 25 | 28 | 31 | **32** | 33 | 34 | 35 | 36 | 36 | 36 | 36 | 36 | 36 |
| 15 | 18 | 21 | 24 | 27 | 30 | **31** | 32 | 33 | 34 | 35 | 35 | 35 | 35 | 35 | 35 |
| 14 | 17 | 20 | 23 | 26 | 29 | **30** | 31 | 32 | 33 | 34 | 34 | 34 | 34 | 34 | 34 |
| 13 | 16 | 19 | 22 | 25 | 28 | **29** | 30 | 31 | 32 | 33 | 33 | 33 | 33 | 33 | 33 |
| 12 | 15 | 18 | 21 | 24 | 27 | **28** | 29 | 30 | 31 | 32 | 32 | 32 | 32 | 32 | 32 |
| 11 | 14 | 17 | 20 | 23 | 26 | **27** | 28 | 29 | 30 | 31 | 31 | 31 | 31 | 31 | 31 |
| 10 | 13 | 16 | 19 | 22 | 25 | **26** | 27 | 28 | 29 | 30 | 30 | 30 | 30 | 30 | 30 |
| 9 | 12 | 15 | 18 | 21 | 24 | **25** | 26 | 27 | 28 | 29 | 29 | 29 | 29 | 29 | 29 |
| 8 | 11 | 14 | 17 | 20 | 23 | **24** | 25 | 26 | 27 | 28 | 28 | 28 | 28 | 28 | 28 |
| 7 | 10 | 13 | 16 | 19 | 22 | **23** | 24 | 25 | 26 | 27 | 27 | 27 | 27 | 27 | 27 |
| 6 | 9 | 12 | 15 | 18 | 21 | **22** | 23 | 24 | 25 | 26 | 26 | 26 | 26 | 26 | 26 |
| 5 | 8 | 11 | 14 | 17 | 20 | **21** | 22 | 23 | 24 | 25 | 25 | 25 | 25 | 25 | 25 |
| 4 | 7 | 10 | 13 | 16 | 19 | **20** | 21 | 22 | 23 | 24 | 24 | 24 | 24 | 24 | 24 |

woods or other panel materials are used, locate the vertical line in Figure A.5-2 corresponding to the surface density scale at the bottom of the Figure.

## A.6 Sound Transmission Class and Impact Insulation Class Calculations

The Sound Transmission Class (STC) is a single-number rating of the effectiveness of a material or construction in retarding the transmission of airborne sound (see Chapter 3, Section 3.2.1). The method for determining the STC (or FSTC and NIC) is a curve-fitting procedure as described in ASTM Designation: E-413, "Classification for Rating Sound Insulation", and is performed as follows (the procedure given in the ASTM standard is different from the method given here):

1. Each STC contour is defined by an row of values given in Table A.6-1.
2. If not already done, round the TL data for which a contour is to be fitted to the nearest integer. Only include one-third octave band TL data from 125 to 4000 Hz.
3. Select a contour (single row of values) such that some TL data lies above and below the contour values.
4. Sum the individual frequency deficiencies of the TL data *below* the contour values. This sum may not exceed 32, and the maximum deficiency may not exceed 8.
5. By comparing adjacent contour values to the TL data as required, determine the contour which just exceeds the rules stated in 4 above. The appropriate contour is the one with the next lower set of values.
6. The STC designation is the contour value at a frequency of 500 Hz.

For example, the TL values for the two curves in Figure A.6-1 (same as Figure 3-3 in Chapter 3), and the values for an STC 46 contour are listed in Table A.6-1. A sample STC calculation for these 2 curves follows.

For the CMU wall, the STC 46 contour was determined by the sum of the deficiencies totaling 27. For the frame wall, the STC 46 contour was determined by the maximum permissible deficiency of 8 at a frequency of 125 Hz. The limitation in defining a TL curve by a single number is apparent here because the two curves have very different characteristics, but have the same STC rating.

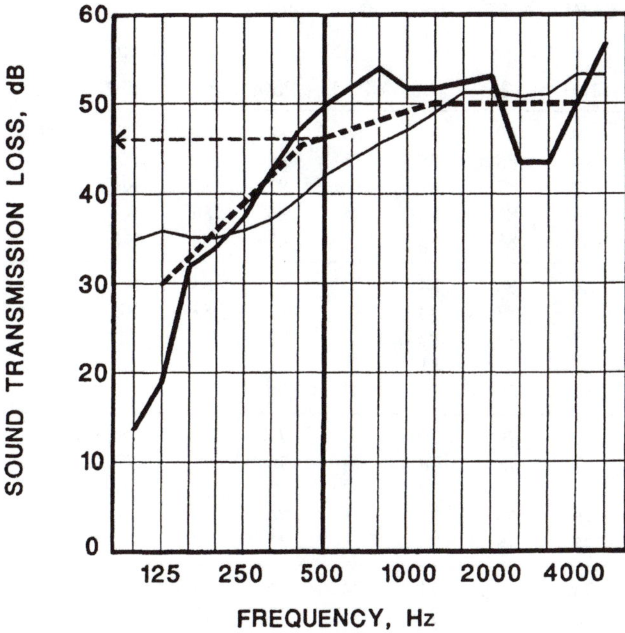


——— Metal stud—gypsum board wall
——— 8 inch concrete block wall
– – – – STC 46 contour

The STC rating is determined by the TL value where the STC contour coincides with the 500 Hz line.

**Figure A.6-1.** Two walls with STC 46 ratings.

The Impact Insulation Class (IIC) is a single-number rating of the effectiveness of a floor/ceiling construction in retarding the transmission of impact sound (see Chapter 3, Section 3.5.1). The method for determining the IIC (or FIIC) is a curve-fitting procedure as described in ASTM Designation: E-989, "Classification for Determination of Impact Insulation Class (IIC)," and is performed as follows (the procedure given in the ASTM standard is different from the method given here):

1. Each IIC contour is defined by a row of values given in Table A.6-2.
2. If not already done, round the NISPL data for which a contour is to be fitted to the nearest integer. Only include one-third octave band TL data from 100 to 3150 Hz.
3. Select a contour (single row of values) such that some NISPL data lies above and below the contour values.

| Frequency, Hz. | 125 | 160 | 200 | 250 | 315 | 400 | 500 | 630 | 800 | 1000 | 1250 | 1600 | 2000 | 2500 | 3150 | 4000 |
|---|---|---|---|---|---|---|---|---|---|---|---|---|---|---|---|---|
| STC 46 Contour | 30 | 33 | 36 | 39 | 42 | 45 | 46 | 47 | 48 | 49 | 50 | 50 | 50 | 50 | 50 | 50 |
| TL, CMU, dB | 36 | 35 | 35 | 36 | 37 | 39 | 42 | 44 | 46 | 47 | 49 | 52 | 52 | 51 | 51 | 53 |
| Deficiencies | 0 | 0 | 1 | 3 | 5 | 6 | 4 | 3 | 2 | 2 | 1 | 0 | 0 | 0 | 0 | 0 |
| SUM OF DEFICIENCIES: 27 | | | | | | | | | | | | | | | | |
| STC 46 Contour | 30 | 33 | 36 | 39 | 42 | 45 | 46 | 47 | 48 | 49 | 50 | 50 | 50 | 50 | 50 | 50 |
| TL, Studs, dB | 22 | 32 | 34 | 38 | 43 | 47 | 50 | 52 | 54 | 52 | 52 | 52 | 53 | 44 | 44 | 50 |
| Deficiencies | 8 | 1 | 2 | 1 | 0 | 0 | 0 | 0 | 0 | 0 | 0 | 0 | 0 | 6 | 6 | 0 |
| MAXIMUM DEFICIENCY: 8 | | | | | | | | | | | | | | | | |

**TABLE A.6-2.** IIC Contour Values

| One–Third Octave Band Center Frequency, Hz | | | | | | | | | | | | | | | | |
|---|---|---|---|---|---|---|---|---|---|---|---|---|---|---|---|---|
| 100 | 125 | 160 | 200 | 250 | 315 | 400 | 500 | 630 | 800 | 1000 | 1250 | 1600 | 2000 | 2500 | 3150 | **IIC** |
| 89 | 89 | 89 | 89 | 89 | 89 | 88 | 87 | 86 | 85 | 84 | 81 | 78 | 75 | 72 | 69 | **23** |
| 88 | 88 | 88 | 88 | 88 | 88 | 87 | 86 | 85 | 84 | 83 | 80 | 77 | 74 | 71 | 68 | **24** |
| 87 | 87 | 87 | 87 | 87 | 87 | 86 | 85 | 84 | 83 | 82 | 79 | 76 | 73 | 70 | 67 | **25** |
| 86 | 86 | 86 | 86 | 86 | 86 | 85 | 84 | 83 | 82 | 81 | 78 | 75 | 72 | 69 | 66 | **26** |
| 85 | 85 | 85 | 85 | 85 | 85 | 84 | 83 | 82 | 81 | 80 | 77 | 74 | 71 | 68 | 65 | **27** |
| 84 | 84 | 84 | 84 | 84 | 84 | 83 | 82 | 81 | 80 | 79 | 76 | 73 | 70 | 67 | 64 | **28** |
| 83 | 83 | 83 | 83 | 83 | 83 | 82 | 81 | 80 | 79 | 78 | 75 | 72 | 69 | 66 | 63 | **29** |
| 82 | 82 | 82 | 82 | 82 | 82 | 81 | 80 | 79 | 78 | 77 | 74 | 71 | 68 | 65 | 62 | **30** |
| 81 | 81 | 81 | 81 | 81 | 81 | 80 | 79 | 78 | 77 | 76 | 73 | 70 | 67 | 64 | 61 | **31** |
| 80 | 80 | 80 | 80 | 80 | 80 | 79 | 78 | 77 | 76 | 75 | 72 | 69 | 66 | 63 | 60 | **32** |
| 79 | 79 | 79 | 79 | 79 | 79 | 78 | 77 | 76 | 75 | 74 | 71 | 68 | 65 | 62 | 59 | **33** |
| 78 | 78 | 78 | 78 | 78 | 78 | 77 | 76 | 75 | 74 | 73 | 70 | 67 | 64 | 61 | 58 | **34** |
| 77 | 77 | 77 | 77 | 77 | 77 | 76 | 75 | 74 | 73 | 72 | 69 | 66 | 63 | 60 | 57 | **35** |
| 76 | 76 | 76 | 76 | 76 | 76 | 75 | 74 | 73 | 72 | 71 | 68 | 65 | 62 | 59 | 56 | **36** |
| 75 | 75 | 75 | 75 | 75 | 75 | 74 | 73 | 72 | 71 | 70 | 67 | 64 | 61 | 58 | 55 | **37** |
| 74 | 74 | 74 | 74 | 74 | 74 | 73 | 72 | 71 | 70 | 69 | 66 | 63 | 60 | 57 | 54 | **38** |
| 73 | 73 | 73 | 73 | 73 | 73 | 72 | 71 | 70 | 69 | 68 | 65 | 62 | 59 | 56 | 53 | **39** |
| 72 | 72 | 72 | 72 | 72 | 72 | 71 | 70 | 69 | 68 | 67 | 64 | 61 | 58 | 55 | 52 | **40** |
| 71 | 71 | 71 | 71 | 71 | 71 | 70 | 69 | 68 | 67 | 66 | 63 | 60 | 57 | 54 | 51 | **41** |
| 70 | 70 | 70 | 70 | 70 | 70 | 69 | 68 | 67 | 66 | 65 | 62 | 59 | 56 | 53 | 50 | **42** |
| 69 | 69 | 69 | 69 | 69 | 69 | 68 | 67 | 66 | 65 | 64 | 61 | 58 | 55 | 52 | 49 | **43** |
| 68 | 68 | 68 | 68 | 68 | 68 | 67 | 66 | 65 | 64 | 63 | 60 | 57 | 54 | 51 | 48 | **44** |
| 67 | 67 | 67 | 67 | 67 | 67 | 66 | 65 | 64 | 63 | 62 | 59 | 56 | 53 | 50 | 47 | **45** |
| 66 | 66 | 66 | 66 | 66 | 66 | 65 | 64 | 63 | 62 | 61 | 58 | 55 | 52 | 49 | 46 | **46** |
| 65 | 65 | 65 | 65 | 65 | 65 | 64 | 63 | 62 | 61 | 60 | 57 | 54 | 51 | 48 | 45 | **47** |
| 64 | 64 | 64 | 64 | 64 | 64 | 63 | 62 | 61 | 60 | 59 | 56 | 53 | 50 | 47 | 44 | **48** |
| 63 | 63 | 63 | 63 | 63 | 63 | 62 | 61 | 60 | 59 | 58 | 55 | 52 | 49 | 46 | 43 | **49** |
| 62 | 62 | 62 | 62 | 62 | 62 | 61 | 60 | 59 | 58 | 57 | 54 | 51 | 48 | 45 | 42 | **50** |
| 61 | 61 | 61 | 61 | 61 | 61 | 60 | 59 | 58 | 57 | 56 | 53 | 50 | 47 | 44 | 41 | **51** |
| 60 | 60 | 60 | 60 | 60 | 60 | 59 | 58 | 57 | 56 | 55 | 52 | 49 | 46 | 43 | 40 | **52** |
| 59 | 59 | 59 | 59 | 59 | 59 | 58 | 57 | 56 | 55 | 54 | 51 | 48 | 45 | 42 | 39 | **53** |
| 58 | 58 | 58 | 58 | 58 | 58 | 57 | 56 | 55 | 54 | 53 | 50 | 47 | 44 | 41 | 38 | **54** |
| 57 | 57 | 57 | 57 | 57 | 57 | 56 | 55 | 54 | 53 | 52 | 49 | 46 | 43 | 40 | 37 | **55** |
| 56 | 56 | 56 | 56 | 56 | 56 | 55 | 54 | 53 | 52 | 51 | 48 | 45 | 42 | 39 | 36 | **56** |
| 55 | 55 | 55 | 55 | 55 | 55 | 54 | 53 | 52 | 51 | 50 | 47 | 44 | 41 | 38 | 35 | **57** |
| 54 | 54 | 54 | 54 | 54 | 54 | 53 | 52 | 51 | 50 | 49 | 46 | 43 | 40 | 37 | 34 | **58** |
| 53 | 53 | 53 | 53 | 53 | 53 | 52 | 51 | 50 | 49 | 48 | 45 | 42 | 39 | 36 | 33 | **59** |
| 52 | 52 | 52 | 52 | 52 | 52 | 51 | 50 | 49 | 48 | 47 | 44 | 41 | 38 | 35 | 32 | **60** |
| 51 | 51 | 51 | 51 | 51 | 51 | 50 | 49 | 48 | 47 | 46 | 43 | 40 | 37 | 34 | 31 | **61** |
| 50 | 50 | 50 | 50 | 50 | 50 | 49 | 48 | 47 | 46 | 45 | 42 | 39 | 36 | 33 | 30 | **62** |
| 49 | 49 | 49 | 49 | 49 | 49 | 48 | 47 | 46 | 45 | 44 | 41 | 38 | 35 | 32 | 29 | **63** |
| 48 | 48 | 48 | 48 | 48 | 48 | 47 | 46 | 45 | 44 | 43 | 40 | 37 | 34 | 31 | 28 | **64** |
| 47 | 47 | 47 | 47 | 47 | 47 | 46 | 45 | 44 | 43 | 42 | 39 | 36 | 33 | 30 | 27 | **65** |
| 46 | 46 | 46 | 46 | 46 | 46 | 45 | 44 | 43 | 42 | 41 | 38 | 35 | 32 | 29 | 26 | **66** |
| 45 | 45 | 45 | 45 | 45 | 45 | 44 | 43 | 42 | 41 | 40 | 37 | 34 | 31 | 28 | 25 | **67** |
| 44 | 44 | 44 | 44 | 44 | 44 | 43 | 42 | 41 | 40 | 39 | 36 | 33 | 30 | 27 | 24 | **68** |
| 43 | 43 | 43 | 43 | 43 | 43 | 42 | 41 | 40 | 39 | 38 | 35 | 32 | 29 | 26 | 23 | **69** |
| 42 | 42 | 42 | 42 | 42 | 42 | 41 | 40 | 39 | 38 | 37 | 34 | 31 | 28 | 25 | 22 | **70** |

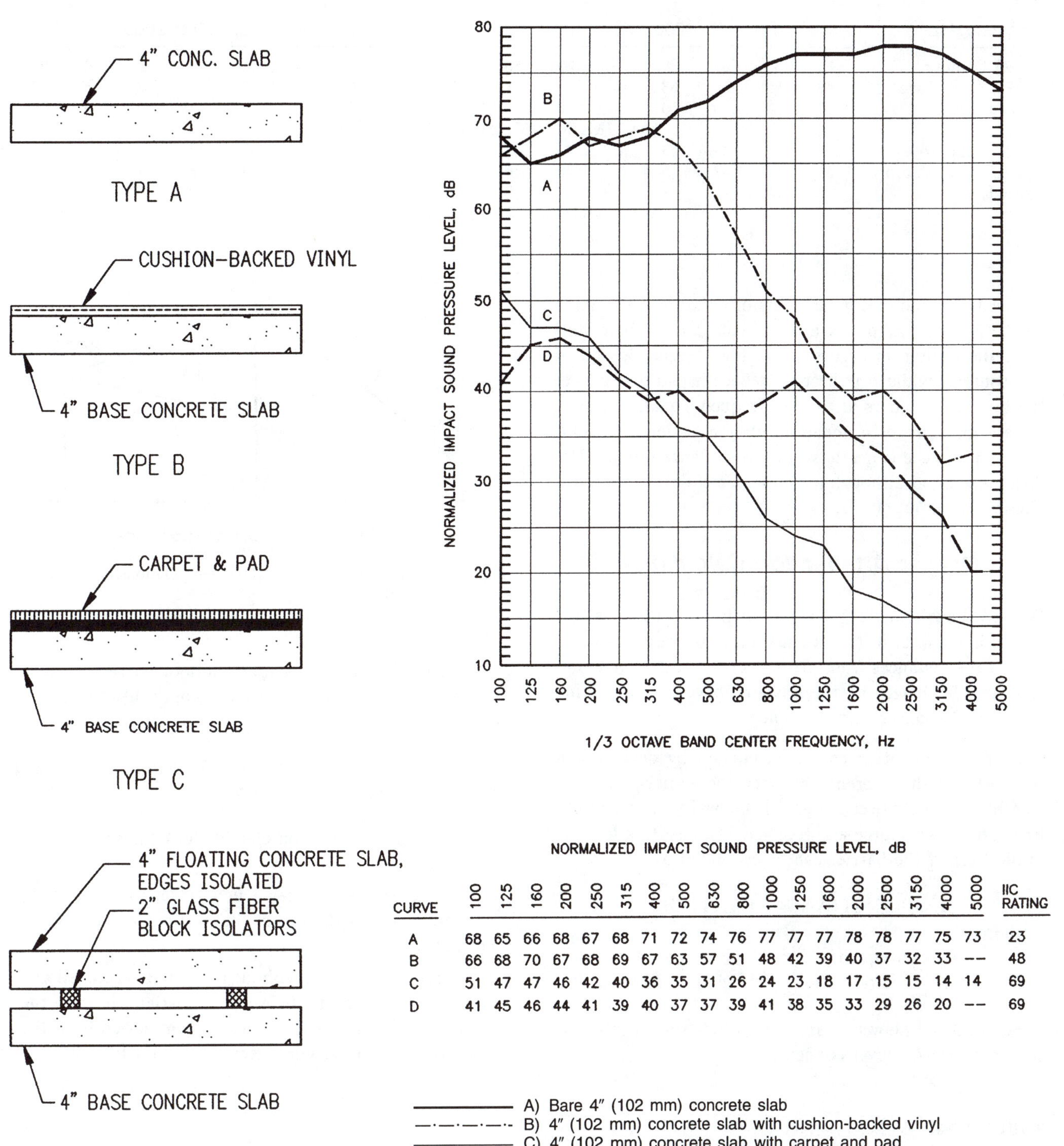


NORMALIZED IMPACT SOUND PRESSURE LEVEL, dB

| CURVE | 100 | 125 | 160 | 200 | 250 | 315 | 400 | 500 | 630 | 800 | 1000 | 1250 | 1600 | 2000 | 2500 | 3150 | 4000 | 5000 | IIC RATING |
|---|---|---|---|---|---|---|---|---|---|---|---|---|---|---|---|---|---|---|---|
| A | 68 | 65 | 66 | 68 | 67 | 68 | 71 | 72 | 74 | 76 | 77 | 77 | 77 | 78 | 78 | 77 | 75 | 73 | 23 |
| B | 66 | 68 | 70 | 67 | 68 | 69 | 67 | 63 | 57 | 51 | 48 | 42 | 39 | 40 | 37 | 32 | 33 | -- | 48 |
| C | 51 | 47 | 47 | 46 | 42 | 40 | 36 | 35 | 31 | 26 | 24 | 23 | 18 | 17 | 15 | 15 | 14 | 14 | 69 |
| D | 41 | 45 | 46 | 44 | 41 | 39 | 40 | 37 | 37 | 39 | 41 | 38 | 35 | 33 | 29 | 26 | 20 | -- | 69 |



**Figure A.6-2.** Impact sound insulation of concrete floors.

4. Sum the individual frequency exceedances of the NISPL data *above* the contour values. This sum may not exceed 32, and the maximum exceedance may not be more than 8.
5. By comparing adjacent contour values to the NISPL data as required, determine the contour which just exceeds the rules stated in 4 above. The appropriate contour is the one with the next higher set of values.
6. The IIC designation is the contour value at a frequency of 500 Hz subtracted from 110.

For example, the NISPL values for curves A (bare concrete) and C (concrete with carpet and pad) in Figure A.6-2 (same as Figure 3-31) are listed below, along with the applicable IIC contours, 23 and 69, respectively. A sample IIC calculation for these 2 curves follows.

| Frequency,Hz. | 100 | 125 | 160 | 200 | 250 | 315 | 400 | 500 | 630 | 800 | 1000 | 1250 | 1600 | 2000 | 2500 | 3150 |
|---|---|---|---|---|---|---|---|---|---|---|---|---|---|---|---|---|
| IIC 23 Contour | 89 | 89 | 89 | 89 | 89 | 89 | 88 | 87 | 86 | 85 | 84 | 81 | 78 | 75 | 72 | 69 |
| NISPL, Bare Conc | 68 | 65 | 66 | 68 | 67 | 68 | 71 | 72 | 74 | 76 | 77 | 77 | 77 | 78 | 78 | 77 |
| Exceedances | 0 | 0 | 0 | 0 | 0 | 0 | 0 | 0 | 0 | 0 | 0 | 0 | 0 | 3 | 6 | 8 |
| MAXIMUM EXCEEDANCE: 8 | | | | | | | | | | | | | | | | |
| IIC 69 Contour | 43 | 43 | 43 | 43 | 43 | 43 | 42 | 41 | 40 | 39 | 38 | 35 | 32 | 29 | 26 | 23 |
| NISPL, Conc w/Carp | 51 | 47 | 47 | 46 | 42 | 40 | 36 | 35 | 31 | 26 | 24 | 23 | 18 | 17 | 15 | 15 |
| Exceedances | 8 | 4 | 4 | 3 | 0 | 0 | 0 | 0 | 0 | 0 | 0 | 0 | 0 | 0 | 0 | 0 |
| MAXIMUM EXCEEDANCE: 8 | | | | | | | | | | | | | | | | |

The IIC contours for both sets of data are determined by the maximum 8 dB exceedance rule. For the bare concrete the limitation occurs at the high frequency end of the contour, while for the concrete with pad and carpet the limitation occurs at the low frequency end. The very large improvement in impact insulation achieved by adding carpet to the concrete surface is evidenced by the IIC difference of 46, although the improvement at the high frequencies is about 60 dB!

## *A.7 Barrier Attenuation Calculation*

Barriers have many applications in noise control as discussed in Chapter 8. The attenuation of barriers is a function of the distances from the barrier for both source and receiver, the height of the barrier above a line joining source and receiver, and frequency.

Barrier attenuation can be calculated by determining the path length difference between the actual path traveled by the sound over the top of the barrier, to a straight line connecting source and receiver. The insertion loss IL, in decibels, of the barrier can be estimated by

$$IL = 10 \log f (A + B - D) - 14 \text{ dB}$$

where

$f$ = frequency, Hz

$A + B - D$ = defracted minus straight line distance, ft

Figure A.7-1 shows these relationships for a particular source/barrier/receiver configuration.

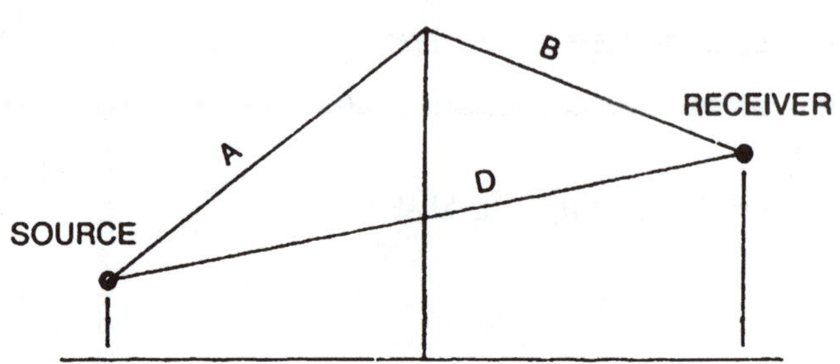


Refer to Table A.7-1 for barrier insertion loss values.

**Figure A.7-1.** Source/Barrier/Receiver relationships affecting barrier insertion loss.

The insertion loss for a solid outdoor barrier is shown in Table A.7-1 for a range of path length differences and frequencies.

The IL values shown in Table A.7-1 will not be realized for a number of conditions as follows:

If the barrier does not completely block the source as viewed from the receiver.

If there are nearby surfaces which reflect sound around the ends or over the top of the barrier into the protected zone.

If the propagation path without the barrier included attenuation from ground cover, the barrier will raise the source to the top of the barrier when viewed from the receiver, and the ground attenuation may be lost.

**TABLE A.7-1.** Barrier Insertion Loss*

| Path Length Difference, Ft | Insertion Loss, dB, at Octave Band Center Frequency, Hz 31.5 | 63 | 125 | 250 | 500 | 1000 | 2000 | 4000 | 8000 |
|---|---|---|---|---|---|---|---|---|---|
| 0.1 | 5 | 5 | 5 | 6 | 7 | 9 | 11 | 13 | 16 |
| 0.5 | 6 | 7 | 9 | 10 | 12 | 15 | 18 | 20 | 22 |
| 1.0 | 7 | 8 | 10 | 12 | 14 | 17 | 20 | 22 | 23 |
| 2.0 | 8 | 10 | 12 | 14 | 17 | 20 | 22 | 23 | 24 |
| 5.0 | 10 | 12 | 14 | 17 | 20 | 22 | 23 | 24 | 24 |
| 10.0 | 12 | 15 | 17 | 20 | 22 | 23 | 24 | 24 | 24 |
| 20.0 | 15 | 18 | 20 | 22 | 23 | 24 | 24 | 24 | 24 |

*Reprinted by permission of the American Society of Heating, Refrigeration and Air-conditioning Engineers, Atlanta, Georgia, from the 1991 *ASHRAE Handbook—Applications.*

If the surface density of the barrier is less than 4 lbs/sq ft (19.5 kg/sq m).

If the barrier has cracks or other sound leaks.

## A.8 Loudness Calculation

Loudness is defined in Chapter 1, Section 1.5.1. Loudness is expressed in sones, and loudness level is expressed in phons. The method for calculating these values for a sound are presented here.

1. Determine the octave band sound pressure levels for the sound of interest.
2. From Table A.8-1, select a band loudness index number for each octave band level.
3. Total all the loudness index numbers ($I_{total}$) *excluding* the band with the highest index number ($I_{max}$).
4. The loudness in sones is determined by adding the highest index number to 0.3 times the sum of all the other index numbers.

$$\text{Sones} = I_{max} + 0.3(I_{total})$$

5. The loudness level in phons is determined by:

$$\text{Phons} = 40 = 33.3 \log \text{sones}$$

For example:

| Frequency, Hz | 31.5 | 63 | 125 | 250 | 500 | 1000 | 2000 | 4000 | 8000 |
|---|---|---|---|---|---|---|---|---|---|
| Band Level, dB | 78 | 76 | 72 | 82 | 81 | 80 | 80 | 73 | 65 |
| Loudness Level | 3.7 | 5.0 | 5.8 | 12.6 | 14.4 | 16.4 | 20.0 | 15.3 | 11.1 |

$$\begin{aligned} \text{Sones} &= 20.0 + 0.3(84.3) \\ &= 45.3 \\ \text{Phons} &= 40 + 33.3 \log \text{sones} \\ &= 40 + 33.3(1.66) \\ &= 95 \end{aligned}$$

**TABLE A.8-1.** Band Loudness Index at Frequency, Hz

| Band Level, dB | 31.5 | 63 | 125 | 250 | 500 | 1000 | 2000 | 4000 | 8000 | Loudness Sones | Loudness Phons |
|---|---|---|---|---|---|---|---|---|---|---|---|
| 20 | – | – | – | – | – | 0.18 | 0.30 | 0.45 | 0.61 | 0.25 | 20 |
| 21 | – | – | – | – | – | 0.22 | 0.35 | 0.50 | 0.67 | 0.27 | 21 |
| 22 | – | – | – | – | 0.07 | 0.26 | 0.40 | 0.55 | 0.73 | 0.29 | 22 |
| 23 | – | – | – | – | 0.12 | 0.30 | 0.45 | 0.61 | 0.80 | 0.31 | 23 |
| 24 | – | – | – | – | 0.16 | 0.35 | 0.50 | 0.67 | 0.87 | 0.33 | 24 |
| 25 | – | – | – | – | 0.21 | 0.40 | 0.55 | 0.73 | 0.94 | 0.35 | 25 |
| 26 | – | – | – | – | 0.26 | 0.45 | 0.61 | 0.80 | 1.02 | 0.38 | 26 |
| 27 | – | – | – | – | 0.31 | 0.50 | 0.67 | 0.87 | 1.10 | 0.41 | 27 |
| 28 | – | – | – | 0.07 | 0.37 | 0.55 | 0.73 | 0.94 | 1.18 | 0.44 | 28 |
| 29 | – | – | – | 0.12 | 0.43 | 0.61 | 0.80 | 1.02 | 1.27 | 0.47 | 29 |
| 30 | – | – | – | 0.16 | 0.49 | 0.67 | .087 | 1.10 | 1.35 | 0.50 | 30 |
| 31 | – | – | – | 0.21 | 0.55 | 0.73 | 0.94 | 1.18 | 1.44 | 0.54 | 31 |
| 32 | – | – | – | 0.26 | 0.61 | 0.80 | 1.02 | 1.27 | 1.54 | 0.57 | 32 |
| 33 | – | – | – | 0.31 | 0.67 | 0.87 | 1.10 | 1.35 | 1.64 | 0.62 | 33 |
| 34 | – | – | 0.07 | 0.37 | 0.73 | 0.94 | 1.18 | 1.44 | 1.75 | 0.66 | 34 |
| 35 | – | – | 0.12 | 0.43 | 0.80 | 1.02 | 1.27 | 1.54 | 1.87 | 0.71 | 35 |
| 36 | – | – | 0.16 | 0.49 | 0.87 | 1.10 | 1.35 | 1.64 | 1.99 | 0.76 | 36 |
| 37 | – | – | 0.21 | 0.55 | 0.94 | 1.18 | 1.44 | 1.75 | 2.11 | 0.81 | 37 |
| 38 | – | – | 0.26 | 0.62 | 1.02 | 1.27 | 1.54 | 1.87 | 2.24 | 0.87 | 38 |
| 39 | – | – | 0.31 | 0.69 | 1.10 | 1.35 | 1.64 | 1.99 | 2.38 | 0.93 | 39 |
| 40 | – | 0.07 | 0.37 | 0.77 | 1.18 | 1.44 | 1.75 | 2.11 | 2.53 | 1.00 | 40 |
| 41 | – | 0.12 | 0.43 | 0.85 | 1.27 | 1.54 | 1.87 | 2.24 | 2.68 | 1.07 | 41 |
| 42 | – | 0.16 | 0.49 | 0.94 | 1.35 | 1.64 | 1.99 | 2.38 | 2.84 | 1.15 | 42 |
| 43 | – | 0.21 | 0.55 | 1.04 | 1.44 | 1.75 | 2.11 | 2.53 | 3.0 | 1.23 | 43 |
| 44 | – | 0.26 | 0.62 | 1.13 | 1.54 | 1.87 | 2.24 | 2.68 | 3.2 | 1.32 | 44 |
| 45 | – | 0.31 | 0.69 | 1.23 | 1.64 | 1.99 | 2.38 | 2.84 | 3.4 | 1.41 | 45 |
| 46 | 0.07 | 0.37 | 0.77 | 1.33 | 1.75 | 2.11 | 2.53 | 3.0 | 3.6 | 1.52 | 46 |
| 47 | 0.12 | 0.43 | 0.85 | 1.44 | 1.87 | 2.24 | 2.68 | 3.2 | 3.8 | 1.62 | 47 |
| 48 | 0.16 | 0.49 | 0.94 | 1.56 | 1.99 | 2.38 | 2.84 | 3.4 | 4.1 | 1.74 | 48 |
| 49 | 0.21 | 0.55 | 1.04 | 1.69 | 2.11 | 2.53 | 3.0 | 3.6 | 4.3 | 1.87 | 49 |
| 50 | 0.26 | 0.62 | 1.13 | 1.82 | 2.24 | 2.68 | 3.2 | 3.8 | 4.6 | 2.00 | 50 |
| 51 | 0.31 | 0.69 | 1.23 | 1.96 | 2.38 | 2.84 | 3.4 | 4.1 | 4.9 | 2.14 | 51 |
| 52 | 0.37 | 0.77 | 1.33 | 2.11 | 2.53 | 3.0 | 3.6 | 4.3 | 5.2 | 2.30 | 52 |
| 53 | 0.43 | 0.85 | 1.44 | 2.24 | 2.68 | 3.2 | 3.8 | 4.6 | 5.5 | 2.46 | 53 |
| 54 | 0.49 | 0.94 | 1.56 | 2.38 | 2.84 | 3.4 | 4.1 | 4.9 | 5.8 | 2.64 | 54 |
| 55 | 0.55 | 1.04 | 1.69 | 2.53 | 3.0 | 3.6 | 4.3 | 5.2 | 6.2 | 2.83 | 55 |
| 56 | 0.62 | 1.13 | 1.82 | 2.68 | 3.2 | 3.8 | 4.6 | 5.5 | 6.6 | 3.03 | 56 |
| 57 | 0.69 | 1.23 | 1.96 | 2.84 | 3.4 | 4.1 | 4.9 | 5.8 | 7.0 | 3.25 | 57 |
| 58 | 0.77 | 1.33 | 2.11 | 3.0 | 3.6 | 4.3 | 5.2 | 6.2 | 7.4 | 3.48 | 58 |
| 59 | 0.85 | 1.44 | 2.27 | 3.2 | 3.8 | 4.6 | 5.5 | 6.6 | 7.8 | 3.73 | 59 |
| 60 | 0.94 | 1.56 | 2.44 | 3.4 | 4.1 | 4.9 | 5.8 | 7.0 | 8.3 | 4.00 | 60 |
| 61 | 1.04 | 1.69 | 2.62 | 3.6 | 4.3 | 5.2 | 6.2 | 7.4 | 8.8 | 4.29 | 61 |
| 62 | 1.13 | 1.82 | 2.81 | 3.8 | 4.6 | 5.5 | 6.6 | 7.8 | 9.3 | 4.59 | 62 |
| 63 | 1.23 | 1.96 | 3.0 | 4.1 | 4.9 | 5.8 | 7.0 | 8.3 | 9.9 | 4.92 | 63 |
| 64 | 1.33 | 2.11 | 3.2 | 4.3 | 5.2 | 6.2 | 7.4 | 8.8 | 10.5 | 5.28 | 64 |
| 65 | 1.44 | 2.27 | 3.5 | 4.6 | 5.5 | 6.6 | 7.8 | 9.3 | 11.1 | 5.66 | 65 |
| 66 | 1.56 | 2.44 | 3.7 | 4.9 | 5.8 | 7.0 | 8.3 | 9.9 | 11.8 | 6.06 | 66 |
| 67 | 1.69 | 2.62 | 4.0 | 5.2 | 6.2 | 7.4 | 8.8 | 10.5 | 12.6 | 6.50 | 67 |
| 68 | 1.82 | 2.81 | 4.3 | 5.5 | 6.6 | 7.8 | 9.3 | 11.1 | 13.5 | 6.96 | 68 |
| 69 | 1.96 | 3.0 | 4.7 | 5.8 | 7.0 | 8.3 | 9.9 | 11.8 | 14.4 | 7.46 | 69 |
| 70 | 2.11 | 3.2 | 5.0 | 6.2 | 7.4 | 8.8 | 10.5 | 12.6 | 15.3 | 8.00 | 70 |
| 71 | 2.27 | 3.5 | 5.4 | 6.6 | 7.8 | 9.3 | 11.1 | 13.5 | 16.4 | 8.6 | 71 |
| 72 | 2.44 | 3.7 | 5.8 | 7.0 | 8.3 | 9.9 | 11.8 | 14.4 | 17.5 | 9.2 | 72 |
| 73 | 2.62 | 4.0 | 6.2 | 7.4 | 8.8 | 10.5 | 12.6 | 15.3 | 18.7 | 9.8 | 73 |
| 74 | 2.81 | 4.3 | 6.6 | 7.8 | 9.3 | 11.1 | 13.5 | 16.4 | 20.0 | 10.6 | 74 |

**TABLE A.8-1.** *(continued)*

| Band Level, dB | 31.5 | 63 | 125 | 250 | 500 | 1000 | 2000 | 4000 | 8000 | Loudness Sones | Loudness Phons |
|---|---|---|---|---|---|---|---|---|---|---|---|
| 75 | 3.0 | 4.7 | 7.0 | 8.3 | 9.9 | 11.8 | 14.4 | 17.5 | 21.4 | 11.3 | 75 |
| 76 | 3.2 | 5.0 | 7.4 | 8.8 | 10.5 | 12.6 | 15.3 | 18.7 | 23.0 | 12.1 | 76 |
| 77 | 3.5 | 5.4 | 7.8 | 9.3 | 11.1 | 13.5 | 16.4 | 20.0 | 24.7 | 13.0 | 77 |
| 78 | 3.7 | 5.8 | 8.3 | 9.9 | 11.8 | 14.4 | 17.5 | 21.4 | 26.5 | 13.9 | 78 |
| 79 | 4.0 | 6.2 | 8.8 | 10.5 | 12.6 | 15.3 | 18.7 | 23.0 | 28.5 | 14.9 | 79 |
| 80 | 4.3 | 6.7 | 9.3 | 11.1 | 13.5 | 16.4 | 20.0 | 24.7 | 30.5 | 16.0 | 80 |
| 81 | 4.7 | 7.2 | 9.9 | 11.8 | 14.4 | 17.5 | 21.4 | 26.5 | 32.9 | 17.1 | 81 |
| 82 | 5.0 | 7.7 | 10.5 | 12 | 6 | 15.3 | 18.7 | 23.0 | 28.5 | 35.3 | 18.4 |
| 82 | 83 | 5.4 | 8.2 | 11.1 | 13.5 | 16.4 | 20.0 | 24.7 | 30.5 | 38 | 19.7 |
| 83 | 84 | 5.8 | 8.8 | 11.8 | 14.4 | 17.5 | 21.4 | 26.5 | 32.9 | 41 | 21.1 |
| 84 | 85 | 6.2 | 9.4 | 12.6 | 15.3 | 18.7 | 23.0 | 28.5 | 35.3 | 44 | 22.6 |
| 85 | 86 | 6.7 | 10.1 | 13.5 | 16.4 | 20.0 | 24.7 | 38.5 | 38 | 48 | 24.3 |
| 86 | 87 | 7.2 | 10.9 | 14.4 | 17.5 | 21.4 | 26.5 | 32.9 | 41 | 52 | 26.0 |
| 87 | 88 | 7.7 | 11.7 | 15.3 | 18.7 | 23.0 | 28.5 | 35.3 | 44 | 56 | 27.9 |
| 88 | 89 | 8.2 | 12.6 | 16.4 | 20.0 | 24.7 | 30.5 | 38 | 48 | 61 | 29.9 |
| 90 | 8.8 | 13.6 | 17.5 | 21.4 | 26.5 | 32.9 | 41 | 52 | 66 | 32.0 | 90 |
| 91 | 9.4 | 14.8 | 18.7 | 23.0 | 28.5 | 35.3 | 44 | 56 | 71 | 34.3 | 91 |
| 92 | 10.1 | 16.0 | 20.0 | 24.7 | 30.5 | 38 | 48 | 61 | 77 | 36.8 | 92 |
| 93 | 10.9 | 17.3 | 21.4 | 26.5 | 32.9 | 41 | 52 | 66 | 83 | 39.4 | 93 |
| 94 | 11.7 | 18.7 | 23.0 | 28.5 | 35.3 | 44 | 56 | 71 | 90 | 42.2 | 94 |
| 95 | 12.6 | 20.0 | 24.7 | 30.5 | 38 | 48 | 61 | 77 | 97 | 45.3 | 95 |
| 96 | 13.6 | 21.4 | 26.5 | 32.9 | 41 | 52 | 66 | 83 | 105 | 48.5 | 96 |
| 97 | 14.8 | 23.0 | 28.5 | 35.3 | 44 | 56 | 71 | 90 | 113 | 52.0 | 97 |
| 98 | 16.0 | 24.7 | 30.5 | 38 | 48 | 61 | 77 | 97 | 121 | 55.7 | 98 |
| 99 | 17.3 | 26.5 | 32.9 | 41 | 52 | 66 | 83 | 105 | 130 | 59.7 | 99 |
| 100 | 18.7 | 28.5 | 35.3 | 44 | 56 | 71 | 90 | 113 | 139 | 64.0 | 100 |
| 101 | 20.3 | 30.5 | 38 | 48 | 61 | 77 | 97 | 121 | 149 | 68.6 | 101 |
| 102 | 22.1 | 32.9 | 41 | 52 | 66 | 83 | 105 | 130 | 160 | 73.5 | 102 |
| 103 | 24.0 | 35.3 | 44 | 56 | 71 | 90 | 113 | 139 | 171 | 78.8 | 103 |
| 104 | 26.1 | 38 | 48 | 61 | 77 | 97 | 121 | 149 | 184 | 84.4 | 104 |
| 105 | 28.5 | 41 | 52 | 66 | 83 | 105 | 130 | 160 | 197 | 90.5 | 105 |
| 106 | 31.0 | 44 | 56 | 71 | 90 | 113 | 139 | 171 | 211 | 97 | 106 |
| 107 | 33.9 | 48 | 61 | 77 | 97 | 121 | 149 | 184 | 226 | 104 | 107 |
| 108 | 36.9 | 52 | 66 | 83 | 105 | 130 | 160 | 197 | 242 | 111 | 108 |
| 109 | 40.3 | 56 | 71 | 90 | 113 | 139 | 171 | 211 | 260 | 119 | 109 |
| 110 | 44 | 61 | 77 | 97 | 121 | 149 | 184 | 226 | 278 | 128 | 110 |
| 111 | 49 | 66 | 83 | 105 | 130 | 160 | 197 | 242 | 298 | 137 | 111 |
| 112 | 54 | 71 | 90 | 113 | 139 | 171 | 211 | 260 | 320 | 147 | 112 |
| 113 | 59 | 77 | 97 | 121 | 149 | 184 | 226 | 278 | 343 | 158 | 113 |
| 114 | 65 | 83 | 105 | 130 | 160 | 197 | 242 | 298 | 367 | 169 | 114 |
| 116 | 77 | 97 | 121 | 149 | 184 | 226 | 278 | 343 | - | 194 | 116 |
| 117 | 83 | 105 | 130 | 160 | 197 | 242 | 298 | 367 | - | 208 | 117 |
| 118 | 90 | 113 | 139 | 171 | 211 | 260 | 320 | - | - | 223 | 118 |
| 119 | 97 | 121 | 149 | 184 | 226 | 278 | 343 | - | - | 239 | 119 |
| 120 | 105 | 130 | 160 | 197 | 242 | 298 | 367 | - | - | 256 | 120 |

## A.9 Acoustical Standards and Regulations

The more frequently referenced acoustical standards relating to architectural or building acoustics are promulgated by organizations listed in Section A.9.1 following. Acoustical standards are also issued by several other professional societies, and by many state, county and municipal regulations and building codes. These standards—including recommended practices and guides—are referred to throughout the handbook and in Appendix A.10, "Acoustical Measurements."

The standards listed were current as of the time of preparation of this handbook. Standards are constantly being revised and updated, and the reader is advised to refer to the issuing agency for the latest version.

### A.9.1 Addresses of Organizations Issuing Acoustical Standards or Guidelines

American Industrial Hygiene Association (AIHA)
66 S. Miller Road
Akron, OH 44313
(216) 425-8333 or (301) 694-5243

American National Standards Institute (ANSI)
11 West 42nd St., 13th Floor
New York, NY 10036-8002
(212) 642-4900
See A.9.4

American Society for Testing and Materials (ASTM)
100 Barr Harbor Drive
West Conshohocken PA 19428
(610) 832-9585
(610) 832-9555 fax
See A.9.3

American Society of Heating, Refrigerating and Air-conditioning Engineers, Inc. (ASHRAE)
1791 Tullie Circle NE
Atlanta, GA 30329
(404) 636-8400

Canadian Construction Materials Centre (CCMC)
Institute for Research in Construction
National Research Council of Canada
Ottawa, ON, Canada K1A 0R6
(613) 993-2463

Canadian Standards Association (CSA)
178 Rexdale Boulevard
Rexdale, ON, Canada M9W 1R3
(416) 747-4000
FAX (416) 747-4149

General Services Administration (GSA)
Public Buildings Service
18th & F Sts. NW
Washington, D.C. 20405
(202) 501-0907

Institute for Research in Construction
National Research Council of Canada
Ottawa, Ontario, Canada K1A 0R6
(613) 993-2463
FAX (613) 952-7673

U.S. Department of Health and Human Services
National Institute for Occupational Safety and Health (NIOSH)
Center for Disease Control
Division of Laboratories and Criteria Development
4676 Columbia Parkway
Cincinnati, OH 45202
(513) 533-8302

U.S. Department of Housing and Urban Development (HUD)
Federal Housing Administration (FHA)
Office of Community Planning and Development
451 7th Street SW
Washington, D.C. 20410
(202) 708-1422
FAX (202) 708-0299

U.S. Department of Labor
Mine Safety and Health Administration (MSHA)
4015 Wilson Blvd.
Arlington, Virginia 22203
(703) 235-1452

Occupational Safety and Health Administration (OSHA)
200 Constitution Avenue
Washington, DC 20210
(202) 219-8148
FAX (202) 219-5986

U.S. Department of Transportation
Federal Aviation Agency (FAA)
800 Independence Avenue SW
Washington, D.C. 20591
(202) 267-3111
FAX (202) 267-5047

Federal Highway Administration (FHWA)
400 7th Street SW
Washington, D.C. 20590
(202) 366-4000

Federal Railway Administration (FRA)
400 7th Street SW
Washington, D.C. 20590
(202) 366-7034
FAX (202) 366-7034

Federal Transit Administration (FTA)
400 7th Street SW
Washington, D.C. 20590

U.S. Environmental Protection Agency (Guidelines only)
401 M Street SW (6202J)
Washington, D.C. 20460
Document Sales:
(202) 260-7751
FAX (202) 260-6257

Noise regulations are also found in other federal government organizations, such as the Departments of Army, Navy, and Air Force.

### A.9.2 Model Code Organizations

AIA Building Performance and Regulations Committee (AIA/BPRC)
1735 New York Avenue NW
Washington, D.C. 20006
(202) 626-7448
FAX (202) 626-7518

Basic Building Code (BBC)
Building Officials and Code Administrators International, Inc. (BOCA)
4051 West Flossmoor Road
Country Club Hills, IL 60478-5795
(708) 799-2300
FAX (708) 799-4981

Construction Specifications Institute (CSI)
601 Madison Street
Alexandria, VA 22314-1791
(800) 689-2900

Council of American Building Officials (CABO)
5203 Leesburg Pike, Suite 708
Falls Church, VA 22041
(703) 931-4533
FAX (703) 379-1546

Factory Mutual Engineering Corp. (FM)
1151 Boston-Providence Tpk.
Norwood, MA 02062
(617) 762-4300
(617) 762-9375

International Conference of Building Officials (ICBO)
5360 Workman Mill Road
Whittier, CA 90601-2298
(310) 699-0541
FAX (310) 699-8031
ICBO Issues the Uniform Building Code (UBC)

National Building Code (NBC)
American Insurance Services Group
85 John Street
New York, NY 10038
(212) 669-0400
FAX (212) 669-0535

National Conference of States on Building Codes and Standards (NCSBCS)
505 Huntmar Park Drive, Suite 210
Herndon, VA 22070
(800) 362-2633
(703) 437-0100
FAX 481-3596

National Building Code of Canada (NBCC)
Associate Committee on the National Building Code, National Research Council of Canada.
Ottawa, Ontario K1A 0R6
(613) 993-2463

Southern Building Code Congress International, Inc. (SBCCI, or SBC)
900 Montclair Road
Birmingham, AL 35213-1206
(205) 591-1853
FAX (205) 592-7001

Underwriters Laboratories (UL)
333 Pfingsten Road
Northbrook, IL 60062
(708) 272-8800
FAX (708) 272-8129

Underwriters Laboratories of Canada (ULC)
7 Crouse Road
Scarborough, ON, Canada M1R 3A9
(416) 757-3611
FAX (416) 757-9540

### A.9.3 ASTM Standards Relating to Environmental Acoustics

*Specifications for:*

| | |
|---|---|
| C 635–94 | Metal Suspension System for Acoustical Tile and Lay-In Panel Ceilings |
| E1289–91 | Reference Specimen for Sound Transmission Loss |
| E1179–87 (1993)$^{\ddot{Y}1}$ | Sound Sources Used for Testing Open Office Components and Systems |

*Test Methods for:*

| | |
|---|---|
| E1414–91 a$^{\ddot{Y}1}$ | Airborne Sound Attenuation Between Rooms Sharing a Common Ceiling Plenum |
| E 336–90 | Airborne Sound Insulation in Buildings, Measurement of |
| E 90–90 | Airborne Sound Transmission Loss of Building Partitions, Laboratory Measurement of |
| C 522–87 (1993)$^{\ddot{Y}1}$ | Airflow Resistance of Acoustical Materials |
| E1503–92 | Conducting Outdoor Sound Measurements Using a Digital Statistical Analysis System |
| E 477–90 | Duct Liner Materials and Prefabricated Silencers, Measuring Acoustical and Airflow Performance of |

| | |
|---|---|
| E1573-93 | Evaluating Masking Sound in Open Offices Using A-Weighted and One-Third Octave Band Sound Pressure Levels |
| E1124-92 | Field Measurement of Sound Power Level by the Two-Surface Method |
| E1007-90 | Field Measurement of Tapping Machine Impact Sound Transmission Through Floor-Ceiling Assemblies and Associated Support Structures. |
| E 492-90 | Impact Sound Transmission Through Floor-Ceiling Assemblies Using the Tapping Machine, Laboratory Measurement of |
| C 384-90 a | Impedance and Absorption of Acoustical Materials by the Impedance Tube Method |
| E1050-90 | Impedance and Absorption of Acoustical Materials Using a Tube, Two Microphones, and a Digital Frequency Analysis System |
| E1222-90 | Insertion Loss of Pipe Lagging Systems, Laboratory Measurement of |
| E1265-90 | Insertion Loss of Pneumatic Exhaust Silencers, Measuring |
| E1111-92 | Interzone Attenuation of Ceiling Systems, Measuring |
| E1375-90 | Interzone Attenuation of Furniture Panels Used as Acoustical Barriers, Measuring |
| E1376-90 | Interzone Attenuation of Sound Reflected by Wall Finishes and Furniture Panels, Measuring |
| E1408-91 | Laboratory Measurement of the Sound Transmission Loss of Door Panels and Door Systems |
| E1574-94 | Measurement of Sound in Residential Spaces |
| E 596-90 | Noise Reduction of Sound-Isolating Enclosures, Laboratory Measurement of |
| C 423-90a | Sound Absorption and Sound Absorption Coefficient by the Reverberation Room Method |

*Test Methods for:*

| | |
|---|---|
| E1130-90 | Speech Privacy in Open Offices Using Articulation Index, Objective Measurement of |
| C 367-78 (1989)$^{\hat{Y}1}$ | Strength Properties of Prefabricated Architectural Acoustical Tile or Lay-in Ceiling Panels |
| E 756-93 | Vibration Damping Properties of Materials, Measuring |

*Classifications for:*

| | |
|---|---|
| E1042-92 | Acoustically Absorptive Materials Applied by Trowel or Spray |
| E 1264-90 | Acoustical Ceiling Products |
| E1110-86 (1990) | Articulation Class, Determination of |
| E 989-89 (1994)$^{\hat{Y}1}$ | Impact Insulation Class (IIC), Determination of |
| E1332-90 | Outdoor-Indoor Transmission Class, Classification of |
| E 413-87 (1994)$^{\hat{Y}1}$ | Rating Sound Insulation |

*Practices for:*

| | |
|---|---|
| E 557-93 | Architectural Application and Installation of Operable Partitions |
| E 580-91 | Ceiling Suspension Systems for Acoustical Tile and Lay-in Panels in Areas Requiring Seismic Restraint, Application of |
| E 497-89 (1994)$^{\hat{Y}1}$ | Installation of Fixed Partitions of Light Frame Type for the Purpose of Conserving Their Sound Insulation Efficiency |
| C 636-92 $^{\hat{Y}1}$ | Installation of Metal Ceiling Suspension Systems for Acoustical Tile and Lay-In Panels |
| E 795-93 | Mounting Test Specimens During Sound Absorption Tests |
| E1123-86 (1990) | Mounting Test Specimens for Sound Transmission Loss Testing of Naval and Marine Ship Bulkhead Treatment Materials |
| E 597-81 (1987) | Single-Number Rating of Airborne Sound Isolation for Use in Multiunit Building Specifications, Determining |

*Terminology Relating to:*

| | |
|---|---|
| C 634-89 | Environmental Acoustics |

*Guides for:*

| | |
|---|---|
| E 717-84 (1989) | Accreditation Annex of Acoustical Test Standards, the Preparation of |
| E 966-92 | Field Measurement of Airborne Sound Insulation of Building Facades and Facade Elements |
| E1014-84 (1990) | Making Measurements of Outdoor A-Weighted Sound Levels |
| E1041-85 (1990) | Measurement of Masking Sound in Open Offices |
| E1374-93 | Open Office Acoustics and Applicable ASTM Standards |
| E 1433-91 | Selection of Standards on Environmental Acoustics |

METRIC PRACTICE

*Standard:*

| | |
|---|---|
| *E 380-93* | *Practice for Use of the International System of Units (SI) (the Modernized Metric System) (Excerpts) (see Related Material Section).* |

*Practice for:*

| | |
|---|---|
| *E 621-84 (1991)*$^{\hat{Y}1}$ | *Metric (SI) Units in Building Design and Construction, Use of (see Vol 04.07)* |

### A.9.4 American National Standards Institute (ANSI) Standards

The Standards Publication Program of the Acoustical Society of America (ASA) is the responsibility of the ASA Committee on Standards (ASACOS) and the ASA Standards Secretariat, headed by its Standards Manager.

The Acoustical Society of America provides the Secretariat for four standards committees accredited by the

American National Standards Institute (ANSI): S1 on Acoustics, S2 on Mechanical Vibration and Shock, S3 on Bioacoustics, and S12 on Noise.

These four accredited standards committees also provide the United States input to various international committees (IEC and ISO). Standards Committees S1 Acoustics and S3 Bioacoustics provide the United States input to ISO/TC 43 Acoustics, and IEC/TC 29 Electroacoustics, as the Technical Advisory Groups. S12 on Noise serves as the U.S. Technical Advisory Group for ISO/TC 43/SC1 Noise, and for ISO/TC 94/SC 12 Hearing Protection. S3 is the U.S. Technical Advisory Group for ISO/TC 108/SC4 Human Exposure to Mechanical Vibration and Shock. S2 serves as the U.S. Technical Advisory Group for ISO/TC 108, Mechanical Vibration and Shock, ISO/TC 108/SC1, Balancing, Including Balancing Machines, ISO/TC 108/SC2, Measurement and Evaluation of Mechanical Vibration and Shock as Applied to Machines, Vehicles, and Structures, ISO/TC108/SC3, Use and Calibration of Vibration and Shock Measuring Instruments, and ISO/TC 108/SC5, Condition Monitoring and Diagnostics of Machines, and ISO/TC 108/SC6, Vibration and Shock Generating Equipment.

ASACOS and the ASA Standards Secretariat provide administration for the U.S. Technical Advisory Groups listed above and administer the international Secretariat for ISO/TC 108 Mechanical Vibration and Shock.

Standards are reaffirmed, revised, or withdrawn every five years. The latest information on current ANSI standards, as well as those under preparation, is available from the Standards Secretariat.

For further information on the ASA Standards Publication Program, please contact:

A. Brenig, Standards Manager
Acoustical Society of America
120 Wall Street, 32nd Floor
New York, NY 10005-3993
(212) 248-0373
(212) 248-0146

Orders for standards may be placed directly with:

Acoustical Society of America
Standards and Publications Fulfillment Center
P.O. Box 1020
Sewickley, PA 15143-9998
(412) 741-1979
(412) 741-0609 FAX

## *A.10 Acoustical Measurements*

It is always useful to perform acoustical measurements to better define an acoustical problem or condition. The instrumentation available today for such measurements is vastly improved over that in use just a few years ago. High speed digital signal processing permits sophisticated measurements to be made in the field with virtually real-time output. Indeed, the ability to define an acoustical problem by measurement has advanced much more rapidly than the technology for solving it. With the possible exception of active noise cancellation, methods of noise control still rely on the use of passive absorbers and barriers.

### A.10.1 Measuring Noise Levels

The sound level meter is the basic sound measuring instrument. The simplest meters will read only linear and frequency-weighted sound levels (see Chapter 1, Section 1.6), while more sophisticated meters will incorporate frequency analyzers and a variety of sound integration modes. The newest meters will store data in digital form for later downloading to a computer or other readout device. Many newer meters also have the capability of calculating various sound descriptors, reverberation times, sound transmission loss data, etc., so that such data are available in the field immediately following a measurement.

For out-of-doors sound measurements, the microphone should be protected by a weather screen to prevent wind noise, provide a rain shield, and discourage birds from using it as a roost. The newer meters are designed in such a way that the instrument has very little effect on the sound field being measured, and is equipped with a small enough microphone to minimize the effects of meter orientation.

Measurement location, duration and meter settings must be selected according to the sound source characteristics, purpose of the measurements, and other sources that may be present. A measurement of short duration is usually adequate for steady sounds such as fans and generators, while longer duration measurements may be required for fluctuating noise such as intermittent traffic and many types of construction equipment. Impulsive noise such as gunshots or pile driving requires the meter to have a fast response time to properly record the maximum level of the impacts.

For a quick assessment of how a noise affects speech interference, a simple A-weighted measurement may be sufficient (see Chapter 1, Figure 1-15). To select the construction of a mechanical room wall next to a classroom, however, a frequency analysis is required to insure that the wall design has adequate TL at the predominant frequencies of the source.

If the sound from a source to be measured is contaminated by sound from other sources that cannot be turned off, it is sometimes possible to erect barriers to shield the measurement location from the unwanted sound. Special sound intensity measurement instrumentation is now available which permits measurement of a particular sound source in the presence of other noise. Since intensity is a vector quantity (meaning that the direction of sound travel is defined), it is possible to measure only that sound coming from the direction of the source of interest.

Sound level meters always require calibration at the time a measurement is performed, usually by a device that fits over the microphone and generates a known sound pressure level. Calibrators themselves must be periodically calibrated at a laboratory specializing in this service.

Older sound level meters were equipped with microphones that were 1″ (25 mm) in diameter or larger, and the response depended on the angle of incident sound. Therefore, orientation of the microphone with respect to direction of the source was critical. Newer meters with smaller microphones are much less sensitive to the angle of arriving sound.

Because of the importance of uniform performance characteristics by sound level meters from different manufacturers, an American National Standards Institute (ANSI) specification has been developed, S 1.4, in which meter grades of 0, 1 and 2 are defined, grade 0 having the strictest accuracy requirements.

### A.10.2 Measuring TL and Impact Sound Insulation

The sound transmission loss (TL—refer to Chapter 3, Sections 3.2 and 3.5 for a definition of terms used in this section, and applicable ASTM testing standards) of a test specimen is measured by locating the specimen between two adjacent rooms. A loud sound source is placed in the room to serve as the source room, in a manner to provide a fairly uniform sound distribution throughout the room, but particularly over the face of the test specimen. The test signal usually consists of broad-band noise covering the frequency range of interest, such as 100 Hz to 5000 Hz to meet ASTM requirements.

The sound pressure level is measured in one-third octave frequency bands in both source and receiving rooms, using a suitable frequency analyzer. A space average of the sound fields in both rooms is usually required because the fields are never perfectly uniform. The difference in averaged levels for each frequency band determines the noise reduction in each band, and these data may be used to determine the Noise Isolation Class (NIC). Further adjustments of the data for test specimen area and receiving room absorption can be made to determine the sound transmission loss (TL), and the Sound Transmission Class (STC), but precautions must be taken to ensure that all the sound transmitted from source to receiving room was through the test specimen.

Impact sound isolation is performed by operating a standard tapping machine at a minimum of three locations on the test floor. The sound produced in the space below is measured in one-third octave bands and then normalized to a standard room absorption. From these data the normalized impact sound pressure level (NISPL) can be determined, and the Impact Insulation Class (IIC). The tapping machine may not yield reliable data at low frequencies for wood frame floors or other flexible floor constructions. It is prudent to perform a live walker test and listen in the room below for creaking, or excessive low frequency thumping.

Flanking sound transmission around the test specimen, and sound leakage should always be investigated. Flanking transmission may be difficult to detect, but the ASTM standard for field TL measurements does describe methods for detecting flanking. (It is assumed that flanking transmission is not significant in laboratory measurements.) Sound leakage can usually be detected with a probe stethoscope or with a sound level meter using a microphone extension.

Most codes or standards specifying minimum performance for airborne and impact sound isolation, such as the UBC and HUD/FHA standards for separation of dwellings in multifamily housing, state that ASTM test methods be employed. For the UBC, the test methods are contained in Chapters 35-1, 35-2 and 35-3 of the manual. The performance requirements are contained in UBC Appendix Chapter 35.

### A.10.3 Measuring Reverberation Times

In Section 4.6 reverberation time (RT) is defined as the time required for a sound to decay 60 dB. Measurement of RT, therefore, involves determining the rate at which sound decays in an enclosed space, at each frequency of interest.

There are several methods for measuring reverberation times, starting with a variety of means for generating the sound. It can be generated by a loudspeaker playing bands of noise, pure tones or some other type of signal that contains the frequencies of interest. Sometimes impulsive sources are used, such as pistols shooting blank cartridges, clap boards or popping balloons. Measurement of the decay begins as soon as the source is discontinued. It is not necessary to measure a full 60 dB decay, as long as the rate of decay can be reliably determined by measuring a smaller range.

An instrument used to determine the rate of decay is the graphic level recorder (or strip chart recorder). More recently available are instruments that record decays by digital means, and can display the decay as a graph or readout directly in RT. Decays are seldom smooth and often times lack linearity, making interpretation difficult. It is always useful to have a graphic display, because the structure of the decay can reveal a lot to an experienced interpreter about the characteristics of the space.

A more refined analysis of a sound decay involves looking at the first 40 to 80 milliseconds of the decay, and comparing this to the remainder of the decay. There has been a considerable amount of research in correlating this "early decay" information with a number of acoustical attributes, particularly for concert halls. In particular, clarity is greatly affected by early decay characteristics.

Another quite common characteristic of reverberation is the double slope decay, where the first part of the decay is at a faster rate than the second part. One possible explanation for this condition is the presence of an acoustically coupled space that has a longer reverberation time than the space in which the measurement is being made. An example would be a reverberant stage house coupled to an auditorium which is less reverberant. The first part of the decay is a measure of the auditorium, then sound feeding back from the stage house dominates the second part of the decay.

Another possible explanation for the double slope decay is the presence of undamped room modes. This condition can occur in a room where most of the acoustical absorption is along one room axis, such as ceiling and floor. The modes in this axis die out quickly, while the modes involving the walls that lack much absorption are more persistent. Both of these conditions indicate problems that require some correction.

### A.10.4 Vibration Measurements

There are many instances where the vibration of a surface, machine or other vibrating body is of interest. For example, such data can provide information on the response of structures to input from some type of forcing function. Other applications could be noise radiation from vibrating surfaces, and unbalanced forces in rotating machinery which could cause damage.

Vibration is usually measured by attaching an accelerometer to a surface such that it moves exactly as the surface moves, and its presence does not significantly affect the surface's vibration. The accelerometer produces an electrical signal with the same characteristics as the acceleration of the surface. From this signal, the other vibration parameters of displacement and velocity can be determined. For some vibration problems, it is necessary to measure the acceleration along 2 or all 3 of the principal axes x, y and z.

Vibration problems often involve rotating machinery which can generate single frequencies that need to be identified. Signal analyses, therefore, require narrow band analyzers, with band widths as small as one Hz. Modern analyzers which operate in the digital domain have this capability.

Vibration data are capable of providing information on the damping of structures, which is important in considering the response of a structure to a specified forcing function. Damping can be measured by determining the rate of vibration decay in a structure, in much the same manner as reverberation time is measured in rooms. Damping can also be measured by determining the "Q" of the vibration at a specified frequency, which is the ratio of the peak vibration at that frequency compared to the bandwidth at some lower vibration level on each side of the peak.

### A.10.5 Time-Energy-Frequency (TEF) Measurements

TEF is a sophisticated measurement technique utilizing special instrumentation which is capable of identifying reflecting surfaces within a space, even in the presence of other noise or reflections. The TEF analyzer generates a sine-wave frequency sweep which is played through a sound system and picked up by a microphone located at a point of interest. The change in frequency of the sweep is linear with time. The microphone signal is fed through a filter that tracks the sweep. The tracking filter is synchronized with the generated frequency sweep, but is offset in time to compensate for the propagation delay of sound traveling from loudspeaker to microphone. By varying the bandwidth and time offset of the tracking filter, it is possible to study the spectrum of the direct sound, or sound reflecting from a surface of interest.

There are other measurement techniques for studying the pattern of reflections in a room, utilizing impulsive type signals and an oscilloscope-type readout. Measurements of this type and TEF measurements are difficult to both perform and interpret, and should be attempted only by an experienced acoustical consultant.

### A.10.6 Acoustical Testing Laboratories

There are a number of laboratories that perform tests of the acoustical characteristics of building materials. All the laboratories listed below perform sound absorption, sound transmission loss and impact insulation tests. Most laboratories performing acoustical tests are accredited by the National Voluntary Laboratory Accreditation Program (NVLAP). These laboratories perform the tests in accordance with applicable ASTM standards. Other laboratories not listed are operated by manufacturers of acoustical products, or specialize in testing of mechanical equipment, duct silencers, etc.

Acoustic Systems
P.O. Box 3610
Austin, TX 78764-3610
(800) 749-1460
(512) 444-1961
FAX (512) 444-2282

Center For Applied Engineering, Inc.
10301 Ninth Street North
St. Petersburg, FL 33716
(813) 578-4316
FAX (813) 578-4280

Clark Laboratory Services
821 East Front Street
Buchanan, MI 49107
(606) 697-8632

Geiger & Hamme, Inc.
P.O. Box 1345
3250 East Morgan Road
Ann Arbor, MI 48106
(313) 971-3033

Maxim Technologies
662 Cromwell Avenue
St. Paul, MN 55114
(612) 659-7310
(612) 659-7348

Noise Unlimited, Inc.
9 Saddle Road
Cedar Knolls, NJ 07927
(908) 713-9300
FAX (908) 713-9001

Riverbank Acoustical Laboratories
P.O. Box 189
1512 South Batavia Avenue
Geneva, IL 60134
(708) 232-0104
FAX (708) 232-0138

University of Alberta
MEANU
6720 30th Street
Edmonton, Alberta, Canada T6P 1J6

Western Electro-Acoustic Laboratory, Inc.
1711 16th Street
Santa Monica, CA 90404
(310) 450-1733

## A.11 Acoustical Consultants

The National Council of Acoustical Consultants (NCAC) issues annually a directory of member acoustical consulting firms. Copies of the directory may be obtained from NCAC Headquarters at 66 Morris Avenue, Suite 1B, Springfield, NJ 07081-1409, (201) 564-5859, FAX (201) 564-7480. The Council offers the following advice for selection of an acoustical consultant:

SELECTION OF AN ACOUSTICAL CONSULTANT

Every project undertaken by a consultant is unique. While many assignments may be similar in nature, no two ever are identical. For this reason, it is essential that a consultant be chosen with deliberate care. In essence, the more experienced and qualified the consultant is to undertake a given project, the more likely the services will be in accord with the goals and objectives of the client. Moreover, provision of consulting services, by definition, implies a close, privileged relationship between consultant and client. To give less than full consideration to the selection/retention process, therefore, would be to jeopardize a successful consultant-client relationship before it begins, thereby jeopardizing the successful outcome of the project at hand.

In the event that you have not already established a relationship with an acoustical consultant, the National Council of Acoustical Consultants recommends for your consideration the following method of selection and retention tested through many years of successful application:

1. Determine to the extent possible the nature and scope of the problem or assignment involved.
2. Through contact with mutual acquaintances who have previously utilized acoustical consultants, or from directories of qualified independent consulting firms provided by an organization such as NCAC, identify one or more acoustical consultants who, by virtue of previous experience, stated capabilities, availability and proximity, as well as other relevant factors, appear to be generally qualified to undertake the project.
3. Provide project details to the consultants so identified and request from each statements of qualification, including a complete description of the firm, previous assignments and clients, names and biographies of persons who would be working on the project, anticipated time schedules involved, and other factors which relate to the quality of work to be performed.
4. After thorough review of applicant firm; credentials and experience data, possibly including direct contact with firm representatives if such can be arranged, identify the firm which appears most qualified to serve your specific requirements.
5. Contact representatives of that firm believed to be most qualified and open negotiations to establish a mutually acceptable consulting fee arrangement and payment methodology. Most consultants are experienced in at least several types of retention agreements, including hourly rate, fixed fee, cost plus fixed fee, percentage of overhead, etc. Usually one of these will be most suited to the type of work involved.
6. If the negotiations prove satisfactory, the client should at this point retain the consultant to ensure availability for the project. If negotiations are not successful, they should be terminated and opened with other qualified firms, one at a time.

It should be noted that NCAC encourages open and frank discussion of financial concerns between the client and consultant. Experience demonstrates that mutually satisfactory client-consultant relationships rest predominantly on the consultant's ability to deliver cost-effective services on time and within the scope of the agreement. In fact, most successful consultants pride themselves on their ability to tailor their efforts to the scope of the project and the budget available for services as well as for implementation of the recommendations resulting from their services. It is urged strongly, however, that discussions of fees be divorced completely from the ranking qualifications to prevent financial considerations from biasing the selection process. True economy results only when services provided are cost-effective in the long term, helping ensure results which satisfy the client's needs from an overall standpoint. A consultant who is fully competent to undertake the work is the one most likely to provide such results.

## A.12 Acoustical Materials Manufacturers

The authors have diligently attempted to contact all manufacturers of acoustical or noise control products. The listing that follows includes the manufacturers that responded to the authors' request for product information. Because new companies are formed, and existing companies move or change names, the authors cannot be responsible for omissions, for information that is out of date, or for lack of response to our inquiry.

(1) Acoustic Systems.
Enclosures, panels, doors, windows.
415 East Saint Elmo Road
Austin, TX 78745
1-800-749-1460
(512) 444-1961
FAX (512) 444-2282

(2) Advanced Equipment Corporation.
Operable walls
2401 West Commonwealth Avenue
Fullerton, CA 92633
(714) 635-5350
FAX (714) 525-6083

(3) Akzo Nobel Geosynthetics Co.
Sound rated floor systems
P.O. Box 7249
318 Ridgefield Business Center
Asheville, NC 28802
(704) 665-5030
FAX (704) 665-5005

(4) Alpro Acoustics
Metal ceiling and wall systems, baffle panels.
600 St. George Avenue, Suite A
New Orleans, LA 70121
(504) 733-3836
FAX (504) 733-3851

(5) Architectural Surfaces, Inc.
SONEX acoustical foam ceilings and walls, Pro-SPEC acoustical materials.
123 Columbia Court North
Chaska, MN 55318
1-800-470-6652
(612) 448-5300
FAX (612) 448-2613

(6) Armstrong World Industries, Inc.
Wall & ceiling acoustical materials.
P.O. Box 3001
Lancaster, PA 17604
1-800-448-1405
FAX (717) 396-3169

(7) AVL Systems, Inc.
Acoustical wall panels.
5540 S. W. 6th Place
Ocala, FL 34474
1-800-228-7842
(904) 854-1170
FAX (904) 854-1278

(8) BHP Steel Building Products USA Inc.
Acustadek
2110 Enterprise Blvd.
West Sacramento, CA 95691
1-800-726-2727
(916) 372-6851
FAX (916) 372-7606

(9) Brejtfus Enterprises
Baffles & panels
410 S Madison Dr., Suite 1
Tempe, AZ 85281
(800) 264-9190
(602) 731-9899
FAX (602) 731-7606

Cafco (See #35)

(10) Capaul
Ceiling and wall panels.
1300 Division Street
Plainfield, IL 60544
1-800-876-5884
(815) 436-8500
FAX (800) 444-5066

(11) The Celotex Corporation
Architectural ceilings.
4010 Boy Scout Blvd.
Tampa, FL 33607
(813) 873-4027
FAX (813) 873-4103

(12) Chicago Metallic
Metallic suspended ceiling systems.
4849 South Austin Avenue
Chicago, IL 60638
(800) 323-7164

(13) CGC Interiors
Panels, tiles, blankets & boards
735 Fourth Line
Oakville, Ontario, Canada L6L 5B7
(800) 387-7920
FAX (800) 771-7460

(14) Conwed
Absorbers, diffusers, panels.
1205 Worden Avenue East
Ladysmith, WI 54848
(800) 932-2388
FAX (800) 833-4798

(15) Decoustics
Fabric panels.
65 Disco Road
Etobicoke, Ontario,
Canada M9W 1M2
(800) 387-3809
(416) 675-3983
FAX (416) 675-5546

(16) Diamond Manufacturing Co.
Perforated metal.
P.O. Box 174
243 West Eighth Street
Wyoming, PA 18644
(800) 233-9601
(717) 693-0300
FAX (717) 693-3500

(17) The E. J. Davis Co.
Sound barriers and sound absorption control
10 Dodge Ave.
P.O. Box 326
North Haven, CT 06473
(203) 239-5391
FAX (203) 234-7724

(18) E-A-R Specialty Composites
Barriers, damping materials.
7911 Zionsville Road
Indianapolis, IN 46268
(317) 872-1111
FAX (317) 692-3111

(19) Eckel Industries, Inc.
Unit absorbers, metal panels.
155 Fawcett St.
Cambridge, MA 02138
(617) 491-3221
FAX (617) 547-2171

(20) Eggers Industries
Acoustical doors.
1918 East River Street
Two Rivers, WI 54241
(414) 793-1351
FAX (414) 793-2958

(21) Empire Acoustical Systems
Acoustical enclosures, noise barriers, sound absorption wall mountings.
89 Park Avenue West
Mansfield, OH 44902
(419) 522-0800
FAX (419) 522-7937

(22) Epic Metals Corporation
Acoustical security wall panels.
Eleven Talbot Avenue
Rankin, PA 15104
(412) 351-3913
FAX (412) 351-2018

(23) ESSI Acoustical Products Company
Walls, ceilings, baffles, banners, masks, pads.
11740 Berea Road
Cleveland, OH 44111
(216) 251-7888
FAX (216) 251-9933

(24) Firedoor Corporation
Firesonic hollow metal acoustic door assemblies.
1350 NW 74th Street
P.O. Box 380878
Miami, FL 33238-0878
(305) 691-1500
FAX (305) 836-4797

(25) Golterman & Sabo, Inc.
Panels, suspended absorbers and ceiling clouds.
5901 Elizabeth
St. Louis, MO 63110
(800) 737-0307
(314) 781-1422
FAX (314) 781-3836

(26) Grace Construction Products
W. R. Grace & Company—Conn.
Spray-on acoustical material.
62 Whittemore Avenue
Cambridge, MA 02140
(617) 498-4316
FAX (617) 498-4419

(27) Guardian Industries, Inc.
Security Glazing Product Center
Sound control windows.
One Belle Ave., Bldg. #35
Lewistown, PA 17044

(717) 242-2571
FAX (717) 242-0450

(28) Guilford Industries, Inc.
Fabrics for acoustical wall panels.
Park 80 West II
Saddle Brook, NJ 07662
(201) 845-3600

(29) Gyp-Crete Corporation
Floor system for noise control.
920 Hamel Road
Hamel, MN 55340
(800) 356-7887
(612) 478-6072
FAX (612) 478-2431

(30) Hacker Industries, Inc.
Gypsum Concrete.
P.O. Box 5918
Newport Beach, CA 92662
(800) 642-3455
(714) 645-8891
FAX (714) 645-5144

(31) Howard Products
Howard Manufacturing Company
Ventwood building specialties.
P.O. Box 1188
Kent, WA 98035
(206) 852-0640
FAX (206) 854-0895

(32) Hunter Douglas
Architectural products, ceiling systems.
11455 Lakefield Dr.
Duluth, GA 30136
(800) 366-4327
(404) 476-8803
FAX (404) 476-5716

(33) Industrial Acoustics Company
Diffusers, attenuators, absorbers for industrial uses.
1160 Commerce Ave.
The Bronx, NY 10462
(718) 931-8000
FAX (718) 863-1138

(34) International Cellulose Corp.
Spray-on insulation.
P.O. Box 450006
12315 Robin Blvd.
Houston, TX 77245
(800) 444-1252
(713) 433-6701
FAX (713) 433-2029

(35) Isolatek International
Spray-on acoustical material.
41 Furnace Street
Stanhope NJ 07874
(80) 631-9600
(201) 347-1200
FAX (201) 347-9170

(36) Jamison Door Company
Acoustical doors.
P.O. Box 70
55 Maple Avenue
Hagerstown, MD 21741
(800) 532-3667
(301) 733-3100
FAX (301) 791-7339

(37) Kinetics Noise Control
Noise & vibration control systems.
6300 Irelan Place
Dublin, OH 43017
1-800-959-1164
(614) 889-0480
FAX (614) 889-0540

(38) Knauf Fiber Glass
Glass fiber insulation.
240 Elizabeth Street
Shelbyville, IN 46176
(800) 825-4434
(317) 398-3675

(39) Krieger Steel Products Co.
Sound-rated doors.
4880 Gregg Road
Pico Rivera, CA 90660
(310) 695-0645
FAX (310) 692-0146

(40) Lamvin, Inc.
Acoustical wall panels.
6260 Mar Industry drive
San Diego, CA 92121
(800) 446-6329
(619) 452-7480
FAX (619) 452-8960

(41) Lencore Acoustics Corp.
Sound masking, reflecting, diffusing and absorbing panels.
2220 Hewlett Avenue
Merrick, NY 11566
(516) 223-4747
FAX (516) 223-4785

(42) Loadmaster Roof Deck Systems
Acoustic roof decks.
P.O. Box 2169
Duluth, GA 30136
(800) 527-4035
(404) 381-6067
FAX (404) 381-1783

(43) Mason Industries, Inc.
Architectural and noise control products.
350 Rabro Drive
Hauppauge, NY 11788
(516) 348-0282
FAX (516) 348-0279

(44) Metal Building Interior Products Company
Wall systems, baffles.
5309 Hamilton Avenue
Cleveland, OH 44114
(216) 431-6400
(216) 431-9000

(45) Milco
Acoustical windows.
P.O. Box 1366
7555 Stewart Avenue
Wausau, WI 54402-1366
(715) 842-0581
FAX (715) 843-4805

(46) Modernfold
Partitions and doors.
17111 I Avenue
New Castle, IN 47362
(317) 529-1450
FAX (317) 521-6204

(47) MPC, Inc.
Noise Control Products Division
Panels, baffles, inserts.
835 Canterbury Road
Westlake, OH 44145
(216) 835-1405
FAX (216) 835-9313

(48) National Guard Products, Inc.
Seals.
540 North Parkway
Memphis, TN 38018
(800) 647-7874
FAX (800) 255-7874

(49) NetWell Noise Control
Foams and barriers.
6125 Blue Circle Drive
Minnetonka, MN 55343
(800) 638-9355
FAX (612) 939-9836

(50) Overly Manufacturing Company
Sound-rated doors.
P.O. Box 70
Greensburg, PA 15601
(412) 834-7300
FAX (412) 830-2871

(51) Owens-Corning
Insulation.
Fiberglas Tower
Toledo, OH 43659
(800) GET-PINK

(52) Panel Solutions, Inc.
Acoustical wall panels.
100 East Diamind Ave.
Hazelton, PA 18201
(800) 523-6671
FAX (717) 459-3499

(53) Panelfold, Inc.
Folding doors.
P.O. Box 680130
Miami, FL 33168
(305) 688-3501
FAX (305) 688-0185

(54) Peer, Inc.
Sintered aluminum absorbing material.
241 West Palatine Road
Wheeling, IL 60090
(800) 433-7337
(708) 870-3300
FAX (708) 870-3337

(55) Peerless Products, Inc.
Sound-rated windows.
3030 South 24th Street
Kansas City, KS 66109
(913) 432-2232
FAX (913) 432-3004

(56) Pemko
Door seals.
P.O. Box 3780
Ventura, CA 93006
(805) 642-2600
FAX (805) 642-4109

(57) Pioneer Industries, Inc.
Sound rated doors.
401 Washington Ave.
Carlstadt, NJ 07072
(201) 933-1900
FAX (201) 933-9580

(58) Proudfoot Company, Inc.
Sound screens, baffles, resonators, wall panels, pipe and duct lagging.
P.O. Box 276
Monroe, CT 06468-0276
(800) 445-0034
(203) 459-0031
FAX (203)-449-0033

(59) Pyrok, Inc.
Acoustical decorative fire protection finishes.
136 Prospect Park West
Brooklyn, NY 11215
(718) 788-1225

(60) RPG Diffusor Systems, Inc.
Diffusor systems.
651-C Commerce Drive
Upper Marlboro, MD 20772
(301) 249-0044
(301) 249-3912

(61) Rubatex Corp.
Gaskets, seals, pipe insulation.
906 Adams St.
Bedford, VA 24523
(800) 782-2839
FAX (703) 587-4228

(62) Security Acoustics
Sound retardant doors and specialty products.
12519 Cerise Avenue
Hawthorne, CA 90250
(213) 772-1171
FAX (310) 978-4762

(63) Snap-Tex Systems, Inc.
Wall and ceiling systems.
84 Resevoir Park Drive
Rockland, MA 02370
(800) 762-7875
(617) 871-4951
FAX (617) 878-8130

(64) Sound Absorption Ltd.,
c/o Tempest Security Systems
Coustone.
P.O. Box 262946
Tampa, FL 33685
(813) 884-4994
(813) 888-9472

(65) Sound Concepts Canada Inc.
Panels.
599 Henry Avenue
Winnipeg, Manitoba, Canada R3A 0V1
(204) 783-6297
FAX (204) 783-7806

(66) Sound Reduction Corporation
Noise absorbing panels.
16601 St. Clair Ave.
Cleveland, OH 44110
(800) 635-3916
(216) 481-1900
FAX (216) 481-2907

(67) Soundcoat
Damping materials, compounds and sheets.
One Burt Drive
Deer Park, NY 11729
(516) 242-2200
FAX (516) 242-2246

(68) SoundSeal, Division of United Process, Inc.
Acoustical materials and components for industrial plants, transportation, equipment and construction.
279 Silver Street
Agawam, MA 01001
(800) 569-1294

(69) Snap-Tex Systems, Inc.
Custom fabric-wrapped acoustical wall panels.
84 Resevoir Park Drive
Rockland, MA 02370
(800) 762-7875
(617) 871-4951
FAX (617) 878-8130

(70) StageRight Corporation
Stage shells, music practice rooms.
495 Holley Drive
P.O. Box 208
Clare, MI 48617
(800) 438-4499
(517) 386-7393
FAX (517) 386-3500

(71) Steel Ceilings
Perforated metal, ceiling clouds, wall panels & absorbers.
P.O. Box 547
Coshocton, OH 43812
(800) 848-0496
(614) 622-4655
FAX (614) 622-8971

(72) Steelcase, Inc.
Wall panels.
P.O. Box 1967
Grand Rapids, MI 49501
(800) 227-2960
(616) 246-4881

(73) Systems Development Group
Acoustical panels, diffusors and wedges.
5744 Industry Lane, Suite J
Frederick, MD 21701
(800) 221-8975
(301) 846-7990
FAX (301) 698-4683

(74) Tamer Industries
Sound control enclosures.
185 Riverside Avenue
Somerset, MA 02725
(800) U-TAME-IT
(508) 677-0900
FAX (508) 677-3037

(75) Tectum, Inc.
Structural acoustic roof deck products.
P.O. Box 3002
Newark, OH 43058
(614) 345-9691
FAX (614) 349-9305

(76) Thermocon, Inc.
Spray-on insulation
2500 Jackson Street
Monroe, LA 71202
(800) 532-6145
(913) 383-0909
FAX (913) 383-3345

(77) Tru Roll, Inc.
Theatrical & entertainment products.
622 Sonora Ave.
Glendale, CA 91201
(818) 240-4835
FAX (818) 240-4855

(78) Unique Concepts, Inc.
Home of Fabri-Trak and Noise Reduction Concepts
Fabri-Trak Wall Systems.
59 Willet Street
Bloomfield, NJ 07003
(201) 748-6331
FAX (201) 748-8250

(79) USG Interiors, Inc.
Gypsum board, ceiling products.
125 South Franklin Street
Chicago, IL 60680-4470
(312) 606-4000

(80) Wall Technology, Inc.
Acoustical wall panels.
2750 Industrial Lane
Broomfield, CO 80020
(303) 466-3700

(81) Wenger Corporation
Music performance and practice areas.
555 Park Drive, Box 448
Owatonna, MN 55060
(800) 733-0393
(507) 455-4100
FAX (507) 455-4258

(82) Whisper Walls
Fabric covered walls and ceiling systems.
10957 East Bethany Drive
Aurora, CO 80014
(800) 527-7817
(303) 671-6696
FAX (303) 671-0606

(83) Working Walls, Inc.
Panels, tiles and clouds.
100 Hayes Drive, Suite B
Cleveland, OH 44131
(216) 749-7850
FAX (216) 749-7855

(84) Zero International
Door and window hardware.
415 Concord Ave.
Bronx, NY 10455-4898
1-800-635-5335
(718) 585-3230
FAX (718) 292-2243

# Bibliography

## *Organizations*

ACOUSTICAL SOCIETY OF AMERICA (ASA)
500 Sunnyside Blvd.
Woodbury, New York 11797-2999
(516) 576-2360 FAX (516) 349-7669

The Society publishes a monthly journal which covers all aspects of acoustics, and is highly technical. (The journal is now available on CD ROM.) The Society also publishes *Echoes*, a less technical newsletter which usually features some unique or unusual acoustical application. There are a number of local chapters of the Society located throughout the United States, often sponsoring programs on popular acoustical subjects.

AMERICAN SOCIETY OF HEATING, REFRIGERATING AND AIR-CONDITIONING ENGINEERS, INC.
1791 Tullie Circle, N.E.
Atlanta, Georgia 30329

The Society publishes handbooks on fundamentals and applications for the design of HVAC and plumbing systems, including chapters on noise and vibration control. Their publications are revised and updated every 3 to 5 years, and are an excellent source of information relating to HVAC and plumbing systems.

INSTITUTE OF NOISE CONTROL ENGINEERING (INCE)
P.O. Box 3206 Arlington Branch
Poughkeepsie, New York 12603
(914) 462-4006 FAX (914) 463-0201

The Institute of Noise Control Engineering is a nonprofit professional organization founded in 1971. A primary objective of the Institute is to advance the technology of noise control with particular emphasis on engineering solutions to environmental noise problems. Full membership in the Institute requires that the applicant pass a 2-part examination similar to that required for registration as a professional engineer. The Institute publishes the *Noise Control Engineering Journal*, issued bimonthly.

NATIONAL COUNCIL OF ACOUSTICAL CONSULTANTS
66 Morris Avenue, Suite 1A
Springfield, New Jersey 07081-1409
(201) 564-5859 FAX (201) 564-7480

The National Council of Acoustical Consultants was established in 1962 when a group of practicing consultants realized that expanding and rapid developments in the relatively new field of acoustics warranted the support of mutual efforts which only a national association could provide.

The purposes of NCAC are as follows:

- Establish and encourage adherence to the highest standards of professional ethics and business practices;
- Inform the public of the existence of acoustical consultants and the services which they provide;
- Provide members with a forum for discussion and exchange of information on matters of common interest;
- Cooperate with representatives of other organizations on matters of mutual interest and concern;
- Preserve and protect the public welfare by encouraging accurate and proven representations concerning acoustical products, materials and services;
- Participate in the development of performance and measurement standards and regulations; and
- Encourage and promote continuing growth and education in the profession.

While there are organizations dedicated to scientific and engineering aspects of the field of acoustics, only NCAC is dedicated to management and related concerns of professional acoustical consulting firms and to safeguarding the interests of the clients and public which they serve.

NCAC members are not associated with the manufacture or sale of any product when such association could jeopardize, tend to jeopardize, or give the impression of jeopardizing their ability to render an independent, unbiased decision regarding product specification or related matters.

NCAC members have established a Speaker's Bureau. For a list of members available to address your organization or group on specific technical or general interest topics in acoustics, contact NCAC headquarters at the address given previously. An NCAC membership directory is also available from the same address. Additional information on NCAC is given in Appendix A.11.

UNITED STATES GYPSUM COMPANY
125 South Franklin Street
Chicago, Illinois 60606-1678
This company has provided useful information to us in a handbook on the SA-100 Construction Selector.

## Periodicals (Also, see under Organizations)

*Airport Noise Report*
43978 Urbancrest Court
Ashburn, VA 22011
(703) 729-4867 FAX (703) 729-4528
A newsletter published 25 times a year covering topics related to airports and aircraft noise.

*Noise Regulation Report*
Business Publishers, Inc.
951 Pershing Drive
Silver Spring, Maryland 20910-4464
(301) 587-6300 FAX (301) 587-1081
A newsletter published biweekly covering general environmental noise topics.

*Sound & Vibration*; The Noise and Vibration Control Magazine
Acoustical Publications, Inc,
27101 E. Oviatt Road
P.O. Box 40416
Bay Village, OH 44140
(216) 835-0101
This magazine is published monthly featuring articles on practical applications of methods of noise and vibration control. Updates on new services and materials are provided.

U.S. Department of Commerce
National Technical Information Service (NTIS)
5285 Port Royal Road
Springfield, VA 22161
(703) 487-4650
NTIS issues periodic bulletins listing a variety of technical reports, including the subjects of acoustics and environmental noise.

*The Wall Journal*
P.O. Box 1217
Lehigh Acres, FL 33970-1217
(813) 369-0178 FAX (813) 369-0451
The journal is published 6 times a year, and features articles on highway noise barriers. A good source of information for barrier design, cost and performance, and occasionally other aspects of highway noise.

Other publications and periodicals are available on subjects such as sound system engineering, mechanical noise and vibration, industrial noise, manufacturers of acoustical instruments and other related subjects. Also, there are a number of useful booklets published by product manufacturer associations listing acoustical performance data on building materials and constructions. Inquire from a local supplier of building materials what booklets are available for their products.

## Classical Texts

Beranek, L. L., *Acoustics*, McGraw-Hill Book Company, Inc., 1954.

Brüel, Per V., Sound Insulation and Room Acoustics, Chapman & Hall, Ltd., 1951.

Harris, C. M., *Acoustical Measurements and Noise Control*, Third Edition, McGraw-Hill, Inc., 1957.

Hunt, Frederick Vinton, *Origins in Acoustics: The Science of Sound from Antiquity to the Age of Newton*, Yale University Press, 1978.

Kinsler, L. E., Frey, A. R., Coppens, A. B., & Sanders, J. V., *Fundamentals of Acoustics*, Third Edition, John Wiley & Sons, Inc., 1982.

Knudsen, V. O., *Architectural Acoustics*, John Wiley & Sons, Inc., 1932.

Knudsen, V. O., & Harris, C. M., *Acoustical Designing in Architecture*, John Wiley & Sons, Inc., 1950.

Morse, P. M., & Ingard, K. U., *Theoretical Acoustics*, McGraw-Hill Book Company, 1968.

Olson, H. F., *Acoustical Engineering*, D. Van Nostrand Company, Inc., 1957.

Sabine, P. E., *Acoustics and Architecture*, McGraw-Hill Book Company, Inc., 1932.

Watson, F. R., *Acoustics of Buildings* including Acoustics of Auditoriums and Soundproofing of Rooms, Second Edition, John Wiley & Sons, Inc., 1930.

## Current References

*A Guide to Airborne, Impact and Structure Borne Noise Control in Multifamily Dwellings*, a HUD publication prepared for the Federal Housing Administration (FHA), 1967.

Beranek, L. L., & Vér, I. L., *Noise and Vibration Control Engineering*, John Wiley & Sons, Inc., 1992.

Berendt, Raymond D. and Corliss, Edith L. R., *Quieting: A Practical Guide to Noise Control*, National Bureau of Standards, Washington, D.C., 1976.

Egan, M. D., *Architectural Acoustics*, McGraw-Hill, Inc., 1988.

Geerdes, Harold P., *Music Facilities: Building, Equipping, and Renovating*, Music Educators National Conference, 1902 Association Drive, Reston, VA, 22091, 1987.

Gypsum Association, *Fire Resistance Design Manual/Sound Control—Gypsum Systems* (reissued with changes every 1–2 years). Contact:

Gypsum Association
810 First Street NE, #510
Washington, D.C. 20002
(202) 289-5440
FAX (202) 289-3707

Harris, C. M., *Noise Control in Buildings*, McGraw-Hill, Inc., 1994.

Hedeen, R. A., *Compendium of Materials for Noise Control*, U.S. Department of Health, Education, and Welfare, 1980.

Kryter, K. D., *The Effects of Noise on Man*, Academic Press, Inc., 1985.

Kuttruff, H., *Room Acoustics*, Third Edition, Elsevier Applied Science, 1991.

Rettinger, M., *Handbook of Architectural Acoustics and Noise Control*, TAB Books, Inc., 1988.

Uniform Building Code, Appendix Chapter 35 (Updated every 2–3 years).

Proceedings of the *Wallace Clement Sabine Centennial Symposium*, Acoustical Society of America, 1994.

# Index